Selected Titles in This Series

234 Theodore P. Hill and Christian Houdré, Editors, Advances in stochastic inequalities, 1999

233 Hanna Nencka, Editor, Low dimensional topology, 1999

232 Krzysztof Jarosz, Editor, Function spaces, 1999

231 Michael Farber, Wolfgang Lück, and Shmuel Weinberger, Editors, Tel Aviv topology conference: Rothenberg Festschrift, 1999

230 Ezra Getzler and Mikhail Kapranov, Editors, Higher category theory, 1998

229 Edward L. Green and Birge Huisgen-Zimmermann, Editors, Trends in the representation theory of finite dimensional algebras, 1998

228 Liming Ge, Huaxin Lin, Zhong-Jin Ruan, Dianzhou Zhang, and Shuang Zhang, Editors, Operator algebras and operator theory, 1999

227 John McCleary, Editor, Higher homotopy structures in topology and mathematical physics, 1999

226 Luis A. Caffarelli and Mario Milman, Editors, Monge Ampère equation: Applications to geometry and optimization, 1999

225 Ronald C. Mullin and Gary L. Mullen, Editors, Finite fields: Theory, applications, and algorithms, 1999

224 Sang Geun Hahn, Hyo Chul Myung, and Efim Zelmanov, Editors, Recent progress in algebra, 1999

223 Bernard Chazelle, Jacob E. Goodman, and Richard Pollack, Editors, Advances in discrete and computational geometry, 1999

222 Kang-Tae Kim and Steven G. Krantz, Editors, Complex geometric analysis in Pohang, 1999

221 J. Robert Dorroh, Giséle Ruiz Goldstein, Jerome A. Goldstein, and Michael Mudi Tom, Editors, Applied analysis, 1999

220 Mark Mahowald and Stewart Priddy, Editors, Homotopy theory via algebraic geometry and group representations, 1998

219 Marc Henneaux, Joseph Krasil'shchik, and Alexandre Vinogradov, Editors, Secondary calculus and cohomological physics, 1998

218 Jan Mandel, Charbel Farhat, and Xiao-Chuan Cai, Editors, Domain decomposition methods 10, 1998

217 Eric Carlen, Evans M. Harrell, and Michael Loss, Editors, Advances in differential equations and mathematical physics, 1998

216 Akram Aldroubi and EnBing Lin, Editors, Wavelets, multiwavelets, and their applications, 1998

215 M. G. Nerurkar, D. P. Dokken, and D. B. Ellis, Editors, Topological dynamics and applications, 1998

214 Lewis A. Coburn and Marc A. Rieffel, Editors, Perspectives on quantization, 1998

213 Farhad Jafari, Barbara D. MacCluer, Carl C. Cowen, and A. Duane Porter, Editors, Studies on composition operators, 1998

212 E. Ramírez de Arellano, N. Salinas, M. V. Shapiro, and N. L. Vasilevski, Editors, Operator theory for complex and hypercomplex analysis, 1998

211 Józef Dodziuk and Linda Keen, Editors, Lipa's legacy: Proceedings from the Bers Colloquium, 1997

210 V. Kumar Murty and Michel Waldschmidt, Editors, Number theory, 1998

209 Steven Cox and Irena Lasiecka, Editors, Optimization methods in partial differential equations, 1997

208 Michel L. Lapidus, Lawrence H. Harper, and Adolfo J. Rumbos, Editors, Harmonic analysis and nonlinear differential equations: A volume in honor of Victor L. Shapiro, 1997

(Continued in the back of this publication)

Advances in Stochastic Inequalities

CONTEMPORARY MATHEMATICS

234

Advances in Stochastic Inequalities

AMS Special Session on
Stochastic Inequalities and Their Applications
October 17–19, 1997
Georgia Institute of Technology

Theodore P. Hill
Christian Houdré
Editors

American Mathematical Society
Providence, Rhode Island

Editorial Board

Dennis DeTurck, managing editor

Andreas Blass   Andy R. Magid   Michael Vogelius

1991 *Mathematics Subject Classification*. Primary 60E15; Secondary 62F15.

Library of Congress Cataloging-in-Publication Data
AMS Special Session on Stochastic Inequalities and their Applications (1997 : Georgia Institute of Technology)
    Advances in stochastic inequalities : AMS Special Session on Stochastic Inequalities and their Applications, October 17–19, 1997, Georgia Institute of Technology / Theodore P. Hill, Christian Houdré, editors.
        p. cm. — (Contemporary mathematics, ISSN 0271-4132 ; 234)
    Includes bibliographical references.
    ISBN 0-8218-1086-3 (alk. paper)
    1. Stochastic inequalities Congresses. I. Hill, Theodore Preston, 1943–. II. Houdré, Christian. III. Title. IV. Series: Contemporary mathematics (American Mathematical Society) ; v. 234.
QA274.223.A68   1997
519.2—dc21   99-22875
                                                                                                                              CIP

    Copying and reprinting. Material in this book may be reproduced by any means for educational and scientific purposes without fee or permission with the exception of reproduction by services that collect fees for delivery of documents and provided that the customary acknowledgment of the source is given. This consent does not extend to other kinds of copying for general distribution, for advertising or promotional purposes, or for resale. Requests for permission for commercial use of material should be addressed to the Assistant to the Publisher, American Mathematical Society, P. O. Box 6248, Providence, Rhode Island 02940-6248. Requests can also be made by e-mail to reprint-permission@ams.org.
    Excluded from these provisions is material in articles for which the author holds copyright. In such cases, requests for permission to use or reprint should be addressed directly to the author(s). (Copyright ownership is indicated in the notice in the lower right-hand corner of the first page of each article.)

    © 1999 by the American Mathematical Society. All rights reserved.
        The American Mathematical Society retains all rights
        except those granted to the United States Government.
              Printed in the United States of America.

    ∞ The paper used in this book is acid-free and falls within the guidelines
        established to ensure permanence and durability.
        Visit the AMS home page at URL: http://www.ams.org/

            10 9 8 7 6 5 4 3 2 1   04 03 02 01 00 99

Contents

Preface	ix
Bounds on the non-convexity of ranges of vector measures with atoms PIETER C. ALLAART	1
The class of Gaussian chaos of order two is closed by taking limits in distribution MIGUEL A. ARCONES	13
Two inequalities and some applications in connection with ρ^*-mixing, a survey RICHARD C. BRADLEY	21
Variance inequalities for functions of multivariate random variables WAN-YING CHANG AND DONALD ST. P. RICHARDS	43
A note on sums of independent random variables PAWEŁ HITCZENKO AND STEPHEN MONTGOMERY-SMITH	69
Exponential integrability of diffusion processes YAOZHONG HU	75
Local dependencies in random fields via a Bonferroni-type inequality ADAM JAKUBOWSKI AND JAN ROSIŃSKI	85
Pricing-differentials and bounds for lookback options, and prophet problems in probability ROBERT P. KERTZ	97
A correlation inequality for stable random vectors ALEXANDER KOLDOBSKY	121
A note on the maximal inequalities for VC classes RAFAŁ LATAŁA	125
Comparison of moments via Poincaré-type inequality KRZYSZTOF OLESZKIEWICZ	135
Fractional sums and integrals of r-concave tails and applications to comparison probability inequalities IOSIF PINELIS	149

Product formula, tails and independence of multiple stable integrals
   J. ROSIŃSKI AND G. SAMORODNITSKY   169

A domination inequality for martingale polynomials
   JERZY SZULGA   195

A log-concavity proof for a Gaussian exponential bound
   RICHARD A. VITALE   209

Preface

This volume contains papers presented at the Special Session on Stochastic Inequalities and Their Applications at the meeting of the American Mathematical Society held October 17–19, 1997 at the Georgia Institute of Technology, Atlanta, Georgia. The goal of this Special Session was to bring together several dozen international experts in stochastic inequalities to exchange ideas and present state-of-the-art results and techniques in this area of basic research in mathematical probability and statistics.

We would like to thank the participants of the Session for their inspiring lectures, and in particular the contributors to this volume and the referees of these articles (each article was sent to at least two referees). Special thanks to Don Richards, who suggested the idea of publishing a proceedings of the Session, to Annette Rohrs for the superb technical assistance with the manuscripts, and Christine Thivierge for her encouragement and patience in supervising the production of this book.

The Editors
Theodore P. Hill, Christian Houdré
Atlanta, Georgia
February 1999

Bounds on the non-convexity of ranges of vector measures with atoms

Pieter C. Allaart

ABSTRACT. Upper bounds are given for the distance between the range, matrix range and partition range of a vector measure to the respective convex hulls of these ranges. The bounds are specified in terms of the maximum atom size, and generalize convexity results of Lyapounov (1940) and Dvoretzky, Wald and Wolfowitz (1951). Applications are given to the bisection problem, the "problem of the Nile", and fair division problems.

1. Introduction

Lyapounov's celebrated convexity theorem of 1940 (e.g. [**3, 10, 14, 15**]) asserts that the range of a finite-dimensional, atomless vector measure is convex and compact. A generalization of Lyapounov's theorem due to Dvoretzky, Wald and Wolfowitz [**6**] says that the same is true for the *matrix-k-range* and the *partition range* (see Definition 2.2 below).

If the vector measure has atoms, then convexity of all three ranges may fail in general, although atomlessness is not a necessary condition. Gouweleeuw [**9**] has given necessary and sufficient conditions for the range (or matrix-k-range) to be convex, as well as non-trivial sufficient conditions for the partition range to be convex.

A different approach was adopted by Elton and Hill [**7**], who proved a bound on how far from convex the range may be, as a function of the maximum atom size. The aim of this paper is to present such non-convexity inequalities for the three types of ranges mentioned above. Some of these are sharp, whereas in other cases the best possible bounds are not known to the author.

The first result is a slightly improved, but sharp, version of Elton and Hill's inequality. The proof presented here is very similar to that of Elton and Hill, with only a few minor adaptations. The original inequality is also included for the sake of comparison.

Next in line are two non-convexity inequalities for the matrix-k-range. These are proved using the improved inequality for the range, and a device of chaining

---

1991 *Mathematics Subject Classification*. Primary 28B05; Secondary 60A10.

Key words and phrases. Range of a vector measure, partition range, matrix range, vector atom, convexity theorems, Hausdorff distance.

© 1999 American Mathematical Society

together vector measures due to Blackwell. It is, however, the author's belief that these inequalities are not very sharp.

The last result is a sharp non-convexity bound for the partition range. Its proof (see [1]) is beyond the scope of this paper and is therefore omitted.

This paper is organized as follows. Section 2 lists the main results described above, accompanied by examples demonstrating their sharpness when applicable. Section 3 contains the necessary preparations for the proofs of the range and matrix-k-range inequalities, which then follow in Section 4. Section 5 gives applications of the main results to some well-known partitioning problems, including the bisection problem, the "problem of the Nile" and the problem of fair division. Section 6, finally, lists two open problems.

2. Non-convexity inequalities

Throughout this paper, $\mu, \mu_1, \ldots, \mu_n$ will always denote finite, non-negative, countably additive measures on a fixed measurable space $(\Omega, \mathcal{F})$. The *vector measure* $\vec{\mu} = (\mu_1, \ldots, \mu_n)$ is defined by

$$\vec{\mu}(A) := (\mu_1(A), \ldots, \mu_n(A)) \in \mathbb{R}^n, \quad A \in \mathcal{F}.$$

A set $E \in \mathcal{F}$ is called a (scalar) *atom* of μ if $\mu(E) > 0$ and for each $F \subset E, F \in \mathcal{F}$: $\mu(F) \in \{0, \mu(E)\}$. Similarly, E is a *vector atom* of $\vec{\mu} = (\mu_1, \ldots, \mu_n)$ if $\vec{\mu}(E) \neq \vec{0}$ and for each $F \subset E, F \in \mathcal{F}$: $\vec{\mu}(F) = \vec{\mu}(E)$ or $\vec{\mu}(F) = \vec{0}$. A (vector) measure is *atomless* if it does not have any atoms. A measure (resp. vector measure) is *purely atomic* if is assigns mass 0 (resp. $\vec{0}$) to the complement of the union of its atoms.

REMARK 2.1. From the definition of vector atom it can be seen that if E is a vector atom of $\vec{\mu}$, then

(i) E is a scalar atom of at least one μ_i;
(ii) for each $i \in \{1, \ldots, n\}$, either E is an atom of μ_i, or $\mu_i(E) = 0$.

Conversely, it follows from Lemma 2.4 (iii) in [9] that if E is a scalar atom of μ_i for some i, then E contains a vector atom F of $\vec{\mu}$ with $\vec{\mu}(F) = \vec{\mu}(E)$.

As a consequence, a vector measure is purely atomic if and only if all its component measures are.

A (measurable) *k-partition* is an ordered collection $(A_1, \ldots, A_k)$ of subsets of Ω such that $A_i \in \mathcal{F}$ ($i = 1, \ldots, k$), $A_i \cap A_j = \emptyset$ for all $i \neq j$, and $\bigcup_{i=1}^{k} A_i = \Omega$. Let Π_k denote the collection of all k-partitions of Ω.

In the following definition, $M_{n,k}(\mathbb{R})$ denotes the vector space of all $n \times k$ matrices with real entries.

DEFINITION 2.2. For a vector measure $\vec{\mu} = (\mu_1, \ldots, \mu_n)$,

(i) $\mathcal{R}(\vec{\mu}) := \{\vec{\mu}(A) : A \in \mathcal{F}\} \subset \mathbb{R}^n$ is the *range* of $\vec{\mu}$.
(ii) $\mathcal{MR}_k(\vec{\mu}) := \{(\mu_i(A_j))_{i=1,j=1}^{n,k} : (A_1, \ldots, A_k) \in \Pi_k\} \subset M_{n,k}(\mathbb{R})$ is the *matrix-k-range* of $\vec{\mu}$.
(iii) $\mathcal{PR}(\vec{\mu}) := \{(\mu_1(A_1), \ldots, \mu_n(A_n)) : (A_1, \ldots, A_n) \in \Pi_n\} \subset \mathbb{R}^n$ is the *partition range* of $\vec{\mu}$.

PROPOSITION 2.3. [Lyapounov (1940)]. $\mathcal{R}(\vec{\mu})$ *is compact, and if $\vec{\mu}$ is atomless, then $\mathcal{R}(\vec{\mu})$ is convex.*

PROPOSITION 2.4. [Dvoretzky, Wald and Wolfowitz (1951)]. *If $\vec{\mu}$ is atomless, then $\mathcal{MR}_k(\vec{\mu})$ is convex and compact.*

Proposition 2.4 was later improved by Dubins and Spanier [5], who proved that $\mathcal{MR}_k(\vec{\mu})$ is always compact.

A direct consequence of Proposition 2.4 is the following:

PROPOSITION 2.5. *If $\vec{\mu}$ is atomless, then $\mathcal{PR}(\vec{\mu})$ is convex and compact.*

The main goal of this paper is to generalize the above convexity results to measures with atoms, as was first done by Elton and Hill (1987). In order to do so, the following notation is needed. Recall that for a vector $x \in \mathbb{R}^n$, the p-norm $\|x\|_p$ of x is defined by

$$\|x\|_p := \begin{cases} \left(\sum_{i=1}^n |x_i|^p\right)^{1/p} & \text{if } 1 \leq p < \infty, \\ \max_{1 \leq i \leq n} |x_i| & \text{if } p = \infty. \end{cases}$$

Note that the norms $(\|.\|_p, p \in [1, \infty])$ are related via the sharp inequalities

(2.1) $\qquad \|x\|_q \leq \|x\|_p \quad \text{and} \quad n^{-1/p}\|x\|_p \leq n^{-1/q}\|x\|_q \quad \text{for } p \leq q < \infty,$

and

(2.2) $\qquad \|x\|_\infty \leq \|x\|_p \leq n^{1/p}\|x\|_\infty \quad \text{for } p < \infty.$

(See, for example, Theorems 16 and 19 in [11]).

By identifying $M_{n,k}(\mathbb{R})$ with $\mathbb{R}^{nk}$, the norm $\|.\|_p$ can be naturally extended to $M_{n,k}(\mathbb{R})$ as follows:

$$\|(a_{i,j})_{i=1,j=1}^{n,k}\|_p := \begin{cases} \left(\sum_{i=1}^n \sum_{j=1}^k |a_{i,j}|^p\right)^{1/p} & \text{if } 1 \leq p < \infty, \\ \max_{1 \leq i \leq n, 1 \leq j \leq k} |a_{i,j}| & \text{if } p = \infty. \end{cases}$$

If x and y are points in $\mathbb{R}^n$, then $d_p(x, y) = \|x - y\|_p$ denotes the distance between x and y. For a set S in $\mathbb{R}^n$ and a point x in $\mathbb{R}^n$, $d_p(x, S) = \inf_{y \in S} d_p(x, y)$ is the distance from x to S, and $D_p(S)$ denotes the Hausdorff distance from S to its convex hull $\text{co}(S)$:

$$D_p(S) := \sup_{x \in \text{co}(S)} d_p(x, S).$$

For $M_{n,k}(\mathbb{R})$, the distances d_p and D_p are defined similarly.

DEFINITION 2.6. For $\alpha \geq 0$ and $p \in [1, \infty]$, $\mathcal{P}_{n,p}(\alpha)$ is the collection of all n-dimensional vector measures $\vec{\mu}$ for which $\|\vec{\mu}(E)\|_p \leq \alpha$ for each atom E of $\vec{\mu}$.

The following theorem generalizes the convexity statement of Proposition 2.3. A proof is given in Section 4 below.

THEOREM 2.7. *Let $\vec{\mu}$ be a vector measure, and let $1 \leq p \leq 2$.*
(i) *If $\vec{\mu} \in \mathcal{P}_{n,\infty}(\alpha)$, then $D_2(\mathcal{R}(\vec{\mu})) \leq \alpha n/2$.*
(ii) *If $\vec{\mu} \in \mathcal{P}_{n,p}(\alpha)$, then $D_p(\mathcal{R}(\vec{\mu})) \leq \frac{1}{2}\alpha n^{1/p}$.*
The bound in (ii) is attained for all $p \in [1, 2]$. The bound in (i) is of the correct order of magnitude in n.

Theorem 2.7 (i) is the original generalization of Lyapounov's theorem by Elton and Hill. Note that (ii) implies (i), as follows easily by substituting $p = 2$ in (ii), and using (2.2). As a consequence, Elton and Hill's inequality holds under more general conditions, namely whenever $\|\vec{\mu}(E)\|_2 \leq \alpha\sqrt{n}$ for each vector atom E of $\vec{\mu}$. The following example shows that the bound in (ii) is attained for all $p \in [1,2]$.

EXAMPLE 2.8. Let $\mu_i = \alpha\delta_{\{i\}}$, $i = 1, \ldots, n$, where δ denotes Dirac measure. Then $\mathcal{R}(\vec{\mu}) = \{0, \alpha\}^n$ and hence $\text{co}(\mathcal{R}(\vec{\mu})) = [0, \alpha]^n$. In particular, $y = (\alpha/2, \ldots, \alpha/2) \in \text{co}(\mathcal{R}(\vec{\mu}))$, and for each $x \in \mathcal{R}(\vec{\mu})$, $\|x - y\|_p = \frac{1}{2}\alpha n^{1/p}$.

Elton and Hill give the following example to show that the bound in Theorem 2.7 (i) is of the correct order of magnitude in n.

EXAMPLE 2.9. Fix $n \in \mathbb{N}$, let $m = 2^k \leq n < 2^{k+1}$, and let $\{w_i\}_{i=1}^{m-1}$ be the $m - 1$ mean-zero *Walsh functions* on m points (see [18]). Then $w_i \in \{-1, 1\}^m$, $w_i \perp w_j$ for $i \neq j$, and $w_i \perp \vec{1}$ for each i, where $\vec{1} = (1, 1, \ldots, 1)$. For example, when $n = 4$ (so $k = 2$ and $m = 4$),

$$w_1 = (1, 1, -1, -1), \quad w_2 = (1, -1, 1, -1), \quad \text{and} \quad w_3 = (1, -1, -1, 1).$$

Let $\Omega = \{1, 2, \ldots, m - 1\}$, and define $\vec{\mu}(\{j\}) = (w_i + \vec{1})/2$, $j = 1, \ldots, m - 1$. Let $y = \vec{\mu}(\Omega)/2 = \vec{\mu}(\emptyset)/2 + \vec{\mu}(\Omega)/2 \in \text{co}(\mathcal{R}(\vec{\mu}))$. It can be shown (see [7]) that $d_2(x, y) \geq m/4$ for each $x \in \mathcal{R}(\vec{\mu})$.

Since $2m = 2^{k+1} > n$, it follows by rescaling that the best possible upper bound in Theorem 2.7 (i) is at least $\alpha n/8$ for general n, and at least $\alpha n/4$ if n is a power of 2.

The next example shows that the statement of Theorem 2.7 (ii) is false for $p > 2$ and large n. No non-trivial inequalities are known to the author for $p > 2$.

EXAMPLE 2.10. Let $m = 2^k \leq n < 2^{k+1}$, and let $\vec{\mu}$ be the same vector measure as in Example 2.9. Then $\|\vec{\mu}(\{j\})\|_p = \left(\frac{m}{2}\right)^{1/p}$ for each j, so $\vec{\mu} \in \mathcal{P}_{m,p}\left(\left(\frac{m}{2}\right)^{1/p}\right)$. From Example 2.9 it follows that $D_2(\mathcal{R}(\vec{\mu})) \geq m/4$, hence using (2.1) it follows that $D_p(\mathcal{R}(\vec{\mu})) \geq m^{1/p} m^{-1/2} m/4 = m^{1/p} m^{1/2}/4$. Since $1/p < 1/2$ it follows that $D_p(\mathcal{R}(\vec{\mu})) > \frac{1}{2}\left(\frac{m}{2}\right)^{1/p} m^{1/p}$ for sufficiently large m.

The following theorem gives upper bounds on the non-convexity of the matrix-k-range. Its proof is given in Section 4 below.

THEOREM 2.11. *Let $\vec{\mu}$ be a vector measure and let $k \in \mathbb{N}$. Garbled time*
(i) *If $\vec{\mu} \in \mathcal{P}_{n,\infty}(\alpha)$, then $D_2(\mathcal{MR}_k(\vec{\mu})) \leq \alpha n\sqrt{2k}$.*
(ii) *If $\vec{\mu} \in \mathcal{P}_{n,2}(\alpha)$, then $D_2(\mathcal{MR}_k(\vec{\mu})) \leq \alpha\sqrt{2nk}$.*

The next theorem gives a sharp non-convexity bound for the partition range. Its proof can be found in [1].

THEOREM 2.12. *If $\vec{\mu} \in \mathcal{P}_{n,\infty}(\alpha)$, then*

$$D_\infty(\mathcal{PR}(\vec{\mu})) \leq \frac{n-1}{n}\alpha,$$

and this bound is attained.

EXAMPLE 2.13. (sharpness of Theorem 2.12) Let $\mu_i = \alpha\delta_{\{0\}}, i = 1, \ldots, n$. Then $\mathcal{PR}(\vec{\mu}) = \{\alpha u_i : i = 1, \ldots, n\}$, where u_i denotes the i-th unit vector in $\mathbb{R}^n$ with 1 in the i-th position and zeroes elsewhere. It follows that $\text{co}(\mathcal{PR}(\vec{\mu})) = \{x \in \mathbb{R}^n_+ : \sum_{i=1}^n x_i = \alpha\}$. In particular, $y = (\alpha/n, \ldots, \alpha/n) \in \text{co}(\mathcal{PR}(\vec{\mu}))$, and for each $x \in \mathcal{PR}(\vec{\mu}), \|x - y\|_\infty = \alpha(n-1)/n$.

The following immediate consequence of Theorem 2.12 improves on an earlier result of Hill and Tong ([12], Theorem 3.2).

COROLLARY 2.14. *If $\vec{\mu} \in \mathcal{P}_{n,\infty}(\alpha)$, then*

$$D_2(\mathcal{PR}(\vec{\mu})) \leq \frac{n-1}{\sqrt{n}}\alpha.$$

EXAMPLE 2.15. The bound in Corollary 2.14 is of the correct order of magnitude in n: let $\mu_i = \alpha \delta_{\{i\}}, i = 1, \ldots, n$; then $\mathcal{PR}(\vec{\mu}) = \{0, \alpha\}^n$, hence $\text{co}(\mathcal{PR}(\vec{\mu})) = [0, \alpha]^n$. In particular, $y = (\alpha/2, \ldots, \alpha/2) \in \text{co}(\mathcal{PR}(\vec{\mu}))$, and for each $x \in \mathcal{PR}(\vec{\mu})$, $\|x - y\|_2 = \alpha\sqrt{n}/2$.

3. Preliminaries

The goal of this section and the next is to prove Theorems 2.7 and 2.11. For a proof of Theorem 2.12 the reader is referred to [1].

Most of the definitions and lemmas in this section are taken from Elton and Hill [7]. However, some of the statements are slightly more general than the corresponding statements in [7]. Most of the proofs are short, and are included here in order to make this paper more self-contained.

LEMMA 3.1. *For each $\vec{\mu}$, each $\varepsilon > 0$ and each $q \in [1, \infty]$, there exists a measurable partition $\{B_i\}_{i=1}^N$ of Ω satisfying*

(3.1) $$\forall B \in \mathcal{F}, \exists J \subset \{1, \ldots, N\} : \|\vec{\mu}(B) - \vec{\mu}(\bigcup_{j \in J} B_j)\|_q < \varepsilon.$$

PROOF. Since $\mathcal{R}(\vec{\mu})$ is bounded, there is an ε-net $\{x^{(1)}, \ldots, x^{(m)}\}$ of $\mathcal{R}(\vec{\mu})$; that is $\{x^{(1)}, \ldots, x^{(m)}\} \subset \mathcal{R}(\vec{\mu})$, and for each $x \in \mathcal{R}(\vec{\mu})$ there is an $i \leq m$ such that $\|x - x^{(i)}\|_q < \varepsilon$. Let $\{A_i\}_{i=1}^m$ satisfy $\vec{\mu}(A_i) = x_i$, $i = 1, \ldots, m$, and let $\{B_i\}_{i=1}^N \subset \mathcal{F}$ be a measurable partition of Ω such that $\sigma(B_1, \ldots, B_N) = \sigma(A_1, \ldots, A_m)$. (Such a partition exists because $\sigma(A_1, \ldots, A_m)$ is finite.) It is easily seen that $\{B_i\}_{i=1}^N$ satisfies (3.1). □

The next lemma is stated and proved in [7] for $p = \infty$ only; the more general statement below requires a different proof. It will be used in the next section for $p = 2$.

LEMMA 3.2. *For each $p \in [1, \infty)$, each $\vec{\mu} \in \mathcal{P}_{n,p}(\alpha)$ and each $B \in \mathcal{F}$ there exists a measurable partition $\{B_i\}_{i=1}^k$ of B such that $\|\vec{\mu}(B_i)\|_p \leq \alpha$ for all $i \leq k$.*

PROOF. Let $B \in \mathcal{F}$. By Rényi [17], p.83, each μ_i has at most countably many atoms, hence $\vec{\mu}$ has at most countably many vector atoms. Let A be the union of all the vector atoms of $\vec{\mu}$. Then $A \in \mathcal{F}$. Since μ_1 is atomless on $B \backslash A$, there is a measurable partition $(C_j)_{j=1}^l$ of $B \backslash A$ such that $\mu_1(C_j) \leq \alpha n^{-1/p}$ for all $j \leq l$ (where $1/\infty = 0$). Repeating this argument for μ_2 and each C_j, then for μ_3, etc., yields a partition $(D_j)_{j=1}^L$ of $B \backslash A$ such that $\mu_i(D_j) \leq \alpha n^{-1/p}$ for all $i \leq n$ and $j \leq L$, which implies $\|\vec{\mu}(D_j)\|_p \leq \alpha$ for all $j \leq L$.

The argument for $B \cap A$ is slightly different. If the number of vector atoms of $\vec{\mu}$ is finite, then there is nothing left to prove. Otherwise, let the atoms of $\vec{\mu}$ be

$E_1, E_2, \ldots$ Since

$$\sum_{j=1}^{\infty} \|\vec{\mu}(E_j)\|_p \leq \sum_{j=1}^{\infty} \|\vec{\mu}(E_j)\|_1 = \sum_{j=1}^{\infty} \sum_{i=1}^{n} \mu_i(E_j) = \sum_{i=1}^{n} \sum_{j=1}^{\infty} \mu_i(E_j) = \sum_{i=1}^{n} \mu_i(A) < \infty,$$

there is $j_0 \in \mathbb{N}$ such that

$$\left\| \vec{\mu} \left(\bigcup_{j=j_0+1}^{\infty} E_j \right) \right\|_p \leq \sum_{j=j_0+1}^{\infty} \|\vec{\mu}(E_j)\|_p \leq \alpha.$$

Taking intersections of the sets $E_1, \ldots, E_{j_0}$ and $\bigcup_{j=j_0+1}^{\infty} E_j$ with B completes the proof. □

LEMMA 3.3. *For all $p, q \in [1, \infty], \varepsilon > 0$ and $\vec{\mu} \in \mathcal{P}_{n,p}(\alpha)$, there is a purely atomic vector measure $\vec{\mu}_0 \in \mathcal{P}_{n,p}(\alpha)$ with finitely many atoms, such that*

$$D_q(\mathcal{R}(\vec{\mu})) \leq D_q(\mathcal{R}(\vec{\mu}_0)) + \varepsilon.$$

The idea of the proof of Lemma 3.3 is that Lemma 3.1 and a repeated application of Lemma 3.2 yield a partition $\{B_i\}_{i=1}^{N}$ of Ω satisfying both (3.1) and $\|\vec{\mu}(B_i)\|_p \leq \alpha$ for all $i \leq N$. The restriction $\vec{\mu}_0$ of $\vec{\mu}$ to $\sigma(B_1, \ldots, B_N)$ then has the desired property. (See [7], §3 for the details).

Lemma 3.3 says that it is in fact sufficient to prove Theorem 2.7 for purely atomic measures with a finite number of atoms. Since the range of such a vector measure is a finite set, this reduction turns the problem into one of finite geometry.

For the remainder of this section, V is a finite set of (not necessarily distinct) points in $\mathbb{R}_+^n = \{(r_1, \ldots, r_n) : r_i \in \mathbb{R}, r_i \geq 0 \text{ for all } i \leq n\}$, and $|V|$ denotes the cardinality of V.

DEFINITION 3.4.

$$\Sigma(V) = \left\{ \sum_{x_i \in V} \delta_i x_i : \delta_i \in \{0, 1\} \right\}; \quad C(V) = \left\{ \sum_{x_i \in V} t_i x_i : t_i \in [0, 1] \right\}.$$

LEMMA 3.5. $\mathrm{co}(\Sigma(V)) = C(V)$.

The next lemma states that $C(V)$ can be expressed as the union of translates of subsets of the form $C(\hat{V})$ where $|\hat{V}| \leq n$. Let the vector sum $V_1 \oplus V_2$ of two sets V_1 and V_2 be defined by $V_1 \oplus V_2 = \{v_1 + v_2 : v_1 \in V_1, v_2 \in V_2\}$.

LEMMA 3.6. $C(V) = \bigcup \{\Sigma(V \backslash \hat{V}) \oplus C(\hat{V}) : \hat{V} \subset V, |\hat{V}| \leq n\}$.

A direct algebraic proof of Lemma 3.6 can be found in [7]. Here a different proof is given, based on the Shapley-Folkman lemma from convex geometry.

LEMMA 3.7. (Shapley and Folkman - see [2], §5) *Let $V_1, \ldots, V_k$ be nonempty subsets of $\mathbb{R}^n$. Then for each $y \in \mathrm{co}(V_1) \oplus \cdots \oplus \mathrm{co}(V_k)$ there exists a representation $y = x_1 + \cdots + x_k$, with $x_i \in \mathrm{co}(V_i)$ for all i, but $x_i \notin V_i$ for at most n indices i.*

LEMMA 3.8. $\mathrm{co}(\oplus_{i=1}^{k} V_i) = \oplus_{i=1}^{k} \mathrm{co}(V_i)$ *for all $V_1, \ldots, V_k$.*

PROOF. Straightforward. □

PROOF OF LEMMA 3.6. Clearly $C(V) \supset \bigcup\{\Sigma(V\backslash\hat{V}) \oplus C(\hat{V}) : \hat{V} \subset V, |\hat{V}| \le n\}$. Conversely, let $x_1, \ldots, x_k$ denote the elements of V (counting multiplicities), and let $V_i := \{0, x_i\}$. Then $\Sigma(V) = \oplus_{i=1}^k V_i$, and (using Lemma 3.8) $C(V) = \text{co}(\Sigma(V)) = \oplus_{i=1}^k \text{co}(V_i)$. Now fix $y \in C(V)$; then by Lemma 3.7 there is a representation $y = \sum_{i=1}^k y_i$, where $y_i \in \text{co}(V_i)$ for all i, and $y_i \notin V_i$ for at most n indices i. Let $I := \{i : y_i \notin V_i\}$, and define $\hat{V} := \{x_i\}_{i \in I}$. It is easily checked that $y \in \Sigma(V\backslash\hat{V}) \oplus C(\hat{V})$. □

The next lemma is critical for the proof of Theorem 2.7. Its proof is taken from [7].

LEMMA 3.9. *For all $x, y \in \mathbb{R}^n$ and all $t \in [0, 1]$,*

$$\min\{\|x + (1-t)y\|_2^2, \|x - ty\|_2^2\} \le \|x\|_2^2 + \|y\|_2^2/4.$$

PROOF. Let $\langle x, y \rangle = \sum_{i=1}^n x_i y_i$ denote the standard inner product on $\mathbb{R}^n$. First consider the case where $2\langle x, y \rangle \ge (t^2 - (1-t)^2)\|y\|_2^2$. Then

$$\begin{aligned}\|x - ty\|_2^2 &= \|x\|_2^2 + t^2\|y\|_2^2 - 2t\langle x, y \rangle \\ &\le \|x\|_2^2 + t^2\|y\|_2^2 - t(t^2 - (1-t)^2)\|y\|_2^2 \\ &= \|x\|_2^2 + t(1-t)\|y\|_2^2 \le \|x\|_2^2 + \|y\|_2^2/4.\end{aligned}$$

The case where $2\langle x, y \rangle \le (t^2 - (1-t)^2)\|y\|_2^2$ is similar, yielding $\|x + (1-t)y\|_2^2 \le \|x\|_2^2 + \|y\|_2^2/4$. □

LEMMA 3.10. *Let $|V| = n$. If $\|y\|_2 \le 1$ for all $y \in V$, then $D_2(\Sigma(V)) \le \sqrt{n}/2$.*

PROOF. Let $V = \{x_1, \ldots, x_n\}$ and fix $x = \sum_{i=1}^n t_i x_i \in C(V)$. Applying Lemma 3.9 n times implies the existence of $(\delta_i)_{i=1}^n \in \{0, 1\}^n$ satisfying

$$\left\|\sum_{i=1}^n \delta_i x_i - x\right\|_2^2 = \left\|\sum_{i=1}^n (\delta_i - t_i) x_i\right\|_2^2 \le n \cdot \max\{\|y\|_2^2 : y \in V\}/4 \le n/4.$$

□

4. Proofs of the range and matrix range inequalities

PROOF OF THEOREM 2.7. Since (ii) implies (i), it is enough to prove (ii). Note first that if (ii) holds for $p = 2$, then it holds for each $p \in [1, 2]$. For, suppose that $\vec{\mu} \in \mathcal{P}_{n,p}(\alpha)$; then also $\vec{\mu} \in \mathcal{P}_{n,2}(\alpha)$ in view of (2.1), so $D_2(\mathcal{R}(\vec{\mu})) \le \frac{1}{2}\alpha n^{1/2}$, and a second application of (2.1) yields $D_p(\mathcal{R}(\vec{\mu})) \le \frac{1}{2}\alpha n^{1/p}$.

It is therefore sufficient to prove (ii) for $p = 2$. Let $\varepsilon > 0$ and $\vec{\mu} \in \mathcal{P}_{n,2}(\alpha)$. By Lemma 3.3 there exists a purely atomic vector measure $\vec{\mu}_0 \in \mathcal{P}_{n,2}(\alpha)$ with finitely many atoms such that

(4.1) $$D_2(\mathcal{R}(\vec{\mu})) \le D_2(\mathcal{R}(\vec{\mu}_0)) + \varepsilon.$$

Now $\mathcal{R}(\vec{\mu}_0) = \Sigma(V)$, where $V = \{\vec{\mu}_0(E) : E \text{ is a vector atom of } \vec{\mu}_0\}$, so applying Lemma 3.6, Lemma 3.10 and rescaling yields

$$D_2(\mathcal{R}(\vec{\mu}_0)) \le \alpha\sqrt{n}/2,$$

which together with (4.1) completes the proof of (ii), since ε was arbitrary. □

PROOF OF THEOREM 2.11. Since (ii) implies (i), it suffices to prove (ii). Fix $(A_1, \ldots, A_k)$ and $(B_1, \ldots, B_k)$ in Π_k and $t \in [0,1]$. Following Dubins and Spanier [5], define the $2nk$-dimensional vector measure $\vec{m}$ by

$$\vec{m}(S) = (\mu_i(S \cap A_j), \mu_i(S \cap B_j))_{i=1, j=1}^{n,k}.$$

Note that $\vec{m}$ and $\vec{\mu}$ have the same vector atoms. If E is a vector atom of $\vec{m}$, then, since distinct vector atoms are essentially disjoint, it can be assumed that $E \subset A_{j_1}$ and $E \subset B_{j_2}$ for some j_1 and j_2. Hence $E \cap A_j = \emptyset$ for all $j \neq j_1$ and $E \cap B_j = \emptyset$ for all $j \neq j_2$, so

$$\|\vec{m}(E)\|_2^2 = 2\|\vec{\mu}(E)\|_2^2 \leq 2\alpha^2.$$

Applying Theorem 2.7 (ii) to $\vec{m}$ now yields $D_2(\mathcal{R}(\vec{m})) \leq \alpha\sqrt{2} \cdot \sqrt{2nk}/2 = \alpha\sqrt{nk}$, so there exists a set $S \in \mathcal{F}$ with $\|\vec{m}(S) - t\vec{m}(\Omega)\|_2 \leq \alpha\sqrt{nk}$, that is,

$$(4.2) \quad \sum_{i=1}^{n}\sum_{j=1}^{k}(\mu_i(S \cap A_j) - t\mu_i(A_j))^2 + \sum_{i=1}^{n}\sum_{j=1}^{k}(\mu_i(S \cap B_j) - t\mu_i(B_j))^2 \leq \alpha^2 nk.$$

Since $|\mu_i(S \cap B_j) - t\mu_i(B_j)| = |\mu_i(B_j \setminus S) - (1-t)\mu_i(B_j)|$, it follows from (4.2) that

$$(4.3) \quad \sum_{i=1}^{n}\sum_{j=1}^{k}(\mu_i(S \cap A_j) - t\mu_i(A_j))^2 + \sum_{i=1}^{n}\sum_{j=1}^{k}(\mu_i(B_j \setminus S) - (1-t)\mu_i(B_j))^2 \leq \alpha^2 nk.$$

Letting $C_j = (A_j \cap S) \cup (B_j \setminus S)$ it follows from (4.3) that

$$(4.4) \quad \sum_{i=1}^{n}\sum_{j=1}^{k}(\mu_i(C_j) - t\mu_i(A_j) - (1-t)\mu_i(B_j))^2$$

$$= \sum_{i=1}^{n}\sum_{j=1}^{k}(\mu_i(A_j \cap S) + \mu_i(B_j \setminus S) - t\mu_i(A_j) - (1-t)\mu_i(B_j))^2$$

$$= \|M + N - P - Q\|_2^2 \leq 2(\|M - P\|_2^2 + \|N - Q\|_2^2) \leq 2\alpha^2 nk,$$

where the matrices M, N, P and Q are defined in the obvious manner, i.e. $M = (\mu_i(A_j \cap S))_{i=1,j=1}^{n,k}$, etc. Taking square roots on both sides of (4.4) completes the proof of (ii). $\square$

REMARK 4.1. The device of chaining together vector measures was introduced by Blackwell [4]. It was used by Dubins and Spanier [5] to derive convexity of the matrix range from Lyapounov's theorem in the atomless case, and by Hill and Tong [12] to obtain a non-convexity bound for the partition range from Theorem 2.7 (i). However, the inequality presented in Theorem 2.12 is much stronger than Hill and Tong's result. It is therefore the author's belief that sharper inequalities than those presented in Theorem 2.11 can be found.

5. Applications

Lyapounov's theorem has been applied in a number of areas including optimal stopping theory, control theory and statistical decision theory. In principle, any application of Propositions 2.3, 2.4 or 2.5 can be generalized to measures with atoms using the corresponding generalization from Section 2. The aim of this section is to illustrate this by a few examples, including the bisection problem, the "problem of the Nile" and the problem of fair division.

1. The objective in the *bisection problem* is to find a set $A \in \mathcal{F}$ such that $\mu_i(A) = \frac{1}{2}\mu_i(\Omega)$ for all i. If $\mu_1, \ldots, \mu_n$ are atomless, then the existence of such a set is a direct consequence of Proposition 2.3. The following theorem generalizes this result to measures with atoms:

THEOREM 5.1. *If $\|\vec{\mu}(E)\|_2 \leq \alpha$ for every vector atom E of $\vec{\mu}$, then there exists a set $A \in \mathcal{F}$ satisfying*
$$|\mu_i(A) - \frac{1}{2}\mu_i(\Omega)| \leq \alpha\sqrt{n}/2 \text{ for all } i \leq n.$$

PROOF. Since $(0, \ldots, 0) = \vec{\mu}(\emptyset) \in \mathcal{R}(\vec{\mu})$ and $(\mu_1(\Omega), \ldots, \mu_n(\Omega)) = \vec{\mu}(\Omega) \in \mathcal{R}(\vec{\mu})$, it follows that $\text{co}(\mathcal{R}(\vec{\mu}))$ contains the vector $(\frac{1}{2}\mu_1(\Omega), \ldots, \frac{1}{2}\mu_n(\Omega))$. Apply Theorem 2.7 (ii). □

An approximate bisection result based on Theorem 2.7 (i) can, of course, be stated and proved analogously.

For an application of Theorem 2.7 to the bang-bang principle of control theory, see Elton and Hill [7].

2. Fisher's *"problem of the Nile"* (see Dubins and Spanier [5]). "Each year the Nile would flood, thereby irrigating or perhaps devastating parts of the agricultural land of a predynastic Egyptian village. The value of different portions of the land would depend upon the height of the flood. In question was the possibility of giving to each of the k residents a piece of land whose value would be $1/k$ of the total land value no matter what the height of the flood."

Neyman [16] proved that the problem has a solution under the assumption that there are only a finite number, say n, of possible flood heights. The following theorem generalizes Neyman's result.

THEOREM 5.2. *If $\mu_1, \ldots, \mu_n$ are probability measures and $\|\vec{\mu}(E)\|_2 \leq \alpha$ for every vector atom E of $\vec{\mu} = (\mu_1, \ldots, \mu_n)$, then there exists a k-partition $(A_1, \ldots, A_k)$ of Ω such that*
$$|\mu_i(A_j) - \frac{1}{k}| \leq \alpha\sqrt{2nk} \text{ for all } i \leq n \text{ and } j \leq k.$$

PROOF. For $r = 1, \ldots, k$, let $(A_j^{(r)})_{j=1}^k$ be the partition with $A_r^{(r)} = \Omega$ and $A_j^{(r)} = \emptyset$ if $j \neq r$. Then
$$M_r := (\mu_i(A_j^{(r)}))_{i=1,j=1}^{n,k} \in \mathcal{MR}_k(\vec{\mu})$$
for $r = 1, \ldots, k$. Note that M_r is the matrix with only 1's in the r-th column and 0's elsewhere. It follows that $\frac{1}{k}M_1 + \ldots \frac{1}{k}M_k \in \text{co}(\mathcal{MR}_k(\vec{\mu}))$. Applying Theorem 2.11 (ii) gives the desired result. □

3. In the classical *fair division problem* the objective is to find a partition $(A_1, \ldots, A_n)$ of Ω such that

(5.1) $$\mu_i(A_i) \geq \frac{1}{n} \text{ for all } i \leq n,$$

where $\mu_1, \ldots, \mu_n$ are probability measures. If $\mu_1, \ldots, \mu_n$ are atomless, then Proposition 2.5 guarantees the existence of such a partition, even with equality in (5.1). If, in addition, $\mu_i \neq \mu_j$ for some $i \neq j$, then there is a partition for which (5.1) holds with *strict* inequality (see for example Dubins and Spanier [5]).

If the measures have atoms then in general a partition satisfying (5.1) need not exist. However, if the measures are not all equal and the atoms are sufficiently small, then the differentiation between the measures may compensate for the non-divisibility of the atoms, making fair division still possible. To make this more precise, let

$$M := \sup\{\sum_{i=1}^{n} \mu_i(B_i) \mid (B_i)_{i=1}^{n} \text{ is a partition of } \Omega\}.$$

The following theorem generalizes a result of Elton, Hill and Kertz [8].

THEOREM 5.3. *If $\|\vec{\mu}(E)\|_\infty \leq \alpha$ for every vector atom E of $\vec{\mu}$, then there exists a partition $(A_1, \ldots, A_n)$ of Ω such that*

$$\mu_i(A_i) \geq (n - M + 1)^{-1} - \frac{n-1}{n}\alpha, \ i = 1, \ldots, n.$$

PROOF. As in Legut [13], applying Theorem 2.12 at the place where [13] uses convexity of the partition range. □

COROLLARY 5.4. *If $\|\vec{\mu}(E)\|_\infty \leq (M-1)(n-1)^{-1}(n-M+1)^{-1}$ for every vector atom E of $\vec{\mu}$, then there is a partition $(A_1, \ldots, A_n)$ of Ω satisfying* (5.1).

EXAMPLE 5.5. Let $n = 3$ and suppose that $M = 2$ (note that 2 is a realistic value in this case, since M can assume any value between 1 and 3). If $\|\vec{\mu}(E)\|_\infty \leq \frac{1}{4}$ for every vector atom E of $\vec{\mu}$, then Corollary 5.4 implies the existence of a fair division in the sense of (5.1).

6. Open problems

PROBLEM 1. Find a non-convexity inequality analogous to Theorem 2.7 (ii) for $p = \infty$; in other words, find the best possible (or at least a good) constant $K(n, \alpha)$ such that if $\vec{\mu} \in \mathcal{P}_{n,\infty}(\alpha)$, then $D_\infty(\mathcal{R}(\vec{\mu})) \leq K(n, \alpha)$. An example of the significance of such a sharp bound is that it would yield the best possible upper bound in Theorem 5.1. Note that by Example 2.9, the order of magnitude of $K(n, \alpha)$ must be at least $\sqrt{n}$.

PROBLEM 2. Find a sharp non-convexity inequality for the matrix-k-range. The inequalities given in Theorem 2.11 are probably far from sharp, as was already pointed out in Remark 4.1. In fact, no examples are known to the author of vector measures $\vec{\mu}$ for which $D_2(\mathcal{MR}_k(\vec{\mu}))$ is unbounded in k. Is there an upper bound for the non-convexity of $\mathcal{MR}_k(\vec{\mu})$ that does not depend on k? What about $D_\infty(\mathcal{MR}_k(\vec{\mu}))$?

ACKNOWLEDGEMENTS. The author is grateful to Professors Ted Hill and Christian Houdré for the invitation to present part of this paper in the Special Session on Stochastic Inequalities at the AMS conference in Atlanta, October 1997; and to an anonymous referee for pointing out an error in the original manuscript.

References

1. Allaart, P.C. (1998). A sharp non-convexity bound for partition ranges of vector measures with atoms. *To appear*.
2. Artstein, Z. (1980). Discrete and continuous bang-bang and facial spaces or: look for the extreme points. *SIAM J. Review* **22** No. 2, 172-185.
3. Artstein, Z. (1990). Yet another proof of the Lyapounov convexity theorem. *Proc. Amer. Math. Soc.* **108**, 89-91.
4. Blackwell, D. (1951). On a theorem of Lyapounov. *Ann. Math. Statist.* **22** 112-114.
5. Dubins, L.E. and Spanier, E.H. (1961). How to cut a cake fairly. *Amer. Math. Monthly* **68**, 1-17.
6. Dvoretzky, A., Wald, A., and Wolfowitz, J. (1951). Relations among certain ranges of vector measures. *Pacific J. Math.* **1**, 59-74.
7. Elton, J. and Hill, T.P. (1987). A generalization of Lyapounov's convexity theorem to measures with atoms. *Proc. Amer. Math. Soc.* **296**, 297-304.
8. Elton, J., Hill, T.P. and Kertz, R.P. (1986). Optimal-partitioning inequalities for nonatomic probability measures. *Trans. Amer. Math. Soc.* **296**, 703-725.
9. Gouweleeuw, J.M. (1995). A characterization of measures with convex range. *Proc. London Math. Soc.* (3) **70**, 336-362.
10. Halmos, P.R. (1948). The range of a vector measure. *Bull. Amer. Math. Soc.* **54**, 416-421.
11. Hardy, G. H., Littlewood, J. E. and Pólya, G. (1934). *Inequalities*. Cambridge University Press.
12. Hill, T. P. and Tong, Y. L. (1989). Optimal-partitioning inequalities in classification and multi-hypotheses testing. *Ann. Stat.* **17**, 1325-1334.
13. Legut, J. (1988). Inequalities for α-optimal partitioning of a measurable space. *Proc. Amer. Math. Soc.* **104**, 1249-1251.
14. Lindenstrauss, J. (1966). A short proof of Liapounoff's convexity theorem. *J. Math. Mech.* **15**, 971-972.
15. Lyapounov, A. (1940). Sur les fonctions-vecteurs complètement additives. *Bull. Acad. Sci. URSS* **4**, 465-478.
16. Neyman, J. (1946). Un théorème d'existence. *C.R. Acad. Sci. Paris* **222**, 843-845.
17. Rényi, A. (1970). *Probability theory*, North Holland Publishing Company, Amsterdam-London.
18. Walsh, J. L. (1923). A closed set of normal orthogonal functions. *Amer. J. Math.* **55**, 5-24.

MATHEMATICS DEPARTMENT, UNIVERSITY OF NORTH TEXAS, DENTON, TX 76203-5118
E-mail address: allaart@unt.edu

The class of Gaussian chaos of order two is closed by taking limits in distribution

Miguel A. Arcones

ABSTRACT. It is shown that the limit in distribution of a Gaussian chaos of order two is also a Gaussian chaos of order two. The proof is based on inequalities for the tails of such distributions.

1. Introduction

Let $\{g_j\}_{j=1}^\infty$ be a sequence of i.i.d.r.v.'s with standard normal distribution. A Gaussian chaos of order two is a random variable of the form

(1.1) $$X = c + \sum_{j=1}^\infty (b_j g_j + a_j(g_j^2 - 1)),$$

where $\sum_{j=1}^\infty (b_j^2 + a_j^2) < \infty$. By diagonalization, a r.v. of the form

(1.2) $$X = c + \sum_{j=1}^\infty b_j g_j + \sum_{j,k=1}^\infty a_{j,k} g_j g_k,$$

can be expressed as in (1.1) (with respect to a new sequence $\{g_j\}_{j=1}^\infty$). In other words a Gaussian chaos of order two is a polynomial of degree 2 in a sequence of i.i.d. normal r.v.'s. The random variables in (1.1) appear very often. Bivariate multiple integrals with respect to a Gaussian random measure can be expressed as in (1.1). For more in these distributions see [**KW**] and the collection of papers in [**HP**].

Suppose that $\{X_n\}_{n=1}^\infty$ is a sequence of r.v.'s of the form in (1.1) and $\{X_n\}_{n=1}^\infty$ converges in distribution to a r.v. X. We show that X is also a chaos of order two, i.e. we prove the following:

THEOREM 1.1. *Let $\{g_j\}_{j=0}^\infty$ be a sequence of i.i.d.r.v.'s with standard normal distribution. Let $X_n = c_n + \sum_{j=1}^\infty (b_{n,j} g_j + a_{n,j}(g_j^2 - 1))$, where $\sum_{j=1}^\infty (b_{n,j}^2 + a_{n,j}^2) < \infty$. Suppose that $\{X_n\}_{n=1}^\infty$ r.v. X. Then,*
  (a) For each positive integer k, $\lim_{n\to\infty} E[X_n^k] = E[X^k]$. In particular, c_n converges.

1991 *Mathematics Subject Classification.* Primary 62F05.
Key words and phrases. Gaussian chaos, quadratic forms, isoperimetric inequalities.

© 1999 American Mathematical Society

(b) *For each n, it is possible to reorder the sequence $\{a_{n,j}\}_{j\geq 1}$ so that* $\lim_{n\to\infty} a_{n,j} = a_j$, *for some a_j, and $|a_1| \geq |a_2| \geq \cdots$. If $a_j \neq 0$, then $\lim_{n\to\infty} |b_{n,j}| = b_j$, for some b_j.*

(c) *X has the distribution of $E[X] + sg_0 + \sum_{j=1}^{\infty}(b_j g_j + a_j(g_j^2 - 1))$, where $b_j = 0$, if $a_j = 0$, and $s^2 = \mathrm{Var}(X) - \sum_{j=1}^{\infty}(b_j^2 + 2a_j^2)$.*

In the previous theorem, it does not make sense to ask for the convergence of $b_{n,j}$. We have to ask for the convergence of $|b_{n,j}|$. Observe that $\sum_{j=1}^{\infty}(b_{n,j}g_j + a_{n,j}(g_j^2 - 1))$ has the same distribution as $\sum_{j=1}^{\infty}(\tau_j b_{n,j} g_j + a_{n,j}(g_j^2 - 1))$, where τ_j is either $+1$ or -1. If $a_j = 0$, we may have that $b_{n,j}$ does not converge. For example, if $\sum_{j=1}^{\infty} b_{n,j}^2$ converges, then $\sum_{j=1}^{\infty} b_{n,j} g_j$ converges to a normal distribution, but $\{b_{n,j}\}_{n=1}^{\infty}$ does not necessarily converge.

The previous theorem is not true for homogeneous chaos, if $X_n = \sum_{j=1}^{\infty} a_{n,j}(g_j^2 - 1)$ converges in distribution the limit does not have to be of that form. For example, by he central limit theorem, $n^{-1/2} \sum_{j=1}^{n}(g_j^2 - 1)$ converges in distribution to a normal limit.

In Section 2, the proof of Theorem 1.1 is given. The proof is based in inequalities on the tails of a Gaussian chaos of order two.

The basic idea in the proof of Theorem 1.1 is to notice that if X_n is as in (1.1), then

$$\lim_{t\to\infty} t^{-1}\log(\Pr\{|X_n| \geq t\}) = -(2\sup_{j\geq 1}|a_{n,j}|)^{-1}.$$

This forces $\sup_{j\geq 1}|a_{n,j}|$ to converge.

The motivation for this work was to understand some limit distributions in [**A**].

2. Proofs

We will need some estimates in the tails of Gaussian chaos. They are similar to estimates in [**LT**].

LEMMA 2.1. *Let $M = (2\sum_{j=2}^{\infty}(b_j^2 + 2a_{n,j}^2))^{1/2}$ and let $t > 0$, then*

$$2^{-1}\Pr\{|b_1 g_1 + a_1(g_1^2 - 1)| \geq t + M\} \leq \Pr(|\sum_{j=1}^{\infty}(b_j g_j + a_j(g_j^2 - 1))| \geq t)$$

and

$$2^{-1}\Pr(b_1 g_1 + a_1(g_1^2 - 1) \geq t + M) \leq \Pr(\sum_{j=1}^{\infty}(b_j g_j + a_j(g_j^2 - 1)) \geq t).$$

PROOF. By Chebyshev's inequality

$$\Pr\{|\sum_{j=2}^{\infty}(b_j g_j + a_j(g_j^2 - 1))| \geq M\} \leq 1/2.$$

From this and Fubini's theorem

$$2^{-1}\Pr\left\{|b_1 g_1 + a_1(g_1^2 - 1)| \geq t + M\right\}$$

$$\leq \Pr\left\{|b_1 g_1 + a_1(g_1^2 - 1)| \geq t + M, |\sum_{j=2}^{\infty}(b_j g_j + a_j(g_j^2 - 1))| < M\right\}$$

$$\leq \Pr\left\{ \left| \sum_{j=1}^{\infty}(b_j g_j + a_j(g_j^2 - 1)) \right| \geq t \right\}.$$

The second inequality follows similarly. □

We will use an exponential inequality for chaos processes. For proving this inequality we will use a version of the isoperimetric inequality for the canonical Gaussian measure in $\mathbb{R}^{\mathbb{N}}$, where $\mathbb{N} = \{1, 2, ...\}$. The canonical Gaussian measure γ in $\mathbb{R}^{\mathbb{N}}$ is the measure determined by $\{g_j\}_{j\geq 1}^{\infty}$. By the isoperimetric inequality (see for example Theorem 1.2 in [**LT**]), for any set A with $\gamma(A) \geq 1/2$,

(2.1) $$\gamma\{x \in \mathbb{R}^{\mathbb{N}} : d(x, A) \geq t\} \leq P\{g_1 \geq t\},$$

where $d(x, A)$ is the Euclidean distance in $\mathbb{R}^N$ between x and A. Next inequality is a variation of Lemma 3.8 in [**LT**].

LEMMA 2.2. *Let $M = 2(\sum_{j=1}^{\infty}(b_j^2 + 2a_j^2))^{1/2}$, let $\sigma = \sup_{j \geq 1}|a_j|$ and let $t > 0$. Then*

$$\Pr\{|\sum_{j=1}^{\infty}(b_j g_j + a_j(g_j^2 - 1))| \geq M + tM + t^2\sigma\} \leq P(g_1 \geq t\}.$$

PROOF. Let

$$A_1 = \{x \in \mathbb{R}^{\mathbb{N}} : |\sum_{j=1}^{\infty}(b_j x_j + a_j(x_j^2 - 1))| \leq M\}$$

and let

$$A_2 = \{x \in \mathbb{R}^{\mathbb{N}} : (\sum_{j=1}^{\infty} a_j^2 x_j^2)^{1/2} \leq 2(\sum_{j=1}^{\infty} a_j^2)^{1/2}\}.$$

Then, $P(A_i) \geq 3/4$ for $i = 1, 2$. So, $P(A) \geq 1/2$, where $A = A_1 \cap A_2$. By (2.1),

$$\gamma\{x \in \mathbb{R}^{\mathbb{N}} : d(x, A) \geq t\} \leq P(g_1 \geq t),$$

where $d(x, y)$ is the Euclidean distance in $\mathbb{R}^N$. Now if $d(x, A) \leq t$, then $x = y + th$, where $y \in A$ and $\sum_{j=1}^{\infty} h_j^2 \leq 1$. Then,

$$|\sum_{j=1}^{\infty}(b_j x_j + a_j(x_j^2 - 1))|$$

$$\leq |\sum_{j=1}^{\infty}(b_j y_j + a_j(y_j^2 - 1))| + t\sum_{j=1}^{\infty}|b_j||h_j| + 2t\sum_{j=1}^{\infty}|a_j||y_j||h_j| + t^2\sum_{j=1}^{\infty}|a_j|h_j^2$$

$$\leq M + t(\sum_{j=1}^{\infty}b_j^2)^{1/2}(\sum_{j=1}^{\infty}h_j^2)^{1/2} + 2t(\sum_{j=1}^{\infty}a_j^2 y_j^2)^{1/2}(\sum_{j=1}^{\infty}h_j^2)^{1/2} + +t^2\sigma\sum_{j=1}^{\infty}h_j^2$$

$$\leq M + t(\sum_{j=1}^{\infty}b_j^2)^{1/2} + 2t(\sum_{j=1}^{\infty}a_j^2)^{1/2} + t^2\sigma$$

$$\leq M + tM + t^2\sigma.$$

Thus, the claim follows. □

In the proof of Theorem 1.1, we will use the definition of cumulants that we recall next (see page 42 in [**H**]). Given a r.v. Y, we have the formal identity

$$E[e^{itY}] = \exp\left(\sum_{j=1}^{\infty} \frac{1}{j!} \kappa_j (it)^j\right).$$

κ_j is the j-th cumulant of the r.v. Y. We will use that

(2.2) $\qquad \kappa_1 = E[Y], \kappa_2 = E[(Y - E[Y])^2], \kappa_3 = E[(Y - E[Y])^3].$

It is easy to see that κ_j is polynomial in moments of Y of order less or equal than j.

We also will use that the characteristic function of $(g, g^2 - 1)$ is

(2.3) $\qquad E[\exp(itg + is(g^2 - 1))] = \exp(\sum_{k=2}^{\infty}(2^{k-3}(it)^2(is)^{k-2} + k^{-1}2^{k-1}(is)^k)).$

PROOF OF THEOREM 1.1. We claim that

(2.4) $$\sup_{n \geq 1} E[|X_n|^p] < \infty,$$

for each $0 < p < \infty$. It is well known that for a Gaussian chaos of order two,

(2.5) $\qquad (E[|X_n|^p])^{1/p} \leq (p-1)(E[|X_n|^2])^{1/2}$

(see for example Proposition 6.5.1 in [**KW**]). Observe that, by the Cauchy–Schwarz inequality and (2.5),

$$E[X_n^2] \leq \lambda^2 + E[X_n^2 I_{|X_n|>\lambda}] \leq \lambda^2 + (E[|X_n|^{2p}])^{1/p}(P(|X_n| > \lambda))^{(p-1)/p}.$$

$$\leq \lambda^2 + (2p-1)^2 E[X_n^2](P(|X_n| > \lambda))^{(p-1)/p}.$$

Equation (2.4) follows taking λ such that $(\Pr(|X_n| > \lambda))^{(p-1)/p} \leq 2^{-1}(2p-1)^{-2}$ for each $n \geq 1$. By uniform integrability, $E[|X_n|^p] < \infty$, $E[|X|^p] < \infty$ and

$$\lim_{n \to \infty} E[X_n^p] = E[X^p],$$

for each $p > 0$. So, (a) follows. We may assume that $c_n = 0$ and $\text{Var}(X_n) = \sum_{j=1}^{\infty}(b_{n,j}^2 + 2a_{n,j}^2) = 1$.

We may assume that $|a_{n,1}| \geq |a_{n,2}| \geq \cdots$. We are going to prove by contradiction that $\lim_{n \to \infty} |a_{n,1}|$ exists. Suppose that

$$\tau_0 := \liminf_{n \to \infty} |a_{n,1}| < \limsup_{n \to \infty} |a_{n,1}| =: \tau_1.$$

By lemmas 2.1 and 2.2,

$$2^{-1} \Pr\{|b_{n,1}g_1 + a_{n,1}(g_1^2 - 1)| \geq 4 + 2t + |a_{n,1}|t^2\}$$

$$\leq \Pr\{|X_n| \geq 2 + 2t + |a_{n,1}|t^2) \leq P(g_1 \geq t\}.$$

Taking $t = |a_{n,1}|^{-1}(-1 + (1 + u|a_{n,1}|)^{1/2})$, we get that

$$2^{-1} \Pr\{|b_{n,1}g_1 + a_{n,1}(g_1^2 - 1)| \geq 4 + u\}$$

$$\leq \Pr(|X_n| \geq 2 + u) \leq \Pr\{g_1 \geq |a_{n,1}|^{-1}(-1 + (1 + u|a_{n,1}|)^{1/2})\}.$$

There is a subsequence n_k such that $a_{n_k,1} \to \tau_1$ and $b_{n_k,1} \to b$ for some b. Let $u > 0$ be such that $\Pr\{|X| = u + 2\} = 0$. Then,

$$2^{-1} \Pr\{|bg_1 + \tau_1(g_1^2 - 1)| \geq 4 + u\} \leq \Pr\{|X| \geq u + 2\}.$$

By a similar argument,
$$\Pr\{|X| \geq u+2\} \leq \Pr\{g_1 \geq \tau_0^{-1}(-1+(1+u\tau_0)^{1/2})\}.$$
Hence,
$$2^{-1}\Pr\{|bg_1 + \tau_1(g_1^2 - 1)| \geq 4+u\} \leq \Pr\{g_1 \geq \tau_0^{-1}(-1+(1+u\tau_0)^{1/2})\},$$
for each $u > 0$. Now for $u > 0$
$$\Pr\{g \geq 2^{-1}\tau_1^{-1}b + (1 + b^2 2^{-2}\tau_1^{-2} + \tau_1^{-1}4u)^{1/2}\}$$
$$\leq 2^{-1}\Pr\{|bg_1 + \tau_1(g_1^2 - 1)| \geq 4+u\}.$$
Hence,
$$\tau_0^{-1}(-2 + (4+u\tau_0)^{1/2}) \leq 2^{-1}\tau_1^{-1}b + (1 + b^2 2^{-2}\tau_1^{-2} + \tau_1^{-1}4u)^{1/2}.$$
Letting $u \to \infty$, we get that $\tau_1 \leq \tau_0$, in contradiction. So, $\lim_{n\to\infty} |a_{n,1}| =: \tau$ exists.

Now, we want to show that $\lim_{n\to\infty} a_{n,1}$ exists. If this does not happen, then $-\tau = \liminf_{n\to\infty} a_{n,1}$ and $\tau = \limsup_{n\to\infty} a_{n,1}$. We have that $2ja_{n,j}^2 \leq 1$. So, there exists a $p < \infty$ such that $\limsup_{n\to\infty} |a_{n,p}| = \tau$, but $\limsup_{n\to\infty} |a_{n,p+1}| < \tau$. If $\lim_{n\to\infty} \max(a_{n,1}, \ldots, a_{n,p}) = \tau$, then reordering $\{a_{n,j}\}_{1 \leq j \leq p}$, we can get that $\{a_{n,1}\}$ converges. Assume that $\liminf_{n\to\infty} \max(a_{n,1}, \ldots, a_{n,p}) < \tau$. Then, there exists a $\tau > \epsilon > 0$ such that $\liminf_{n\to\infty} a_{n,l} < \tau - \epsilon$, for $2 \leq l \leq p$ and $\limsup_{n\to\infty} |a_{n,p+1}| < \tau - \epsilon$. By Lemma 2.1,
$$2^{-1}\Pr\{b_{n,1}g_1 + a_{n,1}(g_1^2 - 1) \geq 2^{1/2} + t\} \leq \Pr\{X_n \geq t\}.$$
Taking a subsequence such that $a_{n_k,1} \to \tau$ and $b_{n_k,1} \to b$, for some $b \in \mathbb{R}$, we get that
$$(2.6) \quad 2^{-1}\Pr\{bg_1 + \tau(g_1^2 - 1) \geq 2^{1/2} + t\} \leq \Pr\{X \geq t\},$$
for each $t > 0$ with $\Pr\{X = t\} = 0$. We also have that, for n large enough,
$$\sum_{\substack{j \leq p \\ a_{n,j} \leq 0}} a_{n,j}(g_j^2 - 1) \leq \tau p.$$
So,
$$P(X_n \geq t) \leq \Pr\{\sum_{j=1}^{\infty} b_{n,j}g_j + \sum_{\substack{j \leq p \\ a_{n,j} > 0}} a_{n,j}(g_j^2 - 1) + \sum_{j=p+1}^{\infty} a_{n,j}(g_j^2 - 1) \geq t - \tau p\},$$
For certain sequence n_k, the $|a_{n,j}|$ of the previous expression are bounded in absolute value by $\tau - \epsilon$. From this and Lemma 2.2,
$$(2.7) \quad \Pr\{X \geq 2 + \tau p + 2t + (\tau - \epsilon)t^2\} \leq \Pr\{g_1 > t\},$$
for each $t > 0$. Combining (2.6) and (2.7) we get that
$$2^{-1}\Pr\{bg_1 + \tau(g_1^2 - 1) \geq 4 + \tau p + 2t + (\tau - \epsilon)t^2\} \leq \Pr\{g_1 > t\},$$
which is not possible because
$$2^{-1}\Pr\{bg_1 + \tau(g_1^2 - 1) \geq 4 + \tau p + 2t + (\tau - \epsilon)t^2\}$$
$$= \Pr\{g_1 \geq \tau^{-1/2}(\tau - \epsilon)^{1/2}t + o(t)\}$$
as $t \to \infty$. We got that $\lim_{n\to\infty} a_{n,1}$ exists.

Since $a_{n,1} \to a_1$, $\{b_{n,1}g_1 + \sum_{j=2}^{\infty}(b_{n,j}g_j + a_{n,j}(g_j^2 - 1))\}$ converges in distribution, by repeating the previous argument and induction we get that $\lim_{n\to\infty} a_{n,j}$ exists for each $j \geq 1$. Observe that even when it is not necessarily true that $|a_{n,1}| \geq |a_{n,2}| \geq \cdots$, we are very close to that case. We have that $|a_1| \geq |a_2| \geq \cdots$. Let $p_1, p_2, \ldots$ be the values where $|a_i|$ decrease, i.e. $|a_1| = |a_{p_1}| > |a_{p_1+1}| = |a_{p_2}| > |a_{p_2+1}| = |a_{p_3}| \cdots$. In the construction, we are reordering the original sequence only in the blocks $p_i + 1, \ldots, p_{i+1}$. Using that $\sum_{j=1}^{\infty}(b_{n,j}^2 + 2a_{n,j}^2) = 1$, we get that $\sup_{n\geq 1}\sup_{j\geq p_i+1}|a_{n,j}| \leq (2(p_i + 1))^{-1/2}$. Therefore,

$$(2.8) \qquad \lim_{j_0\to\infty}\sup_{n\geq 1}\sup_{j\geq j_0}|a_{n,j}| = 0.$$

By (2.3),

$$(2.9) \qquad \begin{array}{c} E[\exp(it\sum_{j=1}^{\infty}(b_{n,j}g_j + a_{n,j}(g_j^2 - 1)))] \\ = \exp(\sum_{k=2}^{\infty}\sum_{j=1}^{\infty}(it)^k(2^{k-3}b_{n,j}^2 a_{n,j}^{k-2} + k^{-1}2^{k-1}a_{n,j}^k)) \end{array}$$

From this and (2.2),

$$E[X_n^3] = \sum_{j=1}^{\infty}(6b_{n,j}^2 a_{n,j} + 8a_{n,j}^3).$$

Since we have convergence of moments,

$$\sum_{j=1}^{\infty}(6b_{n,j}^2 a_{n,j} + 8a_{n,j}^3)$$

converges. By (2.8) and the fact that $2\sum_{j=1}^{\infty}a_{n,j}^2 \leq 1$, we get that $\sum_{j=1}^{\infty}a_{n,j}^3$ converges. Therefore,

$$(2.10) \qquad \sum_{j=1}^{\infty}b_{n,j}^2 a_{n,j} \text{ converges.}$$

Given j_0, since a_{n,j_0} converges, $b_{n,j_0}g_{j_0} + \sum_{j\neq j_0}^{\infty}(b_{n,j}g_j + a_{n,j}(g_j^2 - 1))$ converges in distribution. (2.10) for this sequence implies the convergence of $\sum_{j\neq j_0}^{\infty}b_{n,j}^2 a_{n,j}$. The convergence of this sequence and (2.10) implies that $b_{n,j_0}^2 a_{n,j_0}$ converges. This gives that if $a_{j_0} \neq 0$, then $|b_{n,j_0}| \to b_{j_0}$ for some b_{j_0}. So, part (b) follows.

To end the proof we need to prove that (2.9) converges to

$$(2.11) \qquad \exp(2^{-1}(it)^2 s^2 + \sum_{k=2}^{\infty}\sum_{j=1}^{\infty}(it)^k(2^{k-3}b_j^2 a_j^{k-2} + k^{-1}2^{k-1}a_j^k)),$$

which is the characteristic function of $sg_0 + \sum_{j=1}^{\infty}(b_j g_j + a_j(g_j^2 - 1))$, where $s^2 = \text{Var}(X) - \sum_{j=1}^{\infty}(b_j^2 + 2a_j^2)$. The factor of $(it)^2$ in (2.11) is

$$\sum_{j=1}^{\infty}(2^{-1}b_{n,j}^2 + a_{n,j}^2) = 2^{-1}\text{Var}(X_n) \to 2^{-1}\text{Var}(X).$$

The factor of $(it)^2$ in (2.9) is

$$2^{-1}s^2 + \sum_{j=1}^{\infty}(2^{-1}b_j^2 + a_j^2) = 2^{-1}\text{Var}(X).$$

Given $j_0, K_0 < \infty$, using that $\sum_{j=1}^{\infty}(b_{n,j}^2 + 2a_{n,j}^2) = 1$, we have that

$$|\sum_{k=3}^{\infty}\sum_{j=1}^{\infty}(it)^k 2^{k-3}(b_{n,j}^2 a_{n,j}^{k-2} - b_j^2 a_j^{k-2}))|$$

$$\leq 2\sum_{k=K_0+1}^{\infty}|t|^k 2^{k-3} + (\sup_{j\geq j_0+1}\sup_{n\geq 1}|a_{n,j}| + \sup_{j\geq j_0+1}|a_j|)\sum_{k=3}^{K_0+1}|t|^k 2^{k-3}$$

$$+|\sum_{k=3}^{K_0+1}\sum_{j=1}^{j_0}(it)^k 2^{k-3}(b_{n,j}^2 a_{n,j}^{k-2} - b_j^2 a_j^{k-2})|.$$

Hence,

$$|\sum_{k=3}^{\infty}\sum_{j=1}^{\infty}(it)^k 2^{k-3}(b_{n,j}^2 a_{n,j}^{k-2} - b_j^2 a_j^{k-2})| \to 0.$$

A similar argument gives that

$$\sum_{k=3}^{\infty}\sum_{j=1}^{\infty}(it)^k k^{-1} 2^{k-1} a_{n,j}^k \to \sum_{k=3}^{\infty}\sum_{j=1}^{\infty}(it)^k k^{-1} 2^{k-1} a_j^k.$$

So, (2.9) converges to (2.11). $\square$

References

[A] M. A. Arcones, *Distributional limit theorems over a stationary Gaussian sequence of random vectors*. J. Mult. Anal., 1997.

[H] P. Hall, *The Bootstrap and Edgeworth Expansion*. Springer–Verlag, New York, 1992.

[HP] C. Houdré and V. Pérez-Abreu, *Chaos Expansions, Multiple Wiener–Itô Integrals and Their Applications*. CRC Press, Boca Raton, FL, 1994.

[KW] S. Kwapień and W. A. Woyczyński, *Random Series and Stochastic Integrals: Single and Multiple*. Birkhäuser, Boston, 1992.

[LT] M. Ledoux and M. Talagrand, *Probability in Banach Spaces*. Springer–Verlag, New York, 1991.

DEPARTMENT OF MATHEMATICAL SCIENCES, STATE UNIVERSITY OF NEW YORK AT BINGHAMTON, BINGHAMTON, NY 13091

E-mail address: arcones@math.binghamton.edu

Two inequalities and some applications in connection with ρ^*–mixing, a survey

Richard C. Bradley

ABSTRACT. This paper gives a survey of applications of two closely related inequalities (one involving second moments, and the other a Rosenthal inequality) in connection with the ρ^*–mixing condition and related conditions for random sequences. For some of the results, the formulations in this survey are an improvement over the original formulations. A central limit theorem for random fields is also treated.

1. Introduction

This paper will give a survey of applications of two closely related inequalities. The two inequalities make possible the development of central limit theory under certain strong mixing conditions without the use of certain "extra" assumptions such as higher–order moments or conditions on the mixing rates. The two inequalities also allow quite sharp versions of other limit theorems under those mixing conditions. This paper will give the two inequalities and a survey of the applications of them that have been developed so far by various researchers. The statements of a few of the results in this survey will incorporate improvements over the original formulations.

Here in section 1, some definitions and background material will be given. In section 2, one of the two inequalities will be given (in two different forms, in Theorems 2.1 and 2.2), and applications of it will be discussed. In section 3, the other inequality will be given (in Theorem 3.1), and applications of it will be discussed. In section 4, the two inequalities will be combined in the treatment of a central limit theorem for random fields.

Suppose $(\Omega, \mathcal{F}, P)$ is a probability space. For any two σ–fields $\mathcal{A}$ and $\mathcal{B} \subset \mathcal{F}$, define the following measures of dependence:

(1.1) $$\alpha(\mathcal{A}, \mathcal{B}) := \sup_{A \in \mathcal{A}, B \in \mathcal{B}} |P(A \cap B) - P(A)P(B)|;$$

(1.2) $$\rho(\mathcal{A}, \mathcal{B}) := \sup |\text{Corr}(f, g)|,$$

1991 *Mathematics Subject Classification.* Primary 60E15; secondary 60G10.
Key words and phrases. ρ^*-mixing, Rosenthal inequality, stationary sequences.
This work was partially supported by NSF grant DMS-9703712.

where this latter supremum is taken over all pairs of real–valued, square–integrable random variables f and g such that f is $\mathcal{A}$–measurable and g is $\mathcal{B}$–measurable. One has that $\alpha(\mathcal{A}, \mathcal{B}) \leq \rho(\mathcal{A}, \mathcal{B})$. (To see this, consider indicator functions in (1.2).) The quantity $\rho(\mathcal{A}, \mathcal{B})$ is the "maximal correlation" [24], [21] between $\mathcal{A}$ and $\mathcal{B}$.

In what follows, the σ–field generated by a given family $(X_i, i \in I)$ of random variables will be denoted $\sigma(X_i, i \in I)$.

Suppose $X := (X_k, k \in \mathbb{Z})$ is a sequence of real– or complex–valued random variables. For each $n \geq 1$, define the following mixing coefficients:

(1.3) $\quad \alpha(n) = \alpha(X, n) := \sup_{j \in \mathbb{Z}} \alpha(\sigma(X_k, k \leq j), \sigma(X_k, k \geq j + n));$

(1.4) $\quad \rho(n) = \rho(X, n) := \sup_{j \in \mathbb{Z}} \rho(\sigma(X_k, k \leq j), \sigma(X_k, k \geq j + n));$

(1.5) $\quad \alpha^*(n) = \alpha^*(X, n) := \sup \alpha(\sigma(X_k, k \in Q), \sigma(X_k, k \in S));$

(1.6) $\quad \rho^*(n) = \rho^*(X, n) := \sup \rho(\sigma(X_k, k \in Q), \sigma(X_k, k \in S));$

where in each of eqs. (1.5) and (1.6) the supremum is taken over all pairs of nonempty, disjoint sets Q and $S \subset \mathbb{Z}$ such that

(1.7) $\quad \operatorname{dist}(Q, S) := \inf_{q \in Q, s \in S} |q - s| \geq n.$

(The sets Q and S can be "interlaced"; they don't have to be "past" and "future".)

Obviously one has that $\alpha(1) \geq \alpha(2) \geq \alpha(3) \geq \ldots \geq 0$; and the analogous statements hold for the other dependence coefficients here. Also, $\alpha(n) \leq \rho(n) \leq \rho^*(n)$ and $\alpha(n) \leq \alpha^*(n) \leq \rho^*(n)$.

The random sequence X is said to be "strongly mixing" (Rosenblatt [48]) if $\alpha(n) \to 0$ as $n \to \infty$, "ρ–mixing" (Kolmogorov and Rozanov [34]) if $\rho(n) \to 0$ as $n \to \infty$, and "ρ^*–mixing" if $\rho^*(n) \to 0$ as $n \to \infty$. The origin of the ρ^*–mixing condition seems hard to trace; it goes back at least to Stein [51], and its use in central limit theory for random fields goes back to papers such as [20], [22], [23]. Because of the following theorem, the condition $\alpha^*(n) \to 0$ will not be treated separately:

THEOREM 1.1. *Suppose $X := (X_k, k \in \mathbb{Z})$ is a strictly stationary sequence of (real– or complex–valued) random variables. Then X is ρ^*–mixing if and only if $\alpha^*(n) \to 0$ as $n \to \infty$. If X satisfies these two (equivalent) conditions, then $\forall n \geq 1$, $\alpha^*(n) \leq \rho^*(n) \leq 2\pi\alpha^*(n)$.*

Theorem 1.1 was proved by the author [[5], Theorem 1 and Remarks 1, 2, and 3]. In the 1980's, probabilists (faculty and students) at Moscow State University (Moscow, Russia) became aware of such connections between "α-type" and "ρ-type" mixing conditions for strictly stationary random fields. Apparently they never published anything on that. According to Igor Zhurbenko [56], there may have been some uncertainty over statements or proofs. In the case of stationary Gaussian sequences, Theorem 1.1 was pointed out by Rosenblatt [[49], p. 74, Lemma 2], as an application of arguments of Kolmogorov and Rozanov [34].

Much of this survey will deal with limit theorems involving the dependence coefficients $\rho^*(n)$, $n = 1, 2, 3, \ldots$, sometimes in conjunction with the assumption $\alpha(n) \to 0$ (strong mixing). For strictly stationary random sequences, it was shown with a class of examples in [11] that there are in a certain sense "almost" no restrictions on the simultaneous behavior of the dependence coefficients $\alpha(n)$, $\rho(n)$, and $\rho^*(n)$ except for a few unavoidable, elementary inequalities. In particular, for a strictly stationary sequence, one can have $\alpha(n)$, $\rho(n)$, and $\rho^*(n)$ all converging to zero together arbitrarily slowly. For a strictly stationary sequence, one can have $\alpha(n) \to 0$ arbitrarily slowly and (say) $\rho(n) = \rho^*(n) = .997$ for all $n \geq 1$. For a strictly stationary sequence, one can have $\alpha(n) \not\to 0$ and (say) $\rho(n) = \rho^*(n) = .997$ for all $n \geq 1$. Such random sequences as are described here will be dealt with by various limit theorems in this survey.

Let us mention some further examples. A stationary Gaussian sequence with a continuous positive spectral density is ρ^*-mixing. This was shown by Rosenblatt [[49], pp. 73–74, Theorem 7 and Lemma 2] (in the broader context of random fields), with an extension of an argument of Kolmogorov and Rozanov [34]. If a stationary Gaussian sequence has a (not necessarily continuous) spectral density which is bounded between two positive numbers, then it satisfies $\rho^*(1) < 1$ (but not necessarily $\rho^*(n) \to 0$). This seems to be part of the folklore; see e.g. [[15], p. 630, Remark 3]. (Spectral density functions will be treated in section 2 below.) In [8], it was shown that ρ^*-mixing is satisfied by a certain class of (not necessarily stationary) real Markov chains, including all strictly stationary, finite-state, irreducible, aperiodic ones. In [[6], Theorem 3], a class of strictly stationary, ρ^*-mixing random sequences (or random fields) is constructed in which certain special dependence coefficients meet prescribed specifications. These examples (as well as the ones in [11] alluded to above) do not seem to have any close connection to Gaussian random sequences or Markov chains. Finally, if $X := (X_k, k \in \mathbb{Z})$ is a strictly stationary (real) random sequence which satisfies $\rho^*(X, n) \to 0$ resp. $\rho^*(X, 1) < 1$, and $f : \mathbb{R} \to \mathbb{R}$ is a Borel function, then the random sequence $Y := (f(X_k), k \in \mathbb{Z})$ is strictly stationary and satisfies $\rho^*(Y, n) \to 0$ resp. $\rho^*(Y, 1) < 1$ (since $\rho^*(Y, n) \leq \rho^*(X, n)$ for all $n \geq 1$).

Some results in this survey will involve "linear" dependence coefficients and will be formulated most naturally in terms of complex-valued random variables. A complex-valued random variable W will be said to be "centered" if $E|W| < \infty$ and $EW = 0$, and "square-integrable" if $E|W|^2 < \infty$. For any two families $(X_i, i \in I)$ and $(Y_j, j \in J)$ of complex-valued, centered, square-integrable random variables, with the index sets I and J being nonempty and finite, define the following measure of "linear dependence":

$$(1.8) \quad r((X_i, i \in I), (Y_j, j \in J)) := \sup \frac{|E(\sum_{i \in I} a_i X_i)(\overline{\sum_{j \in J} b_j Y_j})|}{\|\sum_{i \in I} a_i X_i\|_2 \cdot \|\sum_{j \in J} b_j Y_j\|_2},$$

where this supremum is taken over all choices of complex numbers a_i, $i \in I$, and b_j, $j \in J$. Here the fraction $0/0$ is interpreted as 0.

One has that

$$r((X_i, i \in I), (Y_j, j \in J)) \leq \rho(\sigma(X_i, i \in I), \sigma(Y_j, j \in J)).$$

This holds because of the elementary fact (see e.g. [[54], p. 512, Theorem 1.1]) that for σ-fields $\mathcal{A}$ and $\mathcal{B}$, the maximal correlation $\rho(\mathcal{A}, \mathcal{B})$ defined in (1.2) satisfies

$$\rho(\mathcal{A}, \mathcal{B}) = \sup |Ef\overline{g}|/(\|f\|_2 \|g\|_2),$$

where this supremum is taken over all pairs of complex-valued, centered, square-integrable random variables f and g such that f is $\mathcal{A}$-measurable and g is $\mathcal{B}$-measurable. (Again $0/0$ is interpreted as 0.)

Suppose $X := (X_k, k \in \mathbb{Z})$ is a (not necessarily stationary) sequence of complex-valued, centered, square-integrable random variables. For each $n \geq 1$ define the following "linear dependence" coefficients:

(1.9) $r(n) = r(X, n) := \sup_{j \in \mathbb{Z}, m \geq 0} r((X_k, j-m \leq k \leq j), (X_k, j+n \leq k \leq j+n+m));$

(1.10) $\quad q^*(n) = q^*(X, n) := \sup \dfrac{|E(\sum_{k \in Q} X_k)(\overline{\sum_{k \in S} X_k})|}{\|\sum_{k \in Q} X_k\|_2 \cdot \|\sum_{k \in S} X_k\|_2};$

(1.11) $\quad r^*(n) = r^*(X, n) := \sup r((X_k, k \in Q), (X_k, k \in S));$

where in each of (1.10) and (1.11), the supremum is taken over all pairs of nonempty, finite, disjoint sets Q and $S \subset \mathbb{Z}$ such that (1.7) holds. (Again, in (1.10), 0/0 is interpreted as 0.)

Obviously, for each $n \geq 1$, one has that $r(n) \leq \rho(n)$, that $r(n) \leq r^*(n)$, and that $q^*(n) \leq r^*(n) \leq \rho^*(n)$. A more subtle relationship is given by the following theorem:

THEOREM 1.2 (Utev). *Suppose* $A := (a(1), a(2), a(3), \ldots)$ *is a nonincreasing sequence of numbers in [0,1] such that* $\sum_{n=0}^{\infty} a(2^n) < \infty$. *Then there exists a nonincreasing sequence* $B = B(A) := (b(1), b(2), b(3), \ldots)$ *of numbers in [0,1] with* $b(n) \to 0$ *as* $n \to \infty$, *such that the following holds:*

If $X := (X_k, k \in \mathbb{Z})$ *is a (not necessarily stationary) sequence of complex-valued, centered, square-integrable random variables such that* $r(X, n) \leq a(n)$ *for all* $n \geq 1$, *then* $r^*(X, n) \leq b(n)$ *for all* $n \geq 1$.

The main point of this theorem is that $\sum_{n=0}^{\infty} r(2^n) < \infty$ implies $r^*(n) \to 0$. In contrast, the condition $\sum_{n=0}^{\infty} \rho(2^n) < \infty$ does *not* imply $\rho^*(n) \to 0$, even under strict stationarity; see e.g. the examples in [**11**].

Theorem 1.2 is due to Sergei Utev [**53**]. His original formulation looked quite different, but the essence of it is captured in the formulation here, and also in a related formulation of his result that is given in [[**13**], Theorem 3]. The argument in this latter reference can be adapted, with just simple modifications, in order to prove the formulation given here in Theorem 1.2. A careful examination of that argument will show more explicitly a connection between the sequences A and B in Theorem 1.2; but that will not be needed here. In two places in section 2, Theorem 1.2 will be used in order to point out immediate corollaries, under $\sum_{n=0}^{\infty} r(2^n) < \infty$, of results involving the condition $r^*(n) \to 0$.

Suppose that one defines a dependence coefficient $q_1(n) = q_1(X, n)$ by [R.H.S. of (1.10)] with the following additional restriction on the sets Q and S: There exists an integer j such that $Q \subset \{k \in \mathbb{Z} : k \leq j\}$ and $S \subset \{k \in \mathbb{Z} : k \geq j+n\}$. Then with essentially the same proof, one has another version of Utev's result: The statement of Theorem 1.2 holds with $r(X, n)$ and $r^*(X, n)$ replaced by $q_1(X, n)$ and $q^*(X, n)$ respectively. Further generalizations of Utev's result can be easily formulated.

2. An inequality on second moments, and applications

A particular inequality will be given in two different forms, in Theorems 2.1 and 2.2. Then the rest of this section will be devoted to applications of that inequality.

THEOREM 2.1. *Suppose $0 \leq r < 1$. Suppose H is a real or complex Hilbert space, with inner product $\langle ., .\rangle$ and norm $\|.\|$. Suppose $x_1, x_2, \ldots, x_n$ are elements of H. Suppose that for each set $S \subset \{1, 2, \ldots, n\}$, defining its complement $S^* := \{1, 2, \ldots, n\} - S$, one has that*

$$(2.1) \quad |\langle \sum_{k \in S} x_k, \sum_{k \in S^*} x_k \rangle| \leq r \cdot \|\sum_{k \in S} x_k\| \cdot \|\sum_{k \in S^*} x_k\|.$$

Then

$$(2.2) \quad \frac{1-r}{1+r} \sum_{k=1}^{n} \|x_k\|^2 \leq \|\sum_{k=1}^{n} x_k\|^2 \leq \frac{1+r}{1-r} \sum_{k=1}^{n} \|x_k\|^2.$$

This theorem reduces (via a simple Hilbert–space isometry) to the version in Theorem 2.2 below, given by the author in [[4], Lemma 1].

The statement of Theorem 2.1 was given by the author [10], in order to correct an error in the formulation in [9], where the modulus signs in the left side of (2.1) were inadvertently omitted. Even in the case where H is a real Hilbert space, if those modulus signs in (2.1) are omitted then the first inequality in (2.2) may fail to hold. (To see why, consider a trivial example with $n = 2$ and $x_2 = -x_1$.)

In the context and notations of Theorem 2.1, if the left side of eq. (2.1) is replaced (for every $S \subset \{1, 2, \ldots, n\}$) by $Re\langle \sum_{k \in S} x_k, \sum_{k \in S^*} x_k\rangle$, then the second inequality in (2.2) will still hold (but the first may fail to hold). If instead the left side of eq. (2.1) is replaced (for every $S \subset \{1, 2, \ldots, n\}$) by $-Re\langle \sum_{k \in S} X_k, \sum_{k \in S^*} X_k\rangle$, then the first inequality in (2.2) will hold (but the second may fail). To verify these facts, one can transcribe to the context of Theorem 2.2 below (via a Hilbert–space isometry), examine its proof in [[4], Lemma 1], and confirm that only minor changes in that argument would be needed. However, Theorems 2.1 and 2.2 in their present form are sufficient for our purposes.

THEOREM 2.2. *Suppose $0 \leq r < 1$. Suppose $X_1, X_2, \ldots, X_n$ are complex–valued, square–integrable (but not necessarily centered) random variables. Suppose that for each set $S \subset \{1, 2, \ldots, n\}$, defining its complement $S^* := \{1, 2, \ldots, n\} - S$, one has that*

$$|E(\sum_{k \in S} X_k)(\overline{\sum_{k \in S^*} X_k})| \leq r \cdot \|\sum_{k \in S} X_k\|_2 \cdot \|\sum_{k \in S^*} X_k\|_2.$$

Then

$$\frac{1-r}{1+r} \sum_{k=1}^{n} E|X_k|^2 \leq E|\sum_{k=1}^{n} X_k|^2 \leq \frac{1+r}{1-r} \sum_{k=1}^{n} E|X_k|^2.$$

This was proved by the author [[4], Lemma 1]. In the formulation there, the hypothesis was technically stronger, but the proof there used only the hypothesis given here.

COROLLARY 2.3. *Suppose $(X_k, k \in \mathbb{Z})$ is a (not necessarily stationary) sequence of complex–valued, centered, square–integrable random variables, N is a positive integer, and $q^*(N) < 1$. Then for any nonempty, finite set $S \subset \mathbb{Z}$, one has that*

$$E|\sum_{k \in S} X_k|^2 \leq N \cdot \frac{1 + q^*(N)}{1 - q^*(N)} \cdot \sum_{k \in S} E|X_k|^2.$$

This is an improved version of a statement by the author [[**4**], Lemma 2]. The formulation there involved weak stationarity and the dependence coefficient $r^*(N)$. The use of the dependence coefficient $q^*(N)$ was suggested by Magda Peligrad [**43**].

PROOF. The proof is simple. For each $j = 1, 2, \ldots, N$, let $S(j)$ denote the set of elements $k \in S$ such that $k \equiv j \bmod N$. In the sums below, omit the values j such that $S(j)$ is empty. By Minkowski's inequality, Theorem 2.2, and Hölder's inequality, one has that

$$\|\sum_{k \in S} X_k\|_2 \leq \sum_{j=1}^{N} \|\sum_{k \in S(j)} X_k\|_2$$

$$\leq \sum_{j=1}^{N} \left[\frac{1 + q^*(N)}{1 - q^*(N)}\right]^{1/2} \left[\sum_{k \in S(j)} E|X_k|^2\right]^{1/2}$$

$$\leq \left[\frac{1 + q^*(N)}{1 - q^*(N)}\right]^{1/2} \cdot N^{1/2} \cdot \left[\sum_{j=1}^{N} \sum_{k \in S(j)} E|X_k|^2\right]^{1/2}$$

$$= N^{1/2} \cdot \left[\frac{1 + q^*(N)}{1 - q^*(N)}\right]^{1/2} \cdot \left[\sum_{k \in S} E|X_k|^2\right]^{1/2}.$$

Taking squares, one obtains the conclusion of Corollary 2.3. □

In the rest of this section, we shall present a few theorems whose proofs were based partly on Theorems 2.1 and 2.2 (sometimes via some version of Corollary 2.3). For most of these theorems, the original formulation involved the broader context of random fields. For simplicity, we shall consider just the narrow context of random sequences. In that context, the statements of some of the theorems here will be in a little more polished form than in their original formulations.

A random sequence $X := (X_k, k \in \mathbb{Z})$ is said to be CCWS ("centered, complex, weakly stationary") if X is a weakly (i.e. second–order) stationary sequence of complex–valued, centered, square–integrable random variables.

Let T denote the unit circle in the complex plane. Let μ denote (one-dimensional) normalized Lebesgue measure on T (i.e. normalized so that $\mu(T) = 1$). A CCWS random sequence $X := (X_k, k \in \mathbb{Z})$ is said to have a spectral density function $f : T \to [0, \infty)$ if f is a Borel, integrable function and

(2.3) $$\forall k \in \mathbb{Z}, \quad EX_k \overline{X_0} = \int_{t \in T} t^k f(t) \mu(dt).$$

If one identifies any given $t \in T$ with the number $\lambda \in (-\pi, \pi]$ such that $t = e^{i\lambda}$, then (2.3) takes the familiar form

(2.4) $$\forall k \in \mathbb{Z}, \quad EX_k \overline{X_0} = \frac{1}{2\pi} \int_{-\pi}^{\pi} e^{ik\lambda} f(e^{i\lambda}) d\lambda.$$

If a CCWS random sequence $(X_k, k \in \mathbb{Z})$ has a continuous spectral density function f on T, then f is given for $t \in T$ by

(2.5) $$f(t) = \lim_{n \to \infty} n^{-1} E|\sum_{k=1}^{n} t^{-k} X_k|^2.$$

This is a well known consequence of Fejer's theorem. (The factor $1/(2\pi)$ in (2.4) is crucial for this.)

For any given random sequence $(X_k, k \in \mathbb{Z})$, the partial sums will be denoted (for $n = 1, 2, 3, \dots$) by $S_n := X_1 + X_2 + \cdots + X_n$.

THEOREM 2.4. *Suppose $X := (X_k, k \in \mathbb{Z})$ is a CCWS random sequence. Then the following six statements hold:*

(1) *If $q^*(n) < 1$ for some $n \geq 1$, then $\sup_{n \geq 1} n^{-1} E|S_n|^2 < \infty$.*

(2) *If $E|S_n|^2 \to \infty$ as $n \to \infty$, and $q^*(n) < 1$ for some $n \geq 1$, then $\inf_{n \geq 1} n^{-1} E|S_n|^2 > 0$.*

(3) *If $q^*(1) < 1$ and $E|X_0|^2 > 0$, then $\inf_{n \geq 1} n^{-1} E|S_n|^2 > 0$.*

(4) *If $q^*(n) \to 0$ as $n \to \infty$ (but not necessarily $q^*(1) < 1$ or $E|S_n|^2 \to \infty$), then $\sigma^2 := \lim_{n \to \infty} n^{-1} E|S_n|^2$ exists in $[0, \infty)$. If also $E|S_n|^2 \to \infty$ as $n \to \infty$, then $\sigma^2 > 0$.*

(5) *If $r^*(n) \to 0$ as $n \to \infty$, then the sequence X has a continuous spectral density function $f : T \to [0, \infty)$, given by (2.5).*

(6) *If the sequence X is nondegenerate, then the following three statements are equivalent:*
   (a) *$r(1) < 1$ and $r^*(n) \to 0$ as $n \to \infty$;*
   (b) *$r^*(1) < 1$ and $r^*(n) \to 0$ as $n \to \infty$;*
   (c) *X has a continuous positive spectral density function on T.*

The use of the dependence coefficient $q^*(n)$ in statements such as (1)–(4) was suggested by Magda Peligrad [43]. The original formulations of (1)–(4) involved $r^*(n)$ or $\rho^*(n)$, but the proofs only needed $q^*(n)$. In statements (5) and (6), $r^*(n)$ apparently cannot be replaced by $q^*(n)$.

Statement (1) follows immediately from Corollary 2.3.

Statement (2) is due to Peligrad [[42], Corollary 2.3]. The formulation there was more restrictive, but her proof (with just trivial changes) yields statement (2) here.

Statement (3) follows immediately from Theorem 2.2.

For statement (4), the first and second parts were proved respectively by the author [[4], Lemma 4 and proof of Theorem 3 (with $\lambda = 0$)]. The formulations there involved the assumption $r^*(n) \to 0$, but the proofs yield statement (4) as given here.

Statement (5) was proved by the author [[4], Theorem 1 (statement and proof)].

Statement (6) developed through several stages: Kolmogorov and Rozanov [[34], p. 207, lines 23-24] showed (in the context of stationary Gaussian sequences) that a continuous positive spectral density implies $r(n) \to 0$. Adapting their arguments, Rosenblatt [[49], p. 73, Theorem 7] showed (in the context of stationary Gaussian random fields) that a continuous positive spectral density implies $r^*(n) \to 0$. The equivalence of (b) and (c) was proved by the author [[4], Theorem 2]. The equivalence of (a) with (b) and (c) was proved by the author and Utev [[13], Theorem 2].

(In the context of random fields further information on statements (5) and (6) is given in [4] and [13].)

REMARK 2.5. (1) Suppose $0 < b < B < \infty$, and $X := (X_k, k \in \mathbb{Z})$ is a CCWS random sequence which has a (not necessarily continuous) spectral density f on

T such that $b \leq f(t) \leq B \ \forall t \in T$; then X satisfies $r^*(1) \leq 1 - b/B < 1$. This fact seems to be (at least in principle) part of the folklore. A proof (in the broader context of random fields) can be found in [[**4**], p. 365, lines -16 to -3].

(2) Ibragimov [[**28**], p. 29] (see also [[**30**], p. 180, Example 2]) constructed a CCWS random sequence $(X_k, k \in \mathbb{Z})$ which satisfies $r(n) \to 0$ as $n \to \infty$, and which has a spectral density f which is bounded between two positive constants but (in a critical way) fails to be continuous. By comment (1) above, this sequence satisfies $r^*(1) < 1$. It shows that in Theorem 2.4, statement (6), statements (a) and (b), the condition $r^*(n) \to 0$ cannot be replaced by the (weaker) condition $r(n) \to 0$.

(3) By Theorem 1.2, one has as a corollary of statement (5) in Theorem 2.4 the following result of Ibragimov [[**27**], Lemma 2]: If $X := (X_k, k \in \mathbb{Z})$ is a CCWS random sequence such that $\sum_{n=0}^{\infty} r(2^n) < \infty$, then X has a continuous spectral density function f on T (given by (2.5)). (For a treatment of this result in the broader context of random fields, see [[**13**], Theorem 4] or [**39**].)

However, Ibragimov [[**28**], Section 5] obtained, explicitly or implicitly, much further information on the spectral density under that assumption $\sum_{n=0}^{\infty} r(2^n) < \infty$ that cannot be derived from statement (5) in Theorem 2.4. For more on that material in [[**28**], Section 5], see [[**30**], Chapter 5] and [[**13**], Theorem 5]. In a closely related paper by the author [**3**] involving the growth of variances of partial sums, the statements and proofs are just a slightly modified version of material implicitly contained in [[**28**], Section 5].

REMARK 2.6. For various theorems in this survey, one can derive analogs for random processes $(X_t, t \in \mathbb{R})$ indexed by $\mathbb{R}$ instead of $\mathbb{Z}$. Curtis Miller [**39**] derived (among other results) an analog of Theorem 2.4, statement (5) for random processes $(X_t, t \in \mathbb{R})$ (and even for random fields indexed by $\mathbb{R}^d$, and also some some random processes $(X_t, t \in \mathbb{R})$ with $E|X_t|^2 = \infty$). We will not elaborate further on that here, but will instead stick with the index set $\mathbb{Z}$.

THEOREM 2.7 (Miller). *Suppose that for each $\ell = 1, 2, 3, \ldots$, $X^{(\ell)} := (X_k^{(\ell)}, k \in \mathbb{Z})$ is a CCWS random sequence. For each $n \geq 1$, define the number $R^*(n) := \sup_{\ell \geq 1} r^*(X^{(\ell)}, n)$. Suppose that*

(2.6) $$R^*(n) \to 0 \text{ as } n \to \infty; \text{ and}$$

(2.7) $$\forall k \in \mathbb{Z}, \quad c_k := \lim_{\ell \to \infty} E X_k^{(\ell)} \overline{X_0^{(\ell)}} \text{ exists in } \mathbb{C}.$$

For each $\ell \geq 1$, let $f_\ell : T \to [0, \infty)$ denote the continuous spectral density function of the random sequence $X^{(\ell)}$ (from eq. (2.3) and Theorem 2.4, statement (5)).

Then there exists a continuous function $f : T \to [0, \infty)$ such that

(2.8) $$\left[\sup_{t \in T} |f^{(\ell)}(t) - f(t)| \right] \to 0 \text{ as } \ell \to \infty.$$

Furthermore, for each $k \in \mathbb{Z}$, the quantity c_k in (2.7) satisfies $c_k = \int_T t^k f(t) \mu(dt) = (2\pi)^{-1} \int_{-\pi}^{\pi} e^{ik\lambda} f(e^{i\lambda}) d\lambda$.

This theorem is due to Curtis Miller [[**37**], Theorem 3.1]. (His result involved the more general context of random fields.)

The last sentence in Theorem 2.7 follows immediately from (2.8), by (2.3)-(2.4) and (2.7).

REMARK 2.8. As an immediate corollary, one obtains the following earlier similar result of Falk [18]: Theorem 2.7 holds with the assumption (2.6) replaced by $\sum_{n=0}^{\infty} R(2^n) < \infty$, where $R(n) := \sup_{\ell \geq 1} r(X^{(\ell)}, n)$. (By Theorem 1.2, the condition $\sum_{n=0}^{\infty} R(2^n) < \infty$ implies (2.6).) The result actually stated and proved by Falk involves an assumption similar to but technically weaker than $\sum_{n=0}^{\infty} R(2^n) < \infty$, and it apparently cannot be derived from Theorem 2.7.

REMARK 2.9. For strictly stationary sequences $(X_k, k \in \mathbb{Z})$ of real-valued random variables such that $EX_0 = 0, EX_0^4 < \infty$, and $\rho^*(M) < 1/128$ for some $M \geq 1$, the author [[4], proofs of Lemmas 5 and 6] implicitly showed that there exists a positive constant C that depends only on M, such that for any nonempty, finite set $S \subset \mathbb{Z}$,

$$(2.9) \quad E\left[\sum_{k \in S} X_k\right]^4 \leq C \cdot \left[(\text{card } S) \cdot EX_0^4 + (\text{card } S)^2 (EX_0^2)^2\right],$$

where (card S) denotes the cardinality of the set S. The proof there involved an application of the Riesz–Thorin interpolation theorem, followed by an application of Theorem 2.2 to (suitable finite collections of) the centered random variables $U_k := X_k^2 - EX_0^2$. Eq. (2.9) played a key role in the proofs of central limit theorems in [4]. (See e.g. Theorem 2.10 below.)

Eq. (2.9) is of course an example of a Rosenthal inequality. A much stronger version of it will be given in Theorem 3.1 in the next section.

THEOREM 2.10. *Suppose $(X_k, k \in \mathbb{Z})$ is a strictly stationary, ρ^*-mixing sequence of real-valued, centered, square-integrable random variables.*
Then $\sigma^2 := \lim_{n \to \infty} n^{-1} ES_n^2$ exists in $[0, \infty)$.
If also $ES_n^2 \to \infty$ as $n \to \infty$, then $\sigma^2 > 0$, and $S_n/(n^{1/2}\sigma)$ converges to $N(0,1)$ in distribution as $n \to \infty$.

Theorem 2.10 was implicitly proved by the author [[4], Theorems 3 and 4]. (To see this, first refer to statements (4) and (5) in Theorem 2.4, and also to eq. (2.5). Of course the quantity σ^2 in Theorem 2.10 is as in statement (4) in Theorem 2.4.)

In Theorem 2.10, the assumption of ρ^*-mixing can be replaced by the assumption $\alpha^*(n) \to 0$ as $n \to \infty$; see Theorem 1.1.

Peligrad [44] extended Theorem 2.10 to a weak invariance principle. See Theorem 3.7 in the next section.

For strictly stationary random sequences (or random fields) satisfying ρ^*-mixing and certain other appropriate assumptions, Miller [37] proved a central limit theorem of Lindeberg type, and used that result in order to prove a central limit theorem for some kernel-type estimators of (marginal) probability density.

In various references, including Rosenblatt [49], Zhurbenko [55], and Ivanov and Leonenko [31], central limit theory is studied for estimators of spectral density for strictly stationary random sequences (or random fields) under various mixing assumptions. For a given strictly stationary sequence $(X_k, k \in \mathbb{Z})$, the estimation of the spectral density (if it exists) is typically based on periodograms, which involve quadratic forms of the X_k's. In order to prove a central limit theorem for estimators of spectral density, one would ordinarily want the periodograms to have finite second moments, and for this one would want the X_k's to have finite fourth moments.

The following is a version of a question posed by Murray Rosenblatt [50]: Under just finite fourth moments and certain mixing assumptions (but with no assumptions on mixing rates), can one prove a central limit theorem for estimators of spectral density? Miller [[38], p. 328, Theorem 2] answered this question affirmatively under the ρ^*-mixing condition, for a certain class of estimators (based on periodograms). This result of Miller will not be stated here. (It would involve considerable extra notation.) Let us just mention that a key role in its proof was played by (a more general version of) the following theorem:

THEOREM 2.11 (Miller). *Suppose $(X_k, k \in \mathbb{Z})$ is a strictly stationary, ρ^*-mixing sequence of complex-valued, centered random variables such that $E|X_0|^4 < \infty$. Then the following three limits exist in $[0, \infty)$, $\mathbb{C}$, and $[0, \infty)$ respectively:*

$$A := \lim_{n \to \infty} n^{-1} E|S_n|^2;$$

$$B := \lim_{n \to \infty} n^{-1} E S_n^2; \text{ and}$$

$$C := \lim_{n \to \infty} n^{-2} E|S_n|^4.$$

Furthermore, these three numbers satisfy

$$C = 2A^2 + |B|^2.$$

This theorem is due to Miller [[38], p. 322, Theorem 1]. (The existence of the number A was already given in Theorem 2.4, statement (4).) Miller's result actually involved a more general context (random fields), a weaker hypothesis, and a more general conclusion (involving the spectral density). It may have broad application in the study of spectral analysis under ρ^*-mixing. Its proof (indirectly) involved Theorem 2.2, somewhat in the spirit of Remark 2.9.

Now let us take a look at the strong law of large numbers.

THEOREM 2.12 (Peligrad and Gut). *Suppose $(X_k, k \in \mathbb{Z})$ is a (not necessarily stationary) sequence of real-valued (or complex-valued), identically distributed random variables such that $E|X_0| < \infty$ and $\rho^*(n) < 1$ for some $n \geq 1$. Define the number $\mu := EX_0$. Then $n^{-1} S_n \to \mu$ almost surely as $n \to \infty$.*

This theorem is due to Peligrad and Gut [[45], [46], Theorem 2.2]. To prove it, they used Etemadi's [17] argument (see e.g. [[2], Theorem 22.1]), but used Corollary 2.3 in the spot where Etemadi used pairwise independence.

THEOREM 2.13 (Bryc and Smolenski). *Suppose $(X_k, k \in \mathbb{Z})$ is a (not necessarily stationary) sequence of real-valued (or complex-valued), centered, square-integrable random variables such that $\sum_{k=1}^{\infty} k^{-3/2} E|X_k|^2 < \infty$ and $\rho^*(n) < 1$ for some $n \geq 1$. Then $n^{-1} S_n \to 0$ almost surely as $n \to \infty$.*

This theorem is due to Bryc and Smolenski [[15], Theorem 2]. Their proof involved combining an argument of Szablowski [52] with (in essence) Corollary 2.3.

REMARK 2.14. In his study of branching processes in connection with population genetics, Jagers [[32], Theorem 5] proved a limit theorem for the rate of growth of branching processes under a dependence condition somewhat analogous to the dependence assumption in Theorem 2.2. The proof involved an application of Theorem 2.2.

3. A Rosenthal inequality, and applications

In Theorem 2.10, a central limit theorem under ρ^*-mixing was given. In Theorem 3.7 below, a similar central limit theorem of Peligrad [42] will be given, involving the (weaker) pair of dependence assumptions that $\rho^*(n) < 1$ for some $n \geq 1$, and $\alpha(n) \to 0$ as $n \to \infty$ (strong mixing). This latter result was based partly on a Rosenthal inequality that was proved by Bryc and Smolenski [15] (and later extended by Peligrad and Gut [45], [46]). Section 3 here is devoted primarily to applications of this Rosenthal inequality. First, here is a statement of the inequality itself:

THEOREM 3.1. *Suppose $q \geq 2$, $0 \leq R < 1$, and N is a positive integer. Then there exists a positive constant $C = C(q, R, N)$ such that the following holds:*

Suppose $X := (X_k, k \in \mathbb{Z})$ is a (not necessarily stationary) sequence of complex-valued, centered random variables such that $\rho^*(N) \leq R$ and $E|X_k|^q < \infty \; \forall k \in \mathbb{Z}$. Then for any nonempty, finite set $S \subset \mathbb{Z}$, one has that

$$(3.1) \qquad E|\sum_{k \in S} X_k|^q \leq C \cdot \left[\sum_{k \in S} E|X_k|^q + \left(\sum_{k \in S} E|X_k|^2 \right)^{q/2} \right].$$

This theorem is stated here for complex-valued random variables, but with just a trivial change in the constant C it reduces easily to the case of real-valued random variables.

Also, again with just a trivial change in the constant C, this theorem reduces easily to the case $N = 1$ (by essentially the same procedure as in the proof of Corollary 2.3 given in section 2).

For $q = 2$, Theorem 3.1 is simply an application of Theorem 2.2 (for $N = 1$) or Corollary 2.3 (for general $N \geq 1$).

In the special case of strict stationarity, $q = 4$, and $R < 1/128$, this theorem was proved by the author [[4], Lemmas 5 and 6]. (See Remark 2.9.)

For $2 < q \leq 4$ and $0 \leq R < 1$, Theorem 3.1 is due to Bryc and Smolenski [[15], Lemma 3]. This was the result that made possible the transition of some central limit theory from ρ^*-mixing (as in Theorem 2.10) to the weaker pair of assumptions $\rho^*(n) < 1$ and $\alpha(n) \to 0$ (as in Theorem 3.7 below). The proof given by Bryc and Smolenski [15] for Theorem 3.1 (under $2 \leq q \leq 4$ and $0 \leq R < 1$) was an ingenious argument that involved, among other things, an application of Khinchin's inequality as well as an application of a theorem of Bryc [14] that deals with the existence of a suitable random variable whose conditional expectations with respect to two given σ-fields are equal (almost surely) to two given random variables.

For $q > 4$ and $0 \leq R < 1$, Theorem 3.1 is due to Peligrad and Gut [[45], [46], Theorem 1.1]. Their result made possible the derivation, under the dependence assumption $\rho^*(n) < 1$, of certain sharp results on rates of convergence in the strong law of large numbers (see Theorem 3.6 below).

In order to derive Theorem 3.1 for $q > 4$ from the (already known) case $2 \leq q \leq 4$, Peligrad and Gut [[45], [46]] used induction on q (and real-valued random variables). For a given $q > 4$, they assumed the theorem with q replaced by $q/2$, and applied that to the (centered, real-valued) random variables $Y_k := X_k^2 - EX_k^2$. To handle one tricky spot in the argument, Peligrad and Gut [45], [46] proved (in greater generality) and then applied the inequality

$$\left[\sum_{k=1}^{n} EX_k^4\right]^{q/4} \leq 2^{(q/4)-1}\left[\sum_{k=1}^{n} E|X_k|^q + \left(\sum_{k=1}^{n} EX_k^2\right)^{q/2}\right].$$

THEOREM 3.2 (Bryc and Smolenski). *Suppose $q \geq 1$, $0 \leq R < 1$, and N is a positive integer. Then there exists a positive constant $D = D(q, R, N)$ such that the following holds:*

Suppose $X := (X_k, k \in \mathbb{Z})$ is a (not necessarily stationary) sequence of complex-valued, centered random variables such that $\rho^*(N) \leq R$ and $E|X_k|^q < \infty$ $\forall k \in \mathbb{Z}$. Then for any nonempty, finite set $S \subset \mathbb{Z}$, one has that

$$E\left|\sum_{k \in S} X_k\right|^q \leq D \cdot E\left[\sum_{k \in S} |X_k|^2\right]^{q/2}.$$

This theorem is due to Bryc and Smolenski [[15], Lemma 2]. As with Theorem 3.1, it easily reduces to the case of real-valued random variables and $N = 1$.

Theorem 3.2 was part of the proof (in [15]) of Theorem 3.1 for the case $2 \leq q \leq 4$. (It was actually in the proof of Theorem 3.2 that Khinchin's inequality and the result of Bryc [14] were used; see the comments after Theorem 3.1.) Of course for $q = 2$, Theorem 3.2 follows from Corollary 2.3.

REMARK 3.3. Houdré [[26], p. 1206] discussed a corollary (in essence) of Theorem 3.2, and used it to show that if a random sequence $X := (X_k, k \in \mathbb{Z})$ satisfies suitable moment conditions and $\rho^*(n) < 1$ for some $n \geq 1$, then the sequence X is "(p, q)-bounded" for certain parameters p and q. (There, Houdré also made an analogous comment with respect to the "logarithmic ρ-mixing" condition $\sum_{n=0}^{\infty} \rho(2^n) < \infty$ frequently used in central limit theory.) For $1 \leq p < \infty$ and $1 \leq q \leq \infty$, a given sequence $X := (X_k, k \in \mathbb{Z})$ of complex-valued random variables satisfying $EX_k = 0$ and $E|X_k|^p < \infty$ for all $k \in \mathbb{Z}$, is "(p, q)-bounded" (see e.g. [[26], p. 1205]) if there exists a positive constant C such that

$$\left(E\left|\sum_{k=-N}^{N} a_k X_k\right|^p\right)^{1/p} \leq C\left(\int_{-\pi}^{\pi} \left|\sum_{k=-N}^{N} a_k e^{ik\theta}\right|^q d\theta\right)^{1/q}$$

holds for every positive integer N and every choice of complex numbers a_k, $-N \leq k \leq N$. (The right hand side is interpreted as $C \cdot \sup_\theta \left|\sum_{k=-N}^{N} a_k e^{ik\theta}\right|$ if $q = \infty$.) The (p, q)-bounded sequences, for various parameters p and q, together form a rather broad class of random sequences (see the discussion in [[26], pp. 1204–1206]); and there has been a considerable development of limit theory for such sequences (see [25], [26] and the references therein). As Houdré pointed out, many of the known results for (p, q)-bounded sequences apply directly (as a special case) to random sequences satisfying $\rho^*(n) < 1$ for some $n \geq 1$.

For example, the following statement is a special case of [[26], Theorem 3.1]:

THEOREM 3.4 (Houdré). *Suppose $X := (X_k, k \in \mathbb{Z})$ is a CCWS random sequence such that $\rho^*(n) < 1$ for some $n \geq 1$, and $(a_1, a_2, a_3, \ldots)$ is a sequence of complex numbers such that $\sum_{k=1}^{\infty} |a_k|^2 (\log k)^2 < \infty$. Then $\sum_{k=1}^{n} a_k X_k$ converges almost surely as $n \to \infty$.*

The following closely related result was proved by Bryc and Smolenski [[15], Theorem 1] with a direct use of Theorem 3.1.

THEOREM 3.5 (Bryc and Smolenski). *Suppose $X := (X_k, k \in \mathbb{Z})$ is a (not necessarily stationary) sequence of real-valued random variables such that $\rho^*(n) < 1$ for some $n \geq 1$, $EX_k = 0$ and $EX_k^2 = 1$ for each $k \in \mathbb{Z}$, and for some $\delta > 0$, $\sup_{k \in \mathbb{Z}} E|X_k|^{2+\delta} < \infty$. If $(a_1, a_2, a_3, \ldots)$ is a sequence of real (or complex) numbers such that $\sum_{k=1}^{\infty} |a_k|^2 < \infty$, then $\sum_{k=1}^n a_k X_k$ converges almost surely as $n \to \infty$.*

Now let us turn to the strong law and then to central limit theory.

THEOREM 3.6 (Peligrad and Gut). *Suppose $\alpha > 1/2$, $p > 0$, and $\alpha p > 1$. Suppose $X := (X_k, k \in \mathbb{Z})$ is a strictly stationary sequence of real-valued (or complex-valued) random variables such that $\rho^*(n) < 1$ for some $n \geq 1$. If $\alpha \leq 1$, assume also that $E|X_0| < \infty$ and $EX_0 = 0$. Then the following two statements are equivalent:*

(a) $E|X_0|^p < \infty$.
(b) *For all $\varepsilon > 0$,* $\sum_{n=1}^{\infty} n^{p\alpha-2} P\left(\max_{1 \leq j \leq n} |S_j| > \varepsilon n^{\alpha}\right) < \infty$.

This theorem is due to Peligrad and Gut [[45], [46], Theorem 2.1]. Their proof adapted classical arguments of Baum and Katz [1] for analogous results on i.i.d. sequences, and it made critical use of Theorem 3.1 with arbitrarily high values of q.

THEOREM 3.7 (Peligrad). *Suppose $X := (X_k, k \in \mathbb{Z})$ is a strictly stationary sequence of real-valued, centered, square-integrable random variables such that $\sigma_n^2 := ES_n^2 \to \infty$ as $n \to \infty$, $\alpha(n) \to 0$ as $n \to \infty$, and $\rho^*(n) < 1$ for some $n \geq 1$. Then the following statements (1)–(4) hold:*
(1) *The sequence of numbers $(n^{-1}\sigma_n^2, n = 1, 2, 3, \ldots)$ is bounded above and below by positive (finite) constants.*
(2) *The normalized partial sum S_n/σ_n converges to $N(0, 1)$ in distribution as $n \to \infty$.*
(3) *The family of random variables $(n^{-1}M_n^2, n = 1, 2, 3, \ldots)$ is uniformly integrable, where $M_n := \max_{1 \leq k \leq n} |S_k|$.*
(4) *As $n \to \infty$, the normalized sample path $(S_{[nt]}/\sigma_n, 0 \leq t \leq 1)$ converges in distribution (on the space $D(0, 1)$) to a standard Wiener process $(W(t), 0 \leq t \leq 1)$, where $[nt]$ denotes the greatest integer $\leq nt$.*

Of course (4) implies (2), and (3) implies the upper bound in (1). Parts (1) and (2) are due to Peligrad [[42], Corollary 2.3], and parts (3) and (4) are due to Peligrad [[44], Proposition 2.1 and Theorem 2.1]. Part (1) was given in Theorem 2.4, statements (1) and (2). In order to prove part (2), Peligrad [42] first used Theorem 3.1 with $q = 4$, together with truncation, in order to show that the random variables $(S_n^2/\sigma_n^2, n = 1, 2, 3, \ldots)$ are uniformly integrable, and then she cited a well known central limit theorem (see [16] or [40]) for strictly stationary, strongly mixing random sequences. The proof (in [44]) of parts (3) and (4) involved a more delicate argument.

REMARK 3.8. (1) As Peligrad [[42], Corollary 2.4] pointed out, in the context of Theorem 3.7, if one assumes $\rho^*(1) < 1$ (and $EX_0^2 > 0$), then one does not need to explicitly assume that $ES_n^2 \to \infty$ as $n \to \infty$; that property would follow automatically (see Theorem 2.4, statement (3)).

(2) In the context of Theorem 3.7, even if $\rho^*(1) < 1$, the quantity $\lim_{n\to\infty} n^{-1} ES_n^2$ need not exist. That was shown by the author [[**12**], Theorem 1], with a stationary Gaussian sequence that was a modification of one studied earlier by Ibragimov [[**28**], p. 29] and Ibragimov and Rozanov [[**30**], p. 180, Example 2]. (See Remark 2.5, part (2).)

(3) In the context of Theorem 3.7, if $\sum_{n=1}^{\infty} |EX_0 X_n| < \infty$, then $\sigma^2 := \lim_{n\to\infty} n^{-1} ES_n^2$ does exist in $(0, \infty)$, and $S_n/(n^{1/2}\sigma)$ converges to $N(0,1)$ in distribution (and a similar restatement of part (4) in Theorem 3.7 holds).

(4) Peligrad [**42**] also obtained a central limit theorem of Lindeberg type under $\rho^*(n) < 1$ and $\alpha(n) \to 0$.

4. A Central Limit Theorem for random fields

Until now, our attention has been restricted to random sequences, even though some of the results that have been stated here were in fact formulated and proved in the broader context of random fields. Section 4 here will be devoted primarily to a central limit theorem for random fields that is based heavily on both Theorem 2.2 and Theorem 3.1. It is given in Theorem 4.3 below. Some further comments on it, possibly relevant to other potential limit theory for random fields, will be given in Remark 4.4. A closely related CLT for random fields along the lines of Theorem 3.7(2)/Remark 3.8(3) was given by Perera [[**47**], Proposition 3].

Suppose d is a positive integer. A given element $k \in \mathbb{Z}^d$ will be denoted by $k := (k_1, k_2, \ldots, k_d)$; and its Euclidean norm will be denoted by $\|k\| := (k_1^2 + k_2^2 + \cdots + k_d^2)^{1/2}$. The "distance" between any two nonempty, disjoint sets Q and $S \subset \mathbb{Z}^d$ is defined by $\text{dist}(Q, S) := \inf_{q \in Q, s \in S} \|q - s\|$. In the case $d = 1$, one has that $\|k\| = |k|$, and $\text{dist}(Q, S)$ coincides with the definition in (1.7).

Suppose $X := (X_k, k \in \mathbb{Z}^d)$ is a random field. The random variables X_k may be complex-valued. For such a random field, we shall deal with the following three sequences of dependence coefficients:

For each $n = 1, 2, 3, \ldots$,

(4.1) $$\alpha(n) := \sup \alpha(\sigma(X_k, k \in Q),\ \sigma(X_k, k \in S))$$

where the supremum is taken over all pairs of sets Q and S of the form $Q = \{k \in \mathbb{Z}^d,\ k_u \leq j\}$ and $S = \{k \in \mathbb{Z}^d : k_u \geq j + n\}$ where $j \in \mathbb{Z}$ and $u \in \{1, 2, \ldots, d\}$. In the case $d = 1$, this coincides with eq. (1.3).

For each $n = 1, 2, 3, \ldots$,

(4.2) $$\rho'(n) := \sup \rho(\sigma(X_k, k \in Q),\ \sigma(X_k, k \in S))$$

where the supremum is taken over all pairs of sets Q and S of the form $Q = \{k \in \mathbb{Z}^d : k_u \in Q_0\}$ and $S = \{k \in \mathbb{Z}^d : k_u \in S_0\}$ where $u \in \{1, 2, \ldots, d\}$, and Q_0 and S_0 are nonempty, disjoint subsets of $\mathbb{Z}$ such that $\text{dist}(Q_0, S_0) \geq n$. (The sets Q_0 and S_0 may be "interlaced.")

For each $n = 1, 2, 3, \ldots$,

(4.3) $$\rho^*(n) := \sup \rho(\sigma(X_k, k \in Q),\ \sigma(X_k, k \in S))$$

where the supremum is taken over all pairs of nonempty, disjoint sets Q and $S \subset \mathbb{Z}^d$ such that $\text{dist}(Q, S) \geq n$. In the case $d = 1$, $\rho'(n)$ and $\rho^*(n)$ each coincide with the dependence coefficient $\rho^*(n)$ in (1.6). For general $d \geq 1$, one of course has for each $n \geq 1$, $\alpha(n) \leq \rho'(n) \leq \rho^*(n)$.

The central limit theorem given in Theorem 4.3 below will involve the pair of dependence assumptions $\rho'(1) < 1$ and $\alpha(n) \to 0$. We first need two lemmas giving analogs of Theorems 2.2 and 3.1 for random fields satisfying $\rho'(1) < 1$.

LEMMA 4.1. *Suppose d is a positive integer. Suppose $(X_k, k \in \mathbb{Z}^d)$ is a (not necessarily stationary) random field, with the random variables X_k being complex-valued, centered, and square-integrable. Suppose $\rho'(1) < 1$. Then for any nonempty, finite set $S \subset \mathbb{Z}^d$, one has that*

$$(4.4) \quad \left[\frac{1-\rho'(1)}{1+\rho'(1)}\right]^d \sum_{k \in S} E|X_k|^2 \leq E\left|\sum_{k \in S} X_k\right|^2 \leq \left[\frac{1+\rho'(1)}{1-\rho'(1)}\right]^d \sum_{k \in S} E|X_k|^2.$$

In Lemma 4.1, the quantity $\rho'(1)$ can be replaced by $q'(1)$, where for any $n \geq 1$, $q'(n) := \sup |Ef\bar{g}|/(\|f\|_2\|g\|_2)$ with the supremum being taken over all pairs of sets Q and S meeting the requirements of (4.2) and all functions f (resp. g) which are the sums of finitely many of the X_k's, $k \in Q$ (resp. $k \in S$). (For $d = 1$, this definition coincides with (1.10).) However, Lemma 4.1 in its present form will be satisfactory for our purposes.

PROOF. This will be done by induction on d. In the context of random fields, such an induction argument is a standard technique; Gaposhkin [20] referred to it as "layering."

For $d = 1$, Lemma 4.1 holds by Theorem 2.2.

Suppose $d \geq 2$, and suppose Lemma 4.1 holds with d replaced by $d-1$. Our task in the induction step is to show that it holds with d itself. Suppose S is a nonempty, finite subset of $\mathbb{Z}^d$. Our task is to verify eq. (4.4).

For each $j \in \mathbb{Z}$, let $S(j)$ denote the set of all elements $k := (k_1, k_2, \ldots, k_d) \in S$ such that $k_1 = j$. The sets $S(j)$ are nonempty for only finitely many $j \in \mathbb{Z}$, and these sets partition S. In the sums below, j ranges over just the values for which the set $S(j)$ is nonempty.

By Theorem 2.2 and our induction assumption,

$$E\left|\sum_{k \in S} X_k\right|^2 \geq \frac{1-\rho'(1)}{1+\rho'(1)} \sum_j E\left|\sum_{k \in S(j)} X_k\right|^2$$

$$\geq \frac{1-\rho'(1)}{1+\rho'(1)} \sum_j \left[\frac{1-\rho'(1)}{1+\rho'(1)}\right]^{d-1} \sum_{k \in S(j)} E|X_k|^2$$

$$= \left[\frac{1-\rho'(1)}{1+\rho'(1)}\right]^d \sum_{k \in S} E|X_k|^2.$$

This gives the first inequality in (4.4). The proof of the second one is exactly analogous. This completes the induction step. Lemma 4.1 holds by induction. □

LEMMA 4.2. *Suppose d is a positive integer, $q \geq 2$, and $0 \leq R < 1$. Then there exists a positive constant $D = D(q, R, d)$ such that the following holds:*

Suppose $(X_k, k \in \mathbb{Z}^d)$ is a random field such that the random variables X_k are complex-valued, centered, and satisfy $E|X_k|^q < \infty$; and suppose $\rho'(1) \leq R$. Then *for any nonempty, finite set $S \subset \mathbb{Z}^d$, one has that*

$$(4.5) \qquad E\Big|\sum_{k \in S} X_k\Big|^q \le D \cdot \left[\sum_{k \in S} E|X_k|^q + \left(\sum_{k \in S} E|X_k|^2\right)^{q/2}\right].$$

Clearly Lemma 4.2 can be extended to the case where, instead of $\rho'(1) \le R$, one has that $\rho'(N) \le R$, where N is a given positive integer. This would entail a simple change in the constant D, which would now depend on q, R, N, and d. The extension can be carried out by (a d-dimensional version of) the procedure that was used in the proof of Corollary 2.3. However, Lemma 4.2 in its present form will be satisfactory for our purposes.

PROOF. For $d = 1$, Lemma 4.2 holds by Theorem 3.1. We shall again use induction on d ("layering").

Suppose $d \ge 2$, and suppose Lemma 4.2 holds with d replaced by $d-1$. Our task in the induction step is to show that it holds with d itself.

Suppose $q \ge 2$ and $0 \le R < 1$. Define the positive constants $A := D(q, R, 1)$ and $B := D(q, R, d-1)$ (in the terminology of Lemma 4.2 with d replaced by 1 and by $d-1$). Define the positive constant

$$(4.6) \qquad D = D(q, R, d) := AB + A \cdot [(1+R)/(1-R)]^{(d-1)q/2}.$$

Now suppose $X := (X_k, k \in \mathbb{Z}^d)$ is a random field, where the random variables X_k are complex-valued and centered and satisfy $E|X_k|^q < \infty \; \forall k \in \mathbb{Z}$. Suppose also that X satisfies $\rho'(1) \le R$. Suppose S is a nonempty, finite subset of $\mathbb{Z}^d$. In order to complete the induction step, it suffices to show that (4.5) holds.

Define the sets $S(j), j \in \mathbb{Z}$, as in the proof of Lemma 4.1. In all sums that follow in this proof, j is restricted to the (finite collection of) integers such that $S(j)$ is nonempty. For each such j, define the random variables $Y_j := \sum_{k \in S(j)} X_k$. Then $\sum_{k \in S} X_k = \sum_j Y_j$.

In the calculations that follow, we shall use the elementary fact that if $a_1, a_2, \ldots, a_m$ are nonnegative numbers and $t \ge 1$, then $\sum_{k=1}^m a_k^t \le (\sum_{k=1}^m a_k)^t$. Now by eq. (4.6), the definitions of A and B (recall our induction assumption), and Lemma 4.1, one has that

$$E\Big|\sum_{k \in S} X_k\Big|^q \le A \cdot \sum_j E|Y_j|^q + A \cdot \left(\sum_j E|Y_j|^2\right)^{q/2}$$

$$\le AB \cdot \sum_j \sum_{k \in S(j)} E|X_k|^q + AB \cdot \sum_j \left(\sum_{k \in S(j)} E|X_k|^2\right)^{q/2}$$

$$+ A \cdot \left(\sum_j [(1+R)/(1-R)]^{d-1} \sum_{k \in S(j)} E|X_k|^2\right)^{q/2}$$

$$\le AB \cdot \sum_{k \in S} E|X_k|^q + AB \cdot \left(\sum_j \sum_{k \in S(j)} E|X_k|^2\right)^{q/2}$$

$$+ A \cdot [(1+R)/(1-R)]^{(d-1)q/2} \left(\sum_{k \in S} E|X_k|^2 \right)^{q/2}$$

$\leq$ [right side of (4.5)].

Thus (4.5) holds. This completes the induction step. Lemma 4.2 holds by induction. □

Now suppose d is a positive integer, and $X := (X_k, k \in \mathbb{Z}^d)$ is a random field. Let $\mathbb{N}$ denote the set of positive integers. For each $L := (l_1, l_2, \ldots, l_d) \in \mathbb{N}^d$ define the "block sum"

$$S(L) = S(X : L) := \sum_k X_k$$

where this sum is taken over all $k := (k_1, k_2, \ldots, k_d) \in \mathbb{N}^d$ such that $\forall u = 1, \ldots, d$, $1 \leq k_u \leq l_u$. Thus $S(L)$ is the sum of $l_1 \cdot l_2 \cdot \ldots \cdot l_d$ of the random variables X_k.

THEOREM 4.3. *Suppose d is a positive integer. Suppose $X := (X_k, k \in \mathbb{Z}^d)$ is a nondegenerate, strictly stationary random field, with the random variables X_k being real-valued and centered. Suppose that*
(a) $H(c) := EX_0^2 I(|X_0| \leq c)$ *is slowly varying as $c \to \infty$, and*
(b) $\rho'(1) < 1$ *and $\alpha(n) \to 0$ as $n \to \infty$.*
Then as $\|L\| \to \infty$, $L \in \mathbb{N}^d$, one has that
(i) $a_L := (\pi/2)^{1/2} E|S(L)| \to \infty$, *and*
(ii) $S(L)/a_L$ *converges to $N(0,1)$ in distribution.*

Refer to assumption (b). If $d = 1$, then $\rho'(1)$ is simply $\rho^*(1)$. If instead $d \geq 2$, then under our assumption of strict stationarity, the condition $\alpha(n) \to 0$ is equivalent to $\rho(n) \to 0$, where $\rho(n)$ is defined by the right side of (4.1) with the symbol α there replaced by ρ. (See [[5], Theorem 1 and Remarks 1 and 2].)

The "tail" condition in (a) allows (barely) infinite variance. As is well known, it implies $E|X_0|^p < \infty \; \forall \; p \in [0, 2)$. It is (equivalent to) the "tail" condition in the classic result of Khinchin [33], Lévy [35], and Feller [19], that characterized the i.i.d. sequences that are in the domain of attraction to a normal distribution.

The author [[4], Theorem 5] proved Theorem 4.3 with assumption (b) replaced by the stronger pair of assumptions $\rho^*(1) < 1$ and $\rho^*(n) \to 0$. After seeing a preprint of that paper, Magda Peligrad [41] pointed out that with just simple modifications, that argument would still work with the assumption $\rho^*(n) \to 0$ replaced by $\alpha(n) \to 0$. She based this comment on Theorem 3.1 with $q = 4$, for which she cited (a preprint of) Bryc and Smolenski [15].

For a related CLT (under $\rho^*(1) < 1$ and $\alpha(n) \to 0$) for random fields with finite second moments, see the result of Perera [[47], Proposition 3].

PROOF. We shall refer to the argument in [4] and just indicate the changes that are needed. Those changes will involve (i) the suggestions made by Peligrad [41] to allow the use of $\alpha(n) \to 0$ instead of $\rho^*(n) \to 0$, and (ii) applications of Lemmas 4.1 and 4.2 to allow the use of $\rho'(1) < 1$ instead of $\rho^*(1) < 1$.

As in [[4], pp. 368–369, Proposition 1], the proof is reduced to the case where $L^{(1)}, L^{(2)}, L^{(3)}, \ldots$ is a sequence of elements of $\mathbb{N}^d$ such that the first coordinate of $L^{(n)}$ is n. The task is to show that the conclusion of Theorem 4.3 holds for this sequence of vectors $L^{(n)}$. The argument is simply the proof of [[4], Proposition 1], with the following changes, using the assumptions $\rho'(1) < 1$ and $\alpha(n) \to 0$:

First, in order to justify [[**4**], p. 370, eq. (3.11)], one uses Lemma 4.1 above, instead of [[**4**], Lemma 1].

Next, in choosing the sequence $m_1, m_2, m_3, \ldots$ of positive integers satisfying [[**4**], p. 370, eqs. (3.14) – (3.18)], one replaces [[**4**], eq. (3.17)] by the following condition:

$$(4.7) \qquad m_n \cdot \alpha(X, q_n) \leq m_n^2 \cdot [\alpha(X, q_n)]^{1/2} \to 0 \text{ as } n \to \infty.$$

Next, in justifying [[**4**], p. 371, eq. (3.21)], one uses Lemma 4.1 above, instead of [[**4**], Lemma 1].

Next one comes to two crucial equations in the proof: [[**4**], p. 371, eqs. (3.22) and (3.23)]. Here these equations will be derived in reverse order.

In order to derive [[**4**], eq. (3.23)], one makes two changes in the argument. First, one uses Lemma 4.2 above, instead of [[**4**], Lemma 6]. Second, in deriving the last "inequality" in [[**4**], eq. (3.23)], one uses [[**4**], eq. (3.10)] and Lemma 4.1 above, instead of [[**4**], eq. (3.22)].

Now in order to derive [[**4**], eq. (3.22)], one first applies (with fourth moments) a well known covariance inequality of Ibragimov in order to obtain

$$(4.8) \qquad |\mathrm{Corr}(W_i^{(n)}, W_j^{(n)})| \leq C \cdot [\alpha(X, q_n)]^{1/2}$$

for all $n \geq 1$ and all pairs of distinct elements $i, j \in \{1, 2, \ldots, m_n\}$, where the constant C does not depend on n, i, j. (For example, apply [[**29**], Theorem 17.2.2] or [[**2**], p. 365, Lemma 3], to the random variables $W_i^{(n)}/\|W_1^{(n)}\|_2$ and $W_j^{(n)}/\|W_1^{(n)}\|_2$, and use [[**4**], eq. (3.23)], which was verified above.) Combining (4.8) and (4.7), one obtains, in the language of [[**4**], pp. 369–372], the equation

$$E|\sum_{k=1}^{m(n)} W_k^{(n)}|^2 \sim m_n E|W_1^{(n)}|^2 \text{ as } n \to \infty.$$

(Here $b_n \sim c_n$ means that $b_n/c_n \to 1$ as $n \to \infty$.) Then using the observation on [[**4**], p. 371, line -10], one obtains [[**4**], eq. (3.22)].

Finally, in justifying [[**4**], p. 372, eq. (3.24)], one uses eq. (4.7) above, instead of [[**4**], eq. (3.17)].

With the changes listed above, the argument in [[**4**], pp. 368–372, proof of Proposition 1] yields Theorem 4.3. This completes the derivation of this theorem. □

REMARK 4.4. Suppose d is a positive integer, and $X := (X_k, k \in \mathbb{Z}^d)$ is a random field, with the random variables X_k being complex–valued, centered, and square–integrable. For $n \geq 1$, let us define $r'(n) := \sup |Ef\bar{g}|/(\|f\|_2\|g\|_2)$, where the supremum is taken over all pairs of functions f (resp. g) which are linear combinations (with complex coefficients) of finitely many X_k's, $k \in Q$ (resp. $k \in S$) where the sets Q and S are as in (4.2). Let us define $r^*(n)$ the same way, but with Q and S as in (4.3).

The author and Utev [**13**] examined the spectral density of weakly stationary random fields under conditions on $r'(n)$ and $r^*(n)$. When the possible use of the dependence coefficients $r'(n)$ and $\rho'(n)$ in limit theory was mentioned to him, Curtis Miller [**36**] noted that his results in (preprints of) [**37**], [**38**] apparently would still hold with $r^*(n)$ and $\rho^*(n)$ replaced by $r'(n)$ and $\rho'(n)$, and that this would apparently entail just elementary modifications of his arguments. For an arbitrary integer $d \geq 2$, the author [7] constructed a strictly stationary random field $(X_k, k \in \mathbb{Z}^d)$ which satisfies $\rho'(2) = 0$ and $\rho^*(n) = 1$ for all $n \geq 1$.

In central limit theory for random fields, the dependence coefficient $\rho^*(n)$ has been used extensively, but the dependence coefficient $\rho'(n)$ seems to have been hardly used at all. Yet Theorems 2.2 and 3.1 seem to provide ample leverage (through technical statements such as Lemmas 4.1 and 4.2) for the possible use of $\rho'(n)$ in central limit theorems (such as Theorem 4.3).

Acknowledgement. The author thanks Christian Houdré and Ted Hill for helpful conversations.

References

[1] E. Baum and M. Katz, *Convergence rates in the law of large numbers*, Trans. Amer. Math. Soc. **120** (1965), 108–123.

[2] P. Billingsley, Probability and Measure, third ed. Wiley, New York (1995).

[3] R.C. Bradley, *A sufficient condition for linear growth of variances in a stationary random sequence*, Proc. Amer. Math. Soc. **83** (1981), 586–589.

[4] _____, *On the spectral density and asymptotic normality of weakly dependent random fields*, J. Theor. Probab. **5** (1992) 355-373.

[5] _____, *Equivalent mixing conditions for random fields*, Ann. Probab. **21** (1993), 1921–1926.

[6] _____, *Some examples of mixing random fields* Rocky Mountain J. Math. **23** (1993), 495–519.

[7] _____, *On regularity conditions for random fields*, Proc. Amer. Math. Soc. **121** (1994), 593–598.

[8] _____, *Every "lower psi-mixing" Markov chain is "interlaced rho-mixing"*, Stochastic Process. Appl. **72** (1997), 221–239.

[9] _____, *Inequalities and applications connected with the ρ^*-mixing condition for stochastic processes*, Abstract 926-60-47, Abstracts of Amer. Math. Soc. **18** (1997), 516.

[10] _____, *Inequalities and applications connected with the ρ^*-mixing condition for stochastic processes*, a talk in the Special Session on Stochastic Inequalities and their Applications, 926th Meeting of the American Mathematical Society, Atlanta, Georgia, Oct. 17–19, 1997.

[11] _____, *On the simultaneous behavior of the dependence coefficients associated with three mixing conditions*, Rocky Mountain J. Math. **28** (1998), 393–415.

[12] _____, *On the growth of variances in a central limit theorem for strongly mixing sequences*, Bernoulli (to appear).

[13] R.C. Bradley and S.A. Utev, *On second-order properties of mixing random sequences and random fields*, Probability Theory and Mathematical Statistics, Proceedings of the Sixth Vilnius Conference (1993) (B. Grigelionis, J. Kubilius, H. Pragarauskas, and V. Statulevicius, eds.) 99–120, VSP Science Publishers, Utrecht, and TEV Publishers Service Group, Vilnius (1994).

[14] W. Bryc, *Conditional expectation with respect to dependent σ-fields*, Proceedings of the Seventh Conference on Probability Theory, Brasov (Romania) 1982, (M. Iosifescu, ed.) 409–411, Editura Academiei, Bucharest, and VNU Science Press, Utrecht (1985).

[15] W. Bryc and W. Smolenski, *Moment conditions for almost sure convergence of weakly correlated random variables*, Proc. Amer. Math. Soc. **119** (1993), 629–635.

[16] M. Denker, *Uniform integrability and the central limit theorem for strongly mixing processes*, Dependence in Probability and Statistics, (E. Eberlein and M.S. Taqqu, eds.) 269–274, Birkhäuser, Boston (1986).

[17] N. Etemadi, *An elementary proof of the strong law of large numbers*, Z. Wahrsch. verw. Gebiete **55** (1981), 119–122.

[18] M. Falk, *On the convergence of spectral densities of arrays of weakly stationary processes*, Ann. Probab. **12** (1984), 918–921.

[19] W. Feller, *Über den zentralen Grenzwertsatz der Wahrscheinlichkeitsrechnung*, Math. Z. **40** (1935), 521–559.

[20] V.F. Gaposhkin, *Moment bounds for integrals of ρ–mixing fields*, Theor. Probab. Appl. **36** (1991), 249–260.

[21] H. Gebelein, *Das Statistische Problem der Korrelation als Variations- und Eigenwertproblem und sein Zusammenhang mit der Ausgleichungsrechnung*, Z. Angew. Math. Mech. **21** (1941), 364–379.

[22] C.M. Goldie and P.E. Greenwood, *Variance of set-indexed sums of mixing random variables and weak convergence of set-indexed processes*, Ann. Probab. **14** (1986), 817–839.

[23] V.V. Gorodetskiĭ, *The central limit theorem and an invariance principle for weakly dependent random fields*, Soviet Math. Dokl. **29** (1984), 529–532.

[24] H.O. Hirschfeld, *A connection between correlation and contingency*, Proc. Camb. Phil. Soc. **31** (1935) 520–524.

[25] C. Houdré, *On the spectral SLLN and pointwise ergodic theorem in L^α*, Ann. Probab. **20** (1992), 1731–1753.

[26] _____, *On the almost sure convergence of series of stationary and related nonstationary variables*, Ann. Probab. **23** (1995), 1204–1218.

[27] I.A. Ibragimov, *Stationary Gaussian sequences that satisfy the strong mixing condition*, Soviet Math. Dokl. **3** (1962), 1799-1801.

[28] _____, *On the spectrum of stationary Gaussian sequences satisfying the strong mixing condition II. Sufficient conditions. Mixing rate*, Theor. Probab. Appl. **15** (1970), 23–36.

[29] I.A. Ibragimov and Yu.V. Linnik, *Independent and Stationary Sequences of Random Variables*, Wolters-Noordhoff, Groningen (1971).

[30] I.A. Ibragimov and Yu.A. Rozanov, *Gaussian Random Processes*, Springer-Verlag, New York (1978).

[31] A.V. Ivanov and N.N. Leonenko, *Statistical Analysis of Random Fields*, Kluwer, Boston (1989).

[32] P. Jagers, *Towards dependence in general branching processes, Classical and Modern Branching Processes*, (K.B. Athreya and P. Jagers, eds.) 127–139, Springer, New York (1997).

[33] A.Y. Khinchin, *Sul dominio di attrazione della legge di Gauss*, Giorn. Ital. Attuari **6** (1935), 371–393.

[34] A.N. Kolmogorov and Yu.A. Rozanov, *On strong mixing conditions for stationary Gaussian processes*, Theor. Probab. Appl. **5** (1960), 204–208.

[35] P. Lévy, *Propriétés asymptotiques des sommes de variables aléatoires indépendentes ou enchaînees*, J. Math. Pures Appl. **14** (1935), 347–402.

[36] C. Miller, Private communication, (1994).

[37] _____, *Three theorems on ρ^*-mixing random fields*, J. Theor. Probab. **7** (1994), 867–882.

[38] _____, *A CLT for the periodograms of a ρ^*-mixing random field*, Stochastic Process. Appl. **60** (1995), 313–330.

[39] _____, *Spectral densities for continuous random fields with ρ^*- or ρ-mixing*, preprint (1997).

[40] T. Mori and K. Yoshihara, *A note on the central limit theorem for stationary strong-mixing sequences*, Yokohama Math. J. **34** (1986), 143–146.

[41] M. Peligrad, Private communication (1992).

[42] _____, *On the asymptotic normality of sequences of weak dependent random variables*, J. Theor. Probab. **9** (1996), 703–715.

[43] _____, A talk at the Conference on Dependence in Probability, Statistics, and Number Theory, Urbana, Illinois, May 31 - June 1, 1997.

[44] _____, *Maximum of partial sums and an invariance principle for a class of weak dependent random variables*, Proc. Amer. Math. Soc. **126** (1998), 1181–1189.

[45] M. Peligrad and A. Gut, *Almost sure results for a class of dependent random variables*, U.U.D.M. Report 1997:3, ISSN 1101-3591, Department of Mathematics, Uppsala University, Uppsala, Sweden (1997).

[46] _____, *Almost sure results for a class of dependent random variables*, J. Theor. Probab. (to appear).

[47] G. Perera, *Geometry of $\mathbb{Z}^d$ and the central limit theorem for weakly dependent random fields*, J. Theor. Probab. **10** (1997), 581–603.

[48] M. Rosenblatt, *A central limit theorem and a strong mixing condition*, Proc. Natl. Acad. Sci. USA **42** (1956), 43–47.

[49] _____, *Stationary Sequences and Random Fields*, Birkhäuser, Boston (1985).

[50] _____, Private communication (1987).

[51] C. Stein, *A bound for the error in the normal approximation to the distribution of a sum of dependent random variables*, Proceedings of the Sixth Berkeley Symposium on Probability and Statistics, University of California Press, Los Angeles **2**, 583–602 (1972).

[52] P.J. Szablowski, *Generalized laws of large numbers and auxiliary results concerning stochastic approximation with dependent disturbances. II*, Comm. Math. Appl. **13** (1987), 973–987.

[53] S.A. Utev, Private communication (1993).

[54] C.S. Withers, *Central limit theorems for dependent random variables I.*, Z. Wahrsch. verw. Gebiete **57** (1981), 509–534.

[55] I.G. Zhurbenko, *The Spectral Analysis of Time Series*, North-Holland, Amsterdam (1986).

[56] _____, Private communication (1995).

E-mail address: bradleyr@indiana.edu

DEPARTMENT OF MATHEMATICS, INDIANA UNIVERSITY, BLOOMINGTON, IN 47405

Variance inequalities for functions of multivariate random variables

Wan-Ying Chang and Donald St. P. Richards

ABSTRACT. We apply orthogonal expansions to derive variance inequalities for functions of multivariate random variables having a variety of distributions. We give a new construction of the generalized Jacobi polynomials which form an orthogonal system with respect to the generalized Dirichlet distributions; then two variance inequalities for the generalized Dirichlet distributions are obtained: one by the method of orthogonal expansions and another by an inductive method. We obtain variance inequalities for symmetric functions of the eigenvalues of the Wishart and multivariate beta random matrices. We extend an inequality of Houdré and Kagan (1995) to the multivariate normal distributions, and provide an application to the calculation of expectations based on Edgeworth approximations.

1. Introduction

This paper was motivated by the article of Houdré and Kagan (1995) which deals with a class of variance inequalities for the normal distribution. Let X be a random variable having a standard normal distribution. The inequality

(1.1) $$\operatorname{Var} g(X) \leq E[g'(X)]^2,$$

which is valid for all absolutely continuous functions $g : \mathbf{R} \to \mathbf{R}$ such that $\operatorname{Var} g(X) < \infty$, has been proved by several methods: Nash (1958) and Chernoff (1981) proved (1.1) using a Hermite polynomial expansion of g; Brascamp and Lieb (1976) proved a Sobolev-type inequality which implies (1.1); Chen (1982) derived a multivariate generalization of (1.1) using the Cauchy-Schwarz inequality; Cacoullos (1982) and Klaassen (1985) obtained lower bounds for $\operatorname{Var} g(X)$ when X is standard normal; and Vitale (1989) obtained analogs of (1.1) when X is infinitely divisible. Recently, Houdré and Kagan (1995) (cf. Houdré, 1995; Houdré, et al., 1995, 1998) generalized (1.1) using higher-order derivatives of g.

1991 *Mathematics Subject Classification*. Primary 60E15, 62E15; Secondary 33C50, 33D80.
Key words and phrases. Compositional data analysis, Edgeworth approximation, generalized binomial coefficients, generalized Dirichlet distribution, Hermite polynomials, Jack polynomials, Laguerre polynomials, Jacobi polynomials, Laplace-Beltrami operator, multivariate beta distribution, normal distribution, orthogonal expansions, Sobolev inequality, Wishart distribution.
The first author is partially supported by NSF grant DMS-9401322, and by NIMH grant P50-MH49173-04A1.
The second author is partially supported by NSF grant DMS-9401322 and DMS-9703705.

We remark also that the method of orthogonal expansions for deriving inequalities of the type (1.1) dates at least to Weyl (1940, pp. 776-778), who utilized an orthogonal basis of trigonometric functions to derive an inequality similar to (1.1) for functions defined on the unit cube in $\mathbf{R}^3$.

In this paper we derive variance inequalities for random vectors from a variety of distributions. Our primary approach, the method of orthogonal expansions, has two advantages. A review of earlier work on (1.1) indicates that, except for this method, all approaches to (1.1) are strongly dependent on the assumption that the random variable X is normally distributed. Moreover, the method has an algebraic flavor which lends itself readily to the study of inequalities similar to (1.1) for other distributions. As we show, the method leads to inequalities reminiscent of (1.1) for a variety of random entities X. We develop these inequalities for cases in which X has a generalized Dirichlet, or multivariate normal, distribution; and also for cases in which X is the vector of eigenvalues of a Wishart or multivariate beta matrix.

In the first step toward deriving these variance inequalities, we find a a complete, orthogonal polynomial basis for the space of functions which are square-integrable with respect to the specified distribution. Next we find a differential operator for which these orthogonal polynomials all are eigenfunctions. An upper bound for the variance, Var $g(X)$, then is obtained by application of the orthogonality and eigenfunction properties of the system of polynomials.

In the case of the generalized Dirichlet distributions we utilize a system of generalized Jacobi polynomials, first constructed by Koornwinder and Schwartz (1997). We provide a new construction of these polynomials using properties of the marginal and conditional distributions. Two types of variance inequalities for the generalized Dirichlet distributions are obtained: one by the method of orthogonal expansions, and another by an inductive method starting with a theorem of Brascamp and Lieb (1976).

We derive variance inequalities for the eigenvalues of the Wishart distributions using the theory of generalized Laguerre polynomials of matrix argument (Muirhead, 1982). We apply the generalized Jacobi polynomials of matrix argument, due to Constantine and James (1974) and Lassalle (1991a), to derive variance inequalities for the eigenvalues of the multivariate beta matrices. We apply the multivariate Hermite polynomials to obtain a variance inequality for the multivariate normal distributions, thereby strengthening a result of Chen (1982) and extending an inequality of Houdré and Kagan (1995). Finally, we make an application to calculating expectations based on Edgeworth approximations.

2. Generalized Dirichlet distributions and generalized Jacobi polynomials

DEFINITION 2.1. A random vector $(X_1, \ldots, X_n)$ on the simplex $\mathcal{S}_n = \{(x_1, \ldots, x_n) : x_1 > 0, \ldots, x_n > 0, \sum_{i=1}^n x_i < 1\}$ is said to have a *generalized Dirichlet distribution*, denoted $(X_1, \ldots, X_n) \sim GD(a_1, \ldots, a_n; b_1, \ldots, b_n)$, if its probability density function (p.d.f.) exists and is of the form

$$(2.1) \qquad f(x_1, \ldots, x_n) = c \prod_{i=1}^n \left[x_i^{a_i - 1} \bigl(1 - \sum_{k=1}^i x_k\bigr)^{b_i - 1} \right],$$

$(x_1, \ldots, x_n) \in \mathcal{S}_n$, where $a_i, b_i > 0$, $i = 1, \ldots, n$.

It may be shown that the normalizing constant in (2.1) is

$$c = \prod_{i=1}^{n} \frac{\Gamma(\sum_{k=i}^{n}(a_k + b_k - 1) + 1)}{\Gamma(a_i)\Gamma(b_i + \sum_{k=i+1}^{n}(a_k + b_k - 1))}.$$

The generalized Dirichlet distributions arise in the study of concepts of independence satisfied by distributions utilized to model compositional data; cf. Aitchison (1986) and Connor and Mosimann (1969).

LEMMA 2.2 (Connor and Mosimann, 1969). *Suppose that the random vector* $(X_1, \ldots, X_n) \sim GD(a_1, \ldots, a_n; b_1, \ldots, b_n)$, *and let* $1 \leq k \leq n-1$.
 (i) *The marginal distribution of* $(X_1, \ldots, X_k)$ *is* $GD(a_1, \ldots, a_k; b_1, \ldots, b_{k-1}, b_k + \sum_{i=k+1}^{n}(a_i + b_i - 1))$.
 (ii) *Let* $U_i = X_i/(1 - X_1 - \cdots - X_k)$ *for* $i = k+1, \ldots, n$. *Then the conditional distribution of* $(U_{k+1}, \ldots, U_n)$, *given* $(X_1, \ldots, X_k) = (x_1, \ldots, x_k)$, *is* $GD(a_{k+1}, \ldots, a_n; b_{k+1}, \ldots, b_n)$.

Note that the second part of Lemma 2.2 is equivalent to the statement that the conditional distribution of $(X_{k+1}, \ldots, X_n)$ given $(X_1, \ldots, X_k)$, after a suitable scaling, is again in the generalized Dirichlet class.

Let us recall now some properties of the classical Jacobi polynomials.

DEFINITION 2.3 (Szegö, 1975). For $a, b > -1$, the classical Jacobi polynomial $P_j^{(a,b)}(x)$, $x \in (-1, 1)$, $j = 0, 1, 2, \ldots$, is defined by the *Rodrigues formula*,

$$(2.2) \qquad P_j^{(a,b)}(x) = \frac{(-1)^j}{j! 2^j}(1-x)^{-a}(1+x)^{-b} \frac{d^j}{dx^j}(1-x)^{a+j}(1+x)^{b+j}.$$

The system of polynomials $\{P_j^{(a,b)}(\cdot) : j = 0, 1, 2, \ldots\}$ is orthogonal with respect to the weight function $w(x) = (1-x)^a(1+x)^b$, $-1 < x < 1$; and the corresponding orthogonality relations are

$$(2.3) \qquad \int_{-1}^{1} P_j^{(a,b)}(x) P_k^{(a,b)}(x) w(x) dx = \delta_{jk} C_j^{(a,b)},$$

where δ_{jk} denotes Kronecker's delta and

$$(2.4) \qquad C_j^{(a,b)} = \frac{2^{a+b+1} \Gamma(a+j+1)\Gamma(b+j+1)}{j!(a+b+2j+1)\Gamma(a+b+j+1)}.$$

The system $\{P_j^{(a,b)}(\cdot) : j = 0, 1, 2, \ldots\}$ is complete for $L^2(w)$. Further, the polynomials $P_j^{(a,b)}(\cdot)$ satisfy the differential equations

$$(2.5) \qquad \frac{d}{dx} P_j^{(a,b)}(x) = \frac{1}{2}(j + a + b + 1) P_{j-1}^{(a+1,b+1)}(x).$$

It follows by a translation and dilation that, in the case of the weight function $(\theta_2 - x)^a (x - \theta_1)^b$ on the finite interval (θ_1, θ_2), the corresponding orthogonal polynomials are $P_j^{(a,b)}\left(2\frac{x-\theta_1}{\theta_2-\theta_1} - 1\right)$, $j = 0, 1, 2, \ldots$. In the case in which $\theta_1 = 0$ and $\theta_2 = 1$, we find that the system of polynomials $\{P_j^{(a,b)}(2x - 1) : j = 0, 1, 2, \ldots\}$ is complete and orthogonal for the complex-valued functions which are square-integrable with respect to the weight function $x^b(1-x)^a$, $0 < x < 1$.

To illustrate our construction of the generalized Jacobi polynomials for the generalized Dirichlet distributions, we first study the two-dimensional case; then the corresponding weight function is

$$w(x_1, x_2) = x_1^{a_1} x_2^{a_2} (1 - x_1)^{b_1} (1 - x_1 - x_2)^{b_2},$$

$(x_1, x_2) \in \mathcal{S}_2$; that is, w is a multiple of the p.d.f. of the random vector $(X_1, X_2) \sim GD(a_1 + 1, a_2 + 1; b_1 + 1, b_2 + 1)$. We will construct the orthogonal polynomials for this distribution as the product of two orthogonal systems: one orthogonal with respect to the marginal distribution of X_1 and the second orthogonal with respect to the conditional distribution of X_2 given X_1.

By Lemma 2.2, both the marginal distribution of X_1 and the conditional distribution of $X_2/(1 - x_1)$, given $X_1 = x_1$, are beta distributions. Therefore the corresponding orthogonal systems may be expressed in terms of the classical Jacobi polynomials. Since the joint p.d.f. of (X_1, X_2) is the product of the marginal p.d.f. of X_1 and the conditional p.d.f. of X_2, given $X_1 = x_1$, then we define the generalized Jacobi polynomial

$$(2.6) \quad P_{j,k}^{(a_1, a_2, b_1, b_2)}(x_1, x_2) := P_{j-k}^{(b_1 + a_2 + b_2 + 2k + 1, a_1)}(2x_1 - 1) \\ \times P_k^{(b_2, a_2)}\left(\frac{2x_2}{1 - x_1} - 1\right)(1 - x_1)^k,$$

$j \geq k \geq 0$. It is not difficult to verify that these polynomials are orthogonal with respect to the joint distribution of (X_1, X_2); further,

$$(2.7) \quad \begin{aligned} C_{j,k}^{(a_1, a_2, b_1, b_2)} &:= E[P_{j,k}^{(a_1, a_2, b_1, b_2)}(X_1, X_2)]^2 \\ &= 2^{-(a_1 + b_1 + 2a_2 + 2b_2 + 2k + 3)} C_k^{(b_2, a_2)} C_{j-k}^{(b_1 + a_2 + b_2 + 2k + 1, a_1)} \end{aligned}$$

where $C_j^{(a,b)}$ is given in (2.4).

The special case of (2.6), in which $b_1 = 0$, is due to Proriol (1957).

Before presenting the details of the higher-dimensional Jacobi polynomials we need to establish the orthogonality and completeness properties of the system (2.6). We will do so in a more abstract setting as follows. For a random entity $\mathbf{X}$, let $L^2(\mathbf{X})$ denote the space of all complex-valued functions $f(\mathbf{X})$ such that $\text{Var } f(\mathbf{X}) < \infty$. The space $L^2(\mathbf{X})$ is a (complex) Hilbert space when equipped with the inner product $(f, g) := E f(\mathbf{X}) \overline{g(\mathbf{X})}$.

THEOREM 2.4. *Let $(\mathbf{X}, \mathbf{Y})$ be a pair of random entities taking values in a space $(\mathcal{X}, \mathcal{Y})$. Suppose that $\{\phi_j(\mathbf{x}) : j \in I\}$ is a complete, orthonormal set of functions in $L^2(\mathbf{X})$; and for each $\mathbf{x} \in \mathcal{X}$, $\{\psi_k(\mathbf{y}|\mathbf{x}) : k \in I\}$ is a complete orthonormal set of functions for $L^2(\mathbf{Y}|\mathbf{X} = \mathbf{x})$. Then $\{\phi_j(\mathbf{x})\psi_k(\mathbf{y}|\mathbf{x}) : j, k \in I\}$ is a complete orthonormal set for $L^2(\mathbf{X}, \mathbf{Y})$.*

PROOF. First we show that the function $\phi_j(\mathbf{x})\psi_k(\mathbf{y}|\mathbf{x}) \in L^2(\mathbf{X}, \mathbf{Y})$ for all $j, k \in I$. Indeed, by Tonelli's theorem, for any $j, k \in I$,

$$E|\phi_j(\mathbf{X})\psi_k(\mathbf{Y}|\mathbf{X})|^2 = E_{\mathbf{X}}|\phi_j(\mathbf{X})|^2 E_{\mathbf{Y}|\mathbf{X}}|\psi_k(\mathbf{Y}|\mathbf{X})|^2 = E|\phi_j(\mathbf{X})|^2 = 1.$$

To establish the orthonormality of the system, choose two pairs of indices, (j_1, k_1) and (j_2, k_2). By Fubini's theorem,

$$E\,\phi_{j_1}(\mathbf{X})\psi_{k_1}(\mathbf{Y}|\mathbf{X})\overline{\phi_{j_2}(\mathbf{X})\psi_{k_2}(\mathbf{Y}|\mathbf{X})} = E_{\mathbf{X}}E_{\mathbf{Y}|\mathbf{X}}\phi_{j_1}(\mathbf{X})\overline{\phi_{j_2}(\mathbf{X})}\psi_{k_1}(\mathbf{Y}|\mathbf{X})\overline{\psi_{k_2}(\mathbf{Y}|\mathbf{X})}$$
$$= E_{\mathbf{X}}\phi_{j_1}(\mathbf{X})\overline{\phi_{j_2}(\mathbf{X})}\delta_{k_1,k_2}$$
$$= \begin{cases} 1, & \text{if } (j_1, k_1) = (j_2, k_2) \\ 0, & \text{otherwise.} \end{cases}$$

This establishes the orthonormality property.

To prove completeness, we choose a function $g \in L^2(\mathbf{X}, \mathbf{Y})$ and define

$$c_{j,k} = E_{\mathbf{X},\mathbf{Y}}g(\mathbf{X},\mathbf{Y})\overline{\phi_j(\mathbf{X})\psi_k(\mathbf{Y}|\mathbf{X})}].$$

By Parseval's theorem, we must verify that $\|g\|^2 = \sum_{j,k \in I} |c_{j,k}|^2$. Since $\infty > E_{\mathbf{X},\mathbf{Y}}|g(\mathbf{X},\mathbf{Y})|^2 = E_{\mathbf{X}}[E_{\mathbf{Y}|\mathbf{X}=\mathbf{x}}|g(\mathbf{X},\mathbf{Y})|^2]$ then $E_{\mathbf{Y}|\mathbf{X}=\mathbf{x}}|g(\mathbf{X},\mathbf{Y})|^2 < \infty$ a.e. $\mathbf{x}$; that is, $g(\mathbf{x}, \mathbf{Y}) \in L^2(\mathbf{Y}|\mathbf{X} = \mathbf{x})$ for a.e. $\mathbf{x}$. By the completeness of the system $\{\psi_k\}$ in $L^2(\mathbf{Y}|\mathbf{X} = \mathbf{x})$, we have for a.e. $\mathbf{x}$,

$$(2.8) \qquad g(\mathbf{x},\mathbf{y}) = \sum_{k \in I} a_k(\mathbf{x})\psi_k(\mathbf{y}|\mathbf{x})$$

for some sequence $\{a_k(\mathbf{x})\}$, where the series (2.8) converges in the norm on $L^2(\mathbf{X})$. Moreover $a_k(\mathbf{x}) = E_{\mathbf{Y}|\mathbf{x}}g(\mathbf{X},\mathbf{Y})\overline{\psi_k(\mathbf{Y}|\mathbf{x})}$ and, by Parseval's theorem,

$$(2.9) \qquad \|g(\mathbf{x},\mathbf{Y})\|^2_{L^2(\mathbf{Y}|\mathbf{x})} = \sum_{k \in I} |a_k(\mathbf{x})|^2.$$

By the completeness of the system $\{\psi_k(\mathbf{y}|\mathbf{x})\}$, Parseval's theorem, and the monotone convergence theorem,

$$\infty > E_{\mathbf{X},\mathbf{Y}}|g(\mathbf{X},\mathbf{Y})|^2 = E_{\mathbf{X}}E_{\mathbf{Y}|\mathbf{X}}|g(\mathbf{X},\mathbf{Y})|^2 = E_{\mathbf{X}}\sum_{k \in I}|a_k(\mathbf{X})|^2 \geq E_{\mathbf{X}}|a_k(\mathbf{X})|^2.$$

Therefore, $a_k(\mathbf{x}) \in L^2(\mathbf{X})$ for each k, and a.e. $\mathbf{x}$. Further,

$$c_{j,k} = E_{\mathbf{X}}\overline{\phi_j(\mathbf{X})}E_{\mathbf{Y}|\mathbf{X}}g(\mathbf{X},\mathbf{Y})\overline{\psi_k(\mathbf{Y}|\mathbf{X})} = E_{\mathbf{X}}a_k(\mathbf{X})\overline{\phi_j(\mathbf{X})};$$

hence $c_{j,k}$ is the jth Fourier coefficient of $a_k(\mathbf{x})$.

By the completeness of the system $\{\phi_j : j \in I\}$ we may write

$$a_k(\mathbf{x}) = \sum_{j \in I} c_{j,k}\phi_j(\mathbf{x}),$$

and it is now clear that

$$\|a_k(\mathbf{x})\|^2 = \sum_{j \in I} |c_{j,k}|^2.$$

By (2.8), (2.9), and Fubini's theorem, we have

$$\|g\|^2 = E_{\mathbf{X}}E_{\mathbf{Y}|\mathbf{X}}|g(\mathbf{X},\mathbf{Y})|^2 = E_{\mathbf{X}}\sum_{k \in I}|a_k(\mathbf{X})|^2 = \sum_{j,k \in I}|c_{j,k}|^2.$$

This completes the proof of the completeness property. $\square$

COROLLARY 2.5. *Suppose* $(X_1, X_2) \sim GD(a_1 + 1, a_2 + 1; b_1 + 1, b_2 + 1)$. *Then the system* $\{P^{(a_1,a_2,b_1,b_2)}_{j,k}(x_1,x_2) : j \geq k \geq 0\}$ *of generalized Jacobi polynomials, defined in* (2.6), *forms a complete orthogonal basis for* $L^2(X_1, X_2)$.

The following result extends, to the generalized Jacobi polynomials, the differential equation (2.5) for the classical Jacobi polynomials. The result will be needed to derive variance inequalities for the generalized Dirichlet distributions.

LEMMA 2.6. *The generalized Jacobi polynomials defined in (2.6) satisfy the partial differential equation*

$$(2.10) \qquad \frac{\partial}{\partial x_2} P_{j,k}^{a_1,a_2,b_1,b_2}(x_1,x_2) = (k+a_2+b_2+1) P_{j-1,k-1}^{a_1,a_2+1,b_1,b_2+1}(x_1,x_2).$$

PROOF. By (2.5) and (2.6),

$$\frac{\partial}{\partial x_2} P_{j,k}^{(a_1,a_2,b_1,b_2)}(x_1,x_2)$$

$$= P_{j-k}^{(b_1+a_2+b_2+2k+1,a_1)}(2x_1-1)(1-x_1)^k \frac{\partial}{\partial x_2} P_k^{(b_2,a_2)}\left(\frac{2x_2}{1-x_1}-1\right)$$

$$= (k+a_2+b_2+1) P_{j-k}^{(b_1+a_2+b_2+2k+1,a_1)}(2x_1-1)$$

$$\times (1-x_1)^{k-1} P_{k-1}^{(b_2+1,a_2+1)}\left(\frac{2x_2}{1-x_1}-1\right)$$

$$\equiv (k+a_2+b_2+1) P_{(j-1)-(k-1)}^{(b_1+(a_2+1)+(b_2+1)+2(k-1)+1,a_1)}(2x_1-1)$$

$$\times P_{k-1}^{(b_2+1,a_2+1)}\left(\frac{2x_2}{1-x_1}-1\right)(1-x_1)^{k-1},$$

which is the right-hand side of (2.10). □

We now construct the generalized Jacobi polynomials by an approach based on Theorem 2.4. These polynomials will form a complete orthogonal basis for the space of functions square-integrable with respect to the n-dimensional generalized Dirichlet distributions.

Let $(X_1, \ldots, X_n) \sim GD(a_1+1, \ldots, a_n+1; b_1+1, \ldots, b_n+1)$. We decompose the joint p.d.f. (2.1), of $(X_1, \ldots, X_n)$, into the product

$$(2.11) \qquad f_{(X_1,\ldots,X_n)}(x_1,\ldots,x_n) = f_{X_1}(x_1) \prod_{j=2}^n f_{X_j|\{X_1=x_1,\ldots,X_{j-1}=x_{j-1}\}}(x_j),$$

where f_{X_1} is the marginal p.d.f. of X_1; and, for $2 \leq j \leq n$, $f_{X_j|X_1=x_1,\ldots,X_{j-1}=x_{j-1}}$ is the conditional p.d.f. of X_j given $\{X_1 = x_1, \ldots, X_{j-1} = x_{j-1}\}$. By Lemma 2.2, X_1 is beta-distributed; and after suitable scaling, $X_j|\{X_1 = x_1, \ldots, X_{j-1} = x_{j-1}\}$ is also beta-distributed, for all $j = 2, \ldots, n$. Corresponding to each density function in the product (2.11) there exists a complete system of orthogonal polynomials which is expressible in terms of the classical Jacobi polynomials.

Therefore we define the generalized Jacobi polynomials for the generalized Dirichlet distribution, $GD(a_1+1, \ldots, a_n+1; b_1+1, \ldots, b_n+1)$, as the product

$$P_{k_1,\ldots,k_n}^{(a_1,\ldots,a_n,b_1,\ldots,b_n)}(x_1,\ldots,x_n)$$

$$(2.12) \qquad := \prod_{i=1}^n \left[P_{k_i}^{(\beta_i,a_i)}\left(\frac{2x_i}{1-\sum_{j=1}^{i-1} x_j}-1\right)\left(1-\sum_{j=1}^{i-1} x_j\right)^{k_i} \right]$$

where $k_i \geq 0$, $i = 1, \ldots, n$;

$$(2.13) \qquad \beta_i := b_i + \sum_{j=i+1}^n (a_j + b_j + 2k_j + 1);$$

and the $P_{k_i}^{(\beta_i,a_i)}$ are the classical Jacobi polynomials defined in (2.2). Further,

$$
\begin{aligned}
(2.14) \quad C_{k_1,\ldots,k_n}^{(a_1,\ldots,a_n,b_1,\ldots,b_n)} &:= E[P_{k_1,\ldots,k_n}^{(a_1,\ldots,a_n,b_1,\ldots,b_n)}(X_1,\ldots,X_n)]^2 \\
&= \prod_{i=1}^{n} 2^{-(\beta_i+a_i+1)} C_{k_i}^{(\beta_i,a_i)}
\end{aligned}
$$

where $C_j^{(a,b)}$ is given in (2.4).

Now the following result follows directly from Theorem 2.4.

THEOREM 2.7 (Koornwinder and Schwartz, 1997). *The system of generalized Jacobi polynomials (2.12) is orthogonal with respect to the density function (2.1). Moreover, the system (2.12) is complete for $L^2(X_1,\ldots,X_n)$, the space of functions $g: \mathbf{R}^n \to \mathbf{R}$ such that $\operatorname{Var} g(X_1,\ldots,X_n) < \infty$.*

An alternative proof of the orthogonality of the system (2.12) can be obtained by induction on n using the orthogonality property of the classical Jacobi polynomials. Also, an alternative proof of completeness can be obtained by noting that the system is *simple*; that is, for each $(k_1,\ldots,k_n)$, the polynomial $P_{k_1,\ldots,k_n}^{(a_1,\ldots,a_n,b_1,\ldots,b_n)}(x_1,\ldots,x_n)$ is of degree k_i in x_i, for all $i = 1,\ldots,n$; then the completeness property follows from a multi-dimensional analog of a result of Higgins (1977), p. 31.

We also remark that the generalization of (2.10) to the n-dimensional generalized Jacobi polynomials is

$$
\begin{aligned}
&\frac{\partial}{\partial x_n} P_{k_1,\ldots,k_n}^{(a_1,\ldots,a_n,b_1,\ldots,b_n)}(x_1,\ldots,x_n) \\
(2.15) \quad &= (a_n + \beta_n + k_n + 1) P_{k_1,\ldots,k_{n-1},k_n-1}^{(a_1,\ldots,a_{n-1},a_n+1,b_1,\ldots,b_{n-1},b_n+1)}(x_1,\ldots,x_n).
\end{aligned}
$$

Further, it follows from the conditional distribution property in Lemma 2.2 (ii), and the product nature of (2.12) that, for all $i = 1,\ldots,n$,

$$
\begin{aligned}
E_{X_{i+1},\ldots,X_n|X_1=x_1,\ldots,X_i=x_i} & P_{k_1,\ldots,k_n}^{(a_1,\ldots,a_n,b_1,\ldots,b_n)}(X_1,\ldots,X_n) \\
(2.16) \quad &= P_{k_1,\ldots,k_i}^{(a_1,\ldots,a_i,b_1,\ldots,b_i)}(x_1,\ldots,x_i).
\end{aligned}
$$

3. Variance inequalities for the generalized Dirichlet distributions

In this section we derive two variance inequalities for the generalized Dirichlet distributions. We obtain the first inequality by induction, starting with a result on the beta distributions. The second variance inequality is obtained by the method of orthogonal expansions based on the generalized Jacobi polynomials.

3.1. Application of an inductive method.

LEMMA 3.1. *Suppose X is a random variable having a beta distribution on the interval (θ_1, θ_2), with density function proportional to $(\theta_2 - x)^a(x - \theta_1)^b$, where $a, b > 0$. Then for any $g \in C^1(\theta_1, \theta_2)$ with $\operatorname{Var} g(X) < \infty$,*

$$\operatorname{Var} g(X) \leq \frac{a+b}{ab(\theta_2 - \theta_1)^2} E[(\theta_2 - X)(X - \theta_1) g'(X)]^2.$$

PROOF. Since f, the p.d.f. of X, satisfies

$$f(x) \propto (\theta_2 - x)^a (x - \theta_1)^b = e^{a \ln(\theta_2 - x) + b \ln(x - \theta_1)}, \quad \theta_1 < x < \theta_2,$$

then
$$(-\ln f(x))'' = \frac{a(x - \theta_1)^2 + b(\theta_2 - x)^2}{(\theta_2 - x)^2 (x - \theta_1)^2} > 0.$$

By Theorem 4.1 of Brascamp and Lieb (1976),

$$(3.1) \quad \operatorname{Var} g(X) \leq E\left[\frac{(g'(X))^2}{(-\ln f)''(X)}\right] = E\left[(g'(X))^2 \frac{(\theta_2 - X)^2 (X - \theta_1)^2}{a(X - \theta_1)^2 + b(\theta_2 - X)^2}\right].$$

Since the function $a(x - \theta_1)^2 + b(\theta_2 - x)^2$, $\theta_1 < x < \theta_2$, has minimum value $ab(\theta_2 - \theta_1)^2/(a+b)$ at $x = (a\theta_1 + b\theta_2)/(a+b)$, then we obtain

$$\operatorname{Var} g(X) \leq \frac{(a+b)}{ab(\theta_2 - \theta_1)^2} E[(g'(X))^2 (\theta_2 - X)^2 (X - \theta_1)^2],$$

which is the desired result. □

REMARK 3.2. Note also that if we replace $a(x - \theta_1)^2 + b(\theta_2 - x)^2$ in (3.1) by $a(x - \theta_1)^2$ or $b(\theta_2 - x)^2$ then we obtain the simpler upper bound,

$$\operatorname{Var} g(X) \leq \min\left\{a^{-1} E[(\theta_2 - X) g'(X)]^2, b^{-1} E[(X - \theta_1) g'(X)]^2\right\}.$$

NOTATION 3.3. Let $(X_1, \ldots, X_n)$ be a random vector.

(i) For $1 \leq i \leq n$ and g a suitable function of $(X_1, \ldots, X_i)$, we denote by $E_{X_1,\ldots,X_i} g$ and $\operatorname{Var}_{X_1,\ldots,X_i} g$ the expectation and variance, respectively, of $g(X_1, \ldots, X_i)$ with respect to the marginal distribution of $(X_1, \ldots, X_i)$.

(ii) For $i = 1, \ldots, n-1$ and g a suitable function of $(X_1, \ldots, X_n)$, we denote by $E_{X_1,\ldots,X_i | X_{i+1},\ldots,X_n} g$ and $\operatorname{Var}_{X_1,\ldots,X_i | X_{i+1},\ldots,X_n} g$ the conditional expectation and variance, respectively, of $g(X_1, \ldots, X_n)$ with respect to the conditional distribution of $(X_1, \ldots, X_i)$ given $(X_{i+1}, \ldots, X_n)$.

Further, if g is differentiable, we denote by g_{x_i} the partial derivative, $\partial g / \partial x_i$, of $g(x_1, \ldots, x_n)$ with respect to the ith variable x_i.

LEMMA 3.4. *Let (X_1, X_2) be a random vector and $g : \mathbf{R}^2 \to \mathbf{R}$ be such that $\operatorname{Var} g(X_1, X_2) < \infty$. Then*

$$\operatorname{Var} g(X_1, X_2) = E_{X_1} \operatorname{Var}_{X_2 | X_1} g(X_1, X_2) + \operatorname{Var}_{X_1} E_{X_2 | X_1} g(X_1, X_2).$$

PROOF. Denoting $g(X_1, X_2)$ by g, then, by direct calculation, we obtain

$$\operatorname{Var} g = E_{X_1} E_{X_2 | X_1} g^2 - (E_{X_1} E_{X_2 | X_1} g)^2$$
$$= E_{X_1} \big(E_{X_2 | X_1} g^2 - (E_{X_2 | X_1} g)^2\big) + E_{X_1} (E_{X_2 | X_1} g)^2 - \big(E_{X_1} E_{X_2 | X_1} g\big)^2$$
$$= E_{X_1} (\operatorname{Var}_{X_2 | X_1} g) + \operatorname{Var}_{X_1} (E_{X_2 | X_1} g),$$

which is the desired result. □

LEMMA 3.5. *Let $(X_1, X_2) \sim GD(a_1+1, a_2+1; b_1+1, b_2+1)$ where $a_1, a_2, b_2 > 0$. If $g : \mathbf{R}^2 \to \mathbf{R}$ and $g \in C^1(\mathbf{R}^2)$ with $\operatorname{Var} g(X_1, X_2) < \infty$ then*

$$\operatorname{Var} g(X_1, X_2) \leq \frac{a_1 + b_1 + a_2 + b_2 + 1}{a_1(b_1 + a_2 + b_2 + 1)} E_{X_1}[X_1(1 - X_1)(E_{X_2 | X_1} g(X_1, X_2))_{x_1}]^2$$
$$+ \frac{a_2 + b_2}{a_2 b_2} E\left[\frac{X_2(1 - X_1 - X_2)}{1 - X_1} g_{x_2}(X_1, X_2)\right]^2.$$

PROOF. By Lemma 2.2, $X_1 \sim B(a_1 + 1, b_1 + a_2 + b_2 + 2)$; and the conditional p.d.f. of X_2, given $X_1 = x_1$, is proportional to $x_2^{a_2}(1 - x_1 - x_2)^{b_2}$, $0 < x_2 < 1 - x_1$. Now we apply Lemma 3.4 and Lemma 3.1 both to X_1 and to $X_2|X_1$, obtaining

$$\text{Var}\, g(X_1, X_2) = E_{X_1} \text{Var}_{X_2|X_1} g(X_1, X_2) + \text{Var}_{X_1} E_{X_2|X_1} g(X_1, X_2)$$

$$\leq E_{X_1} \frac{a_2 + b_2}{a_2 b_2} E_{X_2|X_1} \left[\frac{X_2(1 - X_1 - X_2)}{1 - X_1} g_{x_2}(X_1, X_2) \right]^2$$

$$+ E_{X_1} \frac{a_1 + b_1 + a_2 + b_2 + 1}{a_1(b_1 + a_2 + b_2 + 1)} \left[X_1(1 - X_1)(E_{X_2|X_1} g(X_1, X_2))_{x_1} \right]^2,$$

which is the desired result. $\square$

THEOREM 3.6. *Let* $(X_1, \ldots, X_n) \sim GD(a_1 + 1, \ldots, a_n + 1; b_1 + 1, \ldots, b_n + 1)$, *where* $a_1, \ldots, a_n, b_n > 0$. *If* $g : \mathbf{R}^n \to \mathbf{R}$ *is such that* $g \in C^1(\mathbf{R}^n)$ *and* $\text{Var}\, g(X_1, \ldots, X_n) < \infty$, *then*

$$\text{Var}\, g(X_1, \ldots, X_n) \leq \sum_{i=1}^n c_i E \left[\frac{X_i(1 - X_1 - \cdots - X_i)}{1 - X_1 - \cdots - X_{i-1}} (h_i(X_1, \ldots, X_i))_{x_i} \right]^2$$

where, for $i = 1, \ldots, n$,

$$h_i(X_1, \ldots, X_i) := E_{X_{i+1}, \ldots, X_n | X_1, \ldots, X_i} g(X_1, \ldots, X_n)$$

and

$$c_i = \frac{n - i + \sum_{k=i}^n (a_k + b_k)}{a_i [b_i + n - i + \sum_{k=i+1}^n (a_k + b_k)]},$$

PROOF. For $n = 1, 2$, the result follows from Lemma 3.1 and Lemma 3.5, respectively. The rest of the proof follows by repeated application of Lemma 3.4, and induction on n. $\square$

3.2. Application of the method of orthogonal expansions. Throughout the rest of the paper, we derive bounds for $\text{Var}\, g(X_1, \ldots, X_n)$ by expanding the (*non-constant*) function g in an orthogonal series; applying a differential operator D termwise to this series; and then comparing the expressions for $\text{Var}\, g(X_1, \ldots, X_n)$ and $E[Dg((X_1, \ldots, X_n)]^2$ resulting from the orthogonal expansions. In all instances we assume, and this will hence be referred to as the *standard assumption*, that all interchanges of derivatives, summations, and integrals, are valid. Generally speaking, the standard assumption requires that both the orthogonal series and its term-wise derivative converge sufficiently rapidly. In effect, this assumption requires that the coefficients in the orthogonal expansion of the function g converge to zero sufficiently quickly.

Consider the beta distribution on the interval $(0,1)$, where the weight function is $x^a(1-x)^b$ and the classical Jacobi polynomials $P_j^{(b,a)}(2x-1)$, $j = 0, 1, 2, \ldots$, are the corresponding orthogonal polynomials. Let d/dx be the differential operator D; then, noting the differential recurrence relation (2.5), the standard assumption reduces to the requirement that the orthogonal expansion,

(3.2) $$g(x) = \sum_{j=0}^\infty c_j P_j^{(b,a)}(2x - 1), \quad 0 < x < 1,$$

and its term-wise derived series,

$$(3.3) \quad \sum_{j=0}^{\infty} c_j [P_j^{(b,a)}(2x-1)]' \equiv \sum_{j=1}^{\infty} (j+a+b+1) c_j P_{j-1}^{(b+1,a+1)}(2x-1),$$

both are convergent pointwise; and that the series (3.3) converge to $g'(x)$. Fomin (1993) proved that the condition

$$\sum_{j=1}^{\infty} j^{(2b+3)/2} \left| \frac{c_j}{P_j^{(b,a)}(1)} - \frac{c_{j+1}}{P_{j+1}^{(b,a)}(1)} \right| < \infty$$

is a necessary assumption about the Fourier coefficients c_j.

Although there are criteria, such as the Rademacher-Menchoff theorem (Alexits, 1961, p. 80), which imply the almost everywhere convergence of a general orthogonal series, little is known about conditions under which a term-wise derived orthogonal series for a function g converges to the function Dg; this situation holds both for general orthogonal series on the real line or on more complicated spaces. Therefore, we will retain the standard assumption throughout the rest of the paper.

Now we derive variance inequalities for the generalized Dirichlet distributions.

LEMMA 3.7. *Let $X \sim B(a+1, b+1)$, a beta distribution on $(0,1)$, where $a+b \geq -1$. If $g: \mathbf{R} \to \mathbf{R}$ is such that $g \in C^1(\mathbf{R})$ and $\operatorname{Var} g(X) < \infty$, then*

$$(3.4) \quad \operatorname{Var} g(X) \leq E X(1-X)[g'(X)]^2.$$

Moreover, equality holds in (3.4) if and only if $a+b = -1$ and g is linear in x.

PROOF. It follows from the orthogonal expansion (3.2) that $Eg(X) = c_0$ and

$$(3.5) \quad \operatorname{Var} g(X) = \sum_{j=1}^{\infty} c_j^2 d_j,$$

where

$$(3.6) \quad d_j = E[P_j^{(b,a)}(2X-1)]^2$$
$$= \frac{\Gamma(a+b+2)\Gamma(a+j+1)\Gamma(b+j+1)}{j!\,\Gamma(a+1)\Gamma(b+1)\Gamma(a+b+j+1)\Gamma(a+b+2j+1)}$$

for $j \geq 1$. By (3.3),

$$(3.7) \quad E X(1-X)[g'(X)]^2 = \sum_{j=1}^{\infty} (j+a+b+1)^2 c_j^2 \tilde{d}_{j-1},$$

where

$$(3.8) \quad \tilde{d}_j = E X(1-X) [P_j^{b+1,a+1}(2X-1)]^2$$
$$= \frac{\Gamma(a+b+2)\Gamma(a+j+2)\Gamma(b+j+2)}{j!\,\Gamma(a+1)\Gamma(b+1)\Gamma(a+b+j+3)\Gamma(a+b+2j+3)}$$

for $j \geq 0$. Comparing (3.6) and (3.8), we deduce that

$$(3.9) \quad (j+a+b+1)^2 \tilde{d}_{j-1} = j(j+a+b+1) d_j \geq d_j$$

for $j \geq 1$; therefore the series (3.5) is bounded above by the series (3.7).

To establish the condition for equality, note that if (3.4) is an equality then, by comparing (3.5) and (3.7), we have

$$0 = \sum_{j=1}^{\infty}[(j+a+b+1)^2\tilde{d}_{j-1} - d_j]c_j^2 = \sum_{j=1}^{\infty}[j(j+a+b+1) - 1]c_j^2 d_j,$$

where we have applied (3.9). Each term in this sum is nonnegative; and since $d_j > 0$ for all $j \geq 1$ then we obtain

(3.10) $$[j(j+a+b+1) - 1]c_j^2 = 0, \quad j \geq 1.$$

Since $a + b \geq -1$ then $j(j+a+b+1) - 1 > 0$ for $j \geq 2$; hence $c_j = 0$ for all $j \geq 2$, which implies that g is linear. Since g is nonconstant then $c_1 \neq 0$, and then (3.10) with $j = 1$ reduces to the condition $a + b + 1 = 0$; hence $a + b = -1$. □

THEOREM 3.8. *Let* $(X_1, X_2) \sim GD(a_1+1, a_2+1, b_1+1, b_2+1)$ *where* $a_i + b_i \geq -1$, $i = 1, 2$. *If* $g : \mathbf{R}^2 \to \mathbf{R}$ *is such that* $g \in C^1(\mathbf{R}^2)$ *and* $Var g(X_1, X_2) < \infty$, *then*

(3.11) $$Var g(X_1, X_2) \leq E X_1(1 - X_1)[(E_{X_2|X_1}g(X_1, X_2))_{x_1}]^2 + E X_2(1 - X_1 - X_2)[g_{x_2}(X_1, X_2)]^2.$$

Moreover, equality holds if and only if $a_2 + b_2 = -1$; g *is linear in* x_2 (*i.e.,* $g(x_1, x_2) = g_1(x_1)x_2 + g_2(x_1)$ *for some functions* g_1 *and* g_2); *and*

$$(a_1 + b_1 + 1)EP_{1,0}^{(a_1,a_2,b_1,b_2)}(X_1, X_2)g(X_1, X_2) = 0.$$

PROOF. Since the generalized Jacobi polynomials $\{P_{j,k}^{(a_1,a_2,b_1,b_2)}(x_1, x_2) : j \geq k \geq 0\}$ in (2.6) form a complete orthogonal basis for $L^2(X_1, X_2)$ then

$$g(x_1, x_2) = \sum_{j=0}^{\infty}\sum_{k=0}^{j} c_{j,k} P_{j,k}^{(a_1,a_2,b_1,b_2)}(x_1, x_2).$$

By orthogonality of the polynomials $P_{j,k}^{(a_1,a_2,b_1,b_2)}$, we have

(3.12) $$Var g(X_1, X_2) = \sum_{j=1}^{\infty}\sum_{k=0}^{j} c_{j,k}^2 C_{j,k}^{(a_1,a_2,b_1,b_2)},$$

where $C_{j,k}^{(a_1,a_2,b_1,b_2)} = E[P_{j,k}^{(a_1,a_2,b_1,b_2)}(X_1, X_2)]^2$ is given in (2.7). By Lemma 2.2, (2.5), and an interchange of derivative and summation, we have

$$(E_{X_2|X_1=x_1}g(x_1, X_2))_{x_1}$$

$$= \frac{\partial}{\partial x_1}\sum_{j=0}^{\infty}\sum_{k=0}^{j} c_{j,k}E_{X_2|X_1=x_1}P_{j,k}^{(a_1,a_2,b_1,b_2)}(x_1, X_2)$$

$$= \frac{\partial}{\partial x_1}\sum_{j=0}^{\infty} c_{j,0}P_j^{(b_1+a_2+b_2+1,a_1)}(2x_1 - 1)$$

$$= \sum_{j=1}^{\infty}(j + a_1 + a_2 + b_1 + b_2 + 2)c_{j,0}P_{j-1}^{(b_1+a_2+b_2+2,a_1+1)}(2x_1 - 1).$$

By the orthogonality property of the classical Jacobi polynomials we obtain

$$E_{X_1} X_1(1-X_1)[(E_{X_2|X_1} g(X_1,X_2))_{x_1}]^2$$

(3.13)
$$= \sum_{j=1}^{\infty} j(j + a_1 + a_2 + b_1 + b_2 + 2) c_{j,0}^2 C_{j,0}^{(a_1,a_2,b_1,b_2)}$$

(3.14)
$$\geq \sum_{j=1}^{\infty} c_{j,0}^2 C_{j,0}^{(a_1,a_2,b_1,b_2)},$$

where the inequality follows from the assumption $a_1 + a_2 + b_1 + b_2 \geq -2$.

Next, by (2.10),

$$g_{x_2}(x_1, x_2) = \sum_{j=1}^{\infty} \sum_{k=1}^{j} (k + a_2 + b_2 + 1) c_{j,k} P_{j-1,k-1}^{(a_1,a_2+1,b_1,b_2+1)}(x_1, x_2).$$

Therefore

$$E X_2(1 - X_1 - X_2)[g_{x_2}(X_1, X_2)]^2$$
$$= E X_2(1 - X_1 - X_2)$$
$$\times \Bigg[\sum_{\substack{j,k=1 \\ k \leq j}}^{\infty} (k + a_2 + b_2 + 1) c_{j,k} P_{j-1,k-1}^{(a_1,a_2+1,b_1,b_2+1)}(X_1, X_2) \Bigg]^2$$

(3.15)
$$= \sum_{j=1}^{\infty} \sum_{k=1}^{j} (k + a_2 + b_2 + 1)^2 c_{j,k}^2 C_{j-1,k-1}^{(a_1,a_2+1,b_1,b_2+1)}.$$

It is straightforward from (2.7) to verify that

(3.16)
$$(k + a_2 + b_2 + 1)^2 C_{j-1,k-1}^{(a_1,a_2+1,b_1,b_2+1)} = k(a_2 + b_2 + k + 1) C_{j,k}^{(a_1,a_2,b_1,b_2)} \geq C_{j,k}^{(a_1,a_2,b_1,b_2)};$$

hence

(3.17)
$$E X_2(1 - X_1 - X_2)[g_{x_2}(X_1, X_2)]^2 \geq \sum_{j=1}^{\infty} \sum_{k=1}^{j} c_{j,k}^2 C_{j,k}^{a_1,a_2,b_1,b_2}.$$

By (3.14) and (3.17)

$$E X_1(1-X_1)[(E_{X_2|X_1} g(X_1,X_2))']^2 + E X_2(1 - X_1 - X_2)[g_{x_2}(X_1,X_2)]^2$$
$$\geq \sum_{j=1}^{\infty} c_{j,0}^2 C_{j,0}^{(a_1,a_2,b_1,b_2)} + \sum_{j=1}^{\infty} \sum_{k=1}^{j} c_{j,k}^2 C_{j,k}^{(a_1,a_2,b_1,b_2)}$$
$$= \sum_{j=1}^{\infty} \sum_{k=0}^{j} c_{j,k}^2 C_{j,k}^{(a_1,a_2,b_1,b_2)}$$
$$\equiv \operatorname{Var} g(X_1, X_2).$$

Finally, we establish the condition for equality in (3.11). By (3.12), (3.13) and (3.15), we find that equality holds if and only if

$$0 = \sum_{j=1}^{\infty} j(j + a_1 + a_2 + b_1 + b_2 + 2)c_{j,0}^2 C_{j,0}^{(a_1,a_2,b_1,b_2)}$$

$$+ \sum_{j=1}^{\infty} \sum_{k=1}^{j} (k + a_2 + b_2 + 1)^2 c_{j,k}^2 C_{j-1,k-1}^{(a_1,a_2+1,b_1,b_2+1)} - \sum_{j=1}^{\infty} \sum_{k=0}^{j} c_{j,k}^2 C_{j,k}^{(a_1,a_2,b_1,b_2)}$$

$$= \sum_{j=1}^{\infty} [j(j + a_1 + a_2 + b_1 + b_2 + 2) - 1] c_{j,0}^2 C_{j,0}^{(a_1,a_2,b_1,b_2)}$$

$$+ \sum_{j=1}^{\infty} \sum_{k=1}^{j} [(k + a_2 + b_2 + 1)^2 C_{j-1,k-1}^{(a_1,a_2+1,b_1,b_2+1)} - C_{j,k}^{(a_1,a_2,b_1,b_2)}].$$

Since $j(j + a_1 + a_2 + b_1 + b_2 + 2) \geq 1$ for all $j \geq 1$ then the first summation is nonnegative.

As for the second sum, note that, by (3.16),

$$(k + a_2 + b_2 + 1)^2 C_{j-1,k-1}^{(a_1,a_2+1,b_1,b_2+1)} - C_{j,k}^{(a_1,a_2,b_1,b_2)}$$
$$= [k(k + a_2 + b_2 + 1) - 1] C_{j,k}^{(a_1,a_2,b_1,b_2)},$$

and these terms are strictly positive for all $j \geq k \geq 2$, and are nonnegative for $j \geq k = 1$. Therefore equality in (3.11) implies

(i) $c_{j,k} = 0$, $j \geq k \geq 2$;
(ii) $(a_2 + b_2 + 1)c_{j,1} = 0$, $j \geq 1$;
(iii) $c_{j,0} = 0$, $j \geq 2$; and
(iv) $(a_1 + a_2 + b_1 + b_2 + 2)c_{1,0} = 0$.

Suppose $a_2 + b_2 + 1 > 0$; then by (ii), $c_{j,1} = 0$, $j \geq 1$. Further, it also follows that $a_1 + a_2 + b_1 + b_2 + 2 > 0$ so that, by (iv), $c_{1,0} = 0$. Hence, $c_{j,k} = 0$ for all $j \geq k \geq 1$, which leads to the conclusion $g = c_{0,0}$, a constant. Since it was assumed that g is non-constant, then we have a contradiction; therefore $a_2 + b_2 = -1$, and it follows that g is of the form

$$g(x_1, x_2) = c_{1,0} P_{1,0}^{(a_1,a_2,b_1,b_2)}(x_1, x_2) + \sum_{j=1}^{\infty} c_{j,1} P_{j,1}^{(a_1,a_2,b_1,b_2)}(x_1, x_2),$$

where the $c_{j,1}$ are arbitrary; and $(a_1 + b_1 + 1)c_{1,0} = 0$, equivalently,

$$(a_1 + b_1 + 1) E P_{1,0}^{(a_1,a_2,b_1,b_2)}(X_1, X_2) g(X_1, X_2) = 0.$$

Since $P_{j,k}^{(a_1,a_2,b_1,b_2)}$ is of degree j in x_1 and degree k in x_2 then it follows that $g(x_1, x_2) = g_1(x_1) x_2 + g_2(x_1)$ for some functions g_1 and g_2. □

THEOREM 3.9. *Let* $(X_1, \ldots, X_n) \sim GD(a_1 + 1, \ldots, a_n + 1; b_1 + 1, \ldots, b_n + 1)$ *where* $a_i + b_i \geq -1$, $i = 1, \ldots, n$. *If* $g : \mathbf{R}^n \to \mathbf{R}$ *is such that* $g \in C^1(\mathbf{R}^n)$ *and* $Var\, g(X_1, \ldots, X_n) < \infty$, *then*

$$(3.18) \quad Var\, g(X_1, \ldots, X_n) \leq \sum_{i=1}^{n} E\, X_i \Big(1 - \sum_{j=1}^{i} X_j\Big) [(h_i(X_1, \ldots, X_i))_{x_i}]^2,$$

where $h_i(X_1, \ldots, X_i) := E_{X_{i+1}, \ldots, X_n | X_1, \ldots, X_i} g(X_1, \ldots, X_n)$, $i = 1, \ldots, n$.

Moreover, equality holds if and only if $a_n + b_n = -1$; g is linear in x_n (i.e., $g(x_1, \ldots, x_n) = h_1(x_1, \ldots, x_{n-1})x_n + h_2(x_1, \ldots, x_{n-1})$ for some functions g_1 and g_2); and, for all $i = 1, \ldots, n-1$,

$$(a_i + \beta_i + 1) E P_{k_1,\ldots,k_{i-1},1,0,\ldots,0}^{(a_1,\ldots,a_n,b_1,\ldots,b_n)}(X_1,\ldots,X_n) g(X_1,\ldots,X_n) = 0,$$

where β_i is defined in (2.13).

PROOF. The proof will use induction on n. For $n = 1$, (3.18) is proved in Lemma 3.7; and, for $n = 2$, (3.18) follows immediately from Theorem 3.8.

Throughout, we denote $g(X_1, \ldots, X_n)$ by g. Assume that (3.18) holds for $n = k - 1$ and consider $(X_1, \ldots, X_k) \sim GD(a_1, \ldots, a_k; b_1, \ldots, b_k)$, where $a_i + b_i \geq -1$, $i = 1, \ldots, k$. By Lemma 3.4, for $g \in C^1(\mathbf{R}^k)$ with $\operatorname{Var} g(X_1, \ldots, X_k) < \infty$,

$$(3.19) \qquad \operatorname{Var} g = E_{X_1} \operatorname{Var}_{X_2,\ldots,X_k|X_1} g + \operatorname{Var}_{X_1} E_{X_2,\ldots,X_k|X_1} g.$$

Conditional on $X_1 = x_1$ let $U_i = (1-x_1)^{-1} X_{i+1}$, $i = 1, \ldots, k-1$. By Lemma 2.2, $(U_1, \ldots, U_{k-1})|X_1 \sim GD(a_2, \ldots, a_k; b_2, \ldots, b_k)$ and, by inductive hypothesis,

$$\operatorname{Var}_{U_1,\ldots,U_{k-1}|X_1} g$$

$$(3.20) \qquad \leq \sum_{i=1}^{k-1} E_{U_1,\ldots,U_i|X_1} U_i \left(1 - \sum_{j=1}^{i} U_j\right) [(E_{U_{i+1},\ldots,U_{k-1}|X_1,U_1,\ldots,U_i} g)_{u_i}]^2.$$

By the chain rule,

$$(E_{U_{i+1},\ldots,U_{k-1}|X_1,U_1,\ldots,U_i} g)_{u_i} = \frac{\partial}{\partial x_{i+1}} \frac{\partial x_{i+1}}{\partial u_i} E_{U_{i+1},\ldots,U_{k-1}|X_1,U_1,\ldots,U_i} g$$

$$= (1-x_1)(E_{U_{i+1},\ldots,U_{k-1}|X_1,U_1,\ldots,U_i} g)_{x_{i+1}},$$

for $i = 1, \ldots, k-1$. Therefore by (3.19) and (3.20) we have

$$\operatorname{Var}_{U_1,\ldots,U_{k-1}|X_1} g$$

$$\leq \sum_{i=1}^{k-1} (1-x_1)^2 E_{U_1,\ldots,U_i|X_1} U_i \left(1 - \sum_{j=1}^{i} U_j\right) [(E_{U_{i+1},\ldots,U_{k-1}|X_1,U_1,\ldots,U_i} g)_{x_{i+1}}]^2.$$

It is straightforward to verify that, conditional on $X_1 = x_1$,

$$(1-x_1)^2 U_i \left(1 - \sum_{j=1}^{i} U_j\right) = X_{i+1} \left(1 - x_1 - \sum_{j=2}^{i+1} X_j\right);$$

hence

$$\operatorname{Var}_{U_1,\ldots,U_{k-1}|X_1} g$$

$$\leq \sum_{i=1}^{k-1} E_{U_1,\ldots,U_i|X_1} X_{i+1} \left(1 - x_1 - \sum_{j=2}^{i+1} X_j\right) [(E_{U_{i+1},\ldots,U_{k-1}|X_1,U_1,\ldots,U_i} g)_{x_{i+1}}]^2.$$

Next we note that $\operatorname{Var}_{U_1,\ldots,U_{k-1}|X_1} g \equiv \operatorname{Var}_{X_2,\ldots,X_k|X_1} g$ and, for $i = 1, \ldots, k-1$,

$$(3.21)$$

$$E_{U_1,\ldots,U_i|X_1} X_{i+1} \left(1 - x_1 - \sum_{j=2}^{i+1} X_j\right) [(E_{U_{i+1},\ldots,U_{k-1}|X_1,U_1,\ldots,U_i} g)_{x_{i+1}}]^2$$

$$= E_{X_2,\ldots,X_{i+1}|X_1} X_{i+1} \left(1 - \sum_{j=1}^{i+1} X_j\right) [(E_{X_{i+2},\ldots,X_k|X_1,X_2,\ldots,X_{i+1}} g)_{x_{i+1}}]^2.$$

By (3.20) and (3.21) we have

(3.22)
$$E_{X_1} \text{Var}_{X_2,\ldots,X_k|X_1} g = E_{X_1} \text{Var}_{U_1,\ldots,U_{k-1}|X_1} g$$
$$\leq E_{X_1} \sum_{i=1}^{k-1} E_{X_2,\ldots,X_{i+1}|X_1} X_{i+1}\Big(1 - \sum_{j=1}^{i+1} X_j\Big)[(E_{X_{i+2},\ldots,X_k|X_1,X_2,\ldots,X_{i+1}} g)_{x_{i+1}}]^2$$
$$= \sum_{i=1}^{k-1} E_{X_1,X_2,\ldots,X_{i+1}} X_{i+1}\Big(1 - \sum_{j=1}^{i+1} X_j\Big)[(E_{X_{i+2},\ldots,X_k|X_1,X_2,\ldots,X_{i+1}} g)_{x_{i+1}}]^2$$
$$\equiv \sum_{i=2}^{k} E_{X_1,X_2,\ldots,X_i} X_i\Big(1 - \sum_{j=1}^{i} X_j\Big)[(E_{X_{i+1},\ldots,X_k|X_1,X_2,\ldots,X_i} g)_{x_i}]^2.$$

It remains for us to obtain an upper bound for the second term on the right-hand side of (3.19), viz., $\text{Var}_{X_1} E_{X_2,\ldots,X_k|X_1} g$. By Lemma 2.2 (i), $X_1 \sim B(a_1 + 1, b_1 + 1)$, a beta distribution. Applying Lemma 3.7 to the function $h(x) = E_{X_2,\ldots,X_k|X_1=x} g$, $x \in \mathbf{R}$, we obtain

(3.23)
$$\text{Var}_{X_1} E_{X_2,\ldots,X_k|X_1} g \leq E_{X_1} X_1(1 - X_1)[(E_{X_2,\ldots,X_k|X_1} g)']^2$$
$$\equiv E_{X_1} X_1(1 - X_1)[(E_{X_2,\ldots,X_k|X_1} g)_{x_1}]^2.$$

Combining (3.22) and (3.23) completes the proof of the inductive step, and hence we have (3.18).

Finally, we establish the condition for equality in (3.18). We expand g in terms of the system of generalized Jacobi polynomials in (2.12),

$$g(x_1,\ldots,x_n) = \sum_{k_1=0}^{\infty} \cdots \sum_{k_n=0}^{\infty} c_{k_1,\ldots,k_n} P_{k_1,\ldots,k_n}^{(a_1,\ldots,a_n,b_1,\ldots,b_n)}(x_1,\ldots,x_n),$$

and use the completeness and orthogonality of the polynomials to obtain

$$\text{Var}\, g = \sum_{k_1=0}^{\infty} \cdots \sum_{k_n=0}^{\infty} c_{k_1,\ldots,k_n}^2 C_{k_1,\ldots,k_n}^{(a_1,\ldots,a_n,b_1,\ldots,b_n)} - c_{0,\ldots,0}^2 C_{0,\ldots,0}^{(a_1,\ldots,a_n,b_1,\ldots,b_n)}.$$

By (2.15) and (2.16),

$$(h_i)_{x_i} = \frac{\partial}{\partial x_i} E_{X_{i+1},\ldots,X_n|X_1=x_1,\ldots,X_i=x_i} g$$
$$= \sum_{k_1=0}^{\infty} \cdots \sum_{k_{i-1}=0}^{\infty} \sum_{k_i=1}^{\infty} c_{k_1,\ldots,k_i,0,\ldots,0}(a_i + \beta_i + k_i + 1)$$
$$\times P_{k_1,\ldots,k_{i-1},k_i-1}^{(a_1,\ldots,a_{i-1},a_i+1,b_1,\ldots,b_{i-1},b_i+1)}(x_1,\ldots,x_i);$$

hence,

$$EX_i\Big(1-\sum_{j=1}^{i}X_j\Big)[(h_i(X_1,\ldots,X_i))_{x_i}]^2$$

$$= \sum_{k_1=0}^{\infty}\cdots\sum_{k_{i-1}=0}^{\infty}\sum_{k_i=1}^{\infty} c^2_{k_1,\ldots,k_i,0,\ldots,0}(a_i+\beta_i+k_i+1)^2$$

$$\times C^{(a_1,\ldots,a_{i-1},a_i+1,b_1,\ldots,b_{i-1},b_i+1)}_{k_1,\ldots,k_{i-1},k_i-1}$$

where $C^{(a_1,\ldots,a_n,b_1,\ldots,b_n)}_{k_1,\ldots,k_n}$ is given in (2.14). Using (2.14), we deduce that

$$(a_i+\beta_i+k_i+1)^2 C^{(a_1,\ldots,a_{i-1},a_i+1,b_1,\ldots,b_{i-1},b_i+1)}_{k_1,\ldots,k_{i-1},k_i-1}$$

$$= k_i(a_i+\beta_i+k_i+1) C^{(a_1,\ldots,a_i,b_1,\ldots,b_i)}_{k_1,\ldots,k_i}.$$

Therefore equality in (3.18) implies

(i) $c_{k_1,\ldots,k_i,0,\ldots,0} = 0$, for any $k_i \geq 2$ and $i = 1,\ldots,n$; and
(ii) $(a_i+\beta_i+1)c_{k_1,\ldots,k_{i-1},1,0,\ldots,0} = 0$, for all $i = 1,\ldots,n$.

Suppose $a_n+b_n+1 > 0$; then by an argument similar to the proof of Theorem 3.8, we deduce g is a constant. Since it was assumed that g is non-constant, then we have $a_n+b_n = -1$. Therefore g is linear in x_n; and, for $i = 1,\ldots,n-1$, $(a_i+\beta_i+1)c_{k_1,\ldots,k_{i-1},1,0,\ldots,0} = 0$, equivalently the condition stated in the theorem holds. □

4. Variance inequalities for random eigenvalues

In this section, we apply the method of orthogonal expansions to derive variance inequalities for the eigenvalues of some random matrices. First we consider the multivariate beta distributions, which are well-known to arise in multivariate analysis of variance (Muirhead, 1982). We apply the theory of the generalized Jacobi polynomials of matrix argument (Lassalle, 1991a; Macdonald, 1995) to derive variance inequalities for symmetric functions of the eigenvalues of the multivariate beta-distributed matrices. As in the previous section, we maintain the standard assumption throughout.

As a limiting case of these results for the multivariate beta distributions, we derive variance inequalities for the eigenvalues of the Wishart distributions.

4.1. The multivariate beta distribution.

Let X be an $n \times n$ positive-definite symmetric matrix, denoted $X > 0$; and let I_n denote the $n \times n$ identity matrix. A function g, on the space of positive-definite matrices X, is *orthogonally invariant* if $g(H^{-1}XH) = g(X)$ for all $X > 0$ and all $H \in O(n)$, the group of $n \times n$ orthogonal matrices. It is not difficult to see that if g is orthogonally invariant then $g(X)$ is a symmetric function of $x_1,\ldots,x_n$, the eigenvalues of X.

A random matrix X has a *multivariate beta distribution* with parameters a and b, denoted $X \sim B_n(a,b)$, if the probability density function of X is proportional to

(4.1) $$(\det X)^{a-\frac{1}{2}(n+1)}(\det(I_n - X))^{b-\frac{1}{2}(n+1)},$$

where $X > 0$, $I_n - X > 0$, $a > (n-1)/2$, $b > (n-1)/2$.

Therefore to develop variance inequalities for an orthogonally invariant function g of the random matrix $X \sim B_n(a,b)$, it suffices to restrict our attention to a symmetric function, also denoted by g, of the eigenvalues of X.

It is well-known (cf. Muirhead, 1982, p. 112) that if $X \sim B_n(a,b)$ then the joint p.d.f. of $X_1,\ldots,X_n$, the eigenvalues of X, is proportional to

$$\prod_{i<j}^{n} |x_i - x_j| \prod_{i=1}^{n} x_i^{a-\frac{1}{2}(n+1)} (1-x_i)^{b-\frac{1}{2}(n+1)}, \tag{4.2}$$

where $0 < x_1,\ldots,x_n < 1$.

Instead of working with the eigenvalue p.d.f. (4.2), we shall work in a more general context with random variables $X_1,\ldots,X_n$ which have joint p.d.f. proportional to

$$\prod_{1 \leq i < j \leq n} |x_i - x_j|^{2/\alpha} \prod_{i=1}^{n} x_i^a (1-x_i)^b, \tag{4.3}$$

where $0 < x_1,\ldots,x_n < 1$, $a > -1$, $b > -1$, and $\alpha > 0$. Thus if $\alpha = 2$ then (4.3) reduces to (4.2). It suffices for our purposes to note that the normalizing constant in (4.3) is provided by an integral formula of Selberg (cf. Richards, 1989).

The orthogonal polynomials for the p.d.f. (4.3) have been constructed by Lassalle (1991a). They are described as follows.

A *partition* $\lambda = (\lambda_1,\ldots,\lambda_n)$ is a sequence of nonnegative integers such that $\lambda_1 \geq \cdots \geq \lambda_n$. The integers $\lambda_1,\ldots,\lambda_n$ are called the *parts* of λ; the number of nonzero parts is called the *length* of λ; and $|\lambda| := \lambda_1 + \cdots + \lambda_n$ is called the *weight* of λ. If $\lambda = (\lambda_1,\ldots,\lambda_n)$ and $\mu = (\mu_1,\ldots,\mu_n)$ are partitions then we write $\mu \subseteq \lambda$ if $\mu_i \leq \lambda_i$ for all $i = 1,\ldots,n$; and we write $\mu \subset \lambda$ if $\mu \subseteq \lambda$ and $\mu \neq \lambda$.

For each partition λ, we denote by $J_\lambda^{(\alpha)}(x_1,\ldots,x_n)$ the *Jack polynomial* indexed by λ. A complete account of the algebraic properties of the Jack polynomials is provided by Macdonald (1995). In particular, for the case in which $\alpha = 2$ the Jack polynomial $J_\lambda^{(2)}$ is a constant multiple of the zonal polynomial denoted C_λ by Muirhead (1982).

For our purposes it suffices to note that the Jack polynomials $J_\lambda^{(\alpha)}$ are characterized by the following properties (Macdonald, 1995, p. 379):

(i) The set of Jack polynomials is a basis for the vector space of symmetric polynomials in n variables; more precisely,

$$J_\lambda^{(\alpha)}(x_1,\ldots,x_n) = m_\lambda(x_1,\ldots,x_n) + \sum_{\mu \subset \lambda} c_{\lambda,\mu}^{(\alpha)} m_\mu(x_1,\ldots,x_n)$$

where $m_\lambda(x_1,\ldots,x_n)$ is the *monomial symmetric function* corresponding to the partition λ, and $c_{\lambda,\mu}^{(\alpha)}$ are constants;

(ii) There exists, on the space of symmetric polynomials, an inner product $\langle\,,\,\rangle_\alpha$ such that $\langle J_\lambda^{(\alpha)}, J_\mu^{(\alpha)} \rangle_\alpha = 0$ if $\lambda \neq \mu$.

Following Lassalle (1990), we denote by $\binom{\lambda}{\mu}$ the *generalized binomial coefficient*, which is defined by the binomial expansion

$$\frac{J_\lambda^{(\alpha)}(1+x_1,\ldots,1+x_n)}{J_\lambda^{(\alpha)}(1,\ldots,1)} = \sum_\mu \binom{\lambda}{\mu} \frac{J_\mu^{(\alpha)}(x_1,\ldots,x_n)}{J_\mu^{(\alpha)}(1,\ldots,1)}. \tag{4.4}$$

It is known (cf. Lassalle, 1990) that $\binom{\lambda}{\mu} \neq 0$ if and only if $\mu \subseteq \lambda$.

Define the differential operator

$$\Delta = \sum_{i=1}^{n} x_i(x_i - 1)\frac{\partial^2}{\partial x_i^2} + \frac{2}{\alpha} \sum_{\substack{i,j=1 \\ i \neq j}}^{n} \frac{x_i(x_i - 1)}{x_i - x_j}\frac{\partial}{\partial x_i} + \sum_{i=1}^{n} [(a+b+2)x_i - a - 1]\frac{\partial}{\partial x_i}.$$

The operator -4Δ is called the *Laplace-Beltrami operator*.

By a result of Lassalle (1991a), Theorem 1, there exists a unique symmetric polynomial, denoted by $P_\lambda^{(a,b)}$, such that

$$P_\lambda^{(a,b)}(0,\ldots,0) = 1;$$

$$P_\lambda^{(a,b)}(x_1,\ldots,x_n) = \sum_{\mu \subseteq \lambda} \gamma_{\lambda,\mu} \frac{J_\mu^{(\alpha)}(x_1,\ldots,x_n)}{J_\mu^{(\alpha)}(1^n)}$$

for some sequence of constants $\gamma_{\lambda,\mu}$; and

(4.5) $$\Delta P_\lambda^{(a,b)}(x_1,\ldots,x_n) = \tau_\lambda P_\lambda^{(a,b)}(x_1,\ldots,x_n);$$

where, with $q := 1 + (n-1)\alpha^{-1}$,

(4.6) $$\tau_\lambda = (a + b + 2q)|\lambda| + \sum_{i=1}^{n} \lambda_i(\lambda_i - i).$$

Moreover, the space of polynomials $\{P_\lambda^{(a,b)} : \lambda \text{ is a partition}\}$ is an orthogonal basis for $L^2((0,1)^n)$, the Hilbert space of *symmetric* functions supported on $(0,1)^n$ and square-integrable with respect to the weight function (4.3).

The system $\{P_\lambda^{(a,b)} : \lambda \text{ is a partition}\}$ are the *generalized Jacobi polynomials*. For the case in which $\alpha = 2$ these polynomials were constructed by Constantine and James (1974); for general α, they were constructed by Lassalle (1991a).

THEOREM 4.1. *Let $(X_1,\ldots,X_n)$ be a random vector with p.d.f. proportional to (4.3), where $a+b+2q > n-1$, and let $g : (0,1)^n \to \mathbf{R}$ be a non-constant symmetric function with $g \in C^2((0,1)^n)$ and $\operatorname{Var} g(X_1,\ldots,X_n) < \infty$. Then*

(4.7) $$\operatorname{Var} g(X_1,\ldots,X_n) \leq (a+b+2q-n+1)^{-2} E[\Delta g(X_1,\ldots,X_n)]^2.$$

If equality holds then g is a polynomial of degree at most 1.

PROOF. We expand the function g in terms of the generalized Jacobi polynomials,

$$g(x_1,\ldots,x_n) = \sum_\lambda c_\lambda P_\lambda^{(a,b)}(x_1,\ldots,x_n).$$

By the orthogonality of the generalized Jacobi polynomials, we have

(4.8) $$\operatorname{Var} g(X_1,\ldots,X_n) = \sum_{|\lambda| \geq 1} c_\lambda^2 \|P_\lambda^{(a,b)}\|^2,$$

where $\|P_\lambda^{(a,b)}\|^2 = E[P_\lambda^{(a,b)}(X_1,\ldots,X_n)]^2$. By (4.5) and the orthogonality of the generalized Jacobi polynomials,

$$(4.9) \quad E[\Delta g(X)]^2 = E\left[\Delta \sum_{|\lambda|\geq 0} c_\lambda P_\lambda^{(a,b)}(X_1,\ldots,X_n)\right]^2$$

$$= E\left[\sum_{|\lambda|\geq 1} c_\lambda \tau_\lambda P_\lambda^{(a,b)}(X)\right]^2$$

$$= \sum_{|\lambda|\geq 1} c_\lambda^2 \tau_\lambda^2 \|P_\lambda^{(a,b)}\|^2,$$

where, by (4.6),

$$\tau_\lambda = (a+b+2q)|\lambda| + \sum_{i=1}^n \lambda_i(\lambda_i - i) = \sum_{i=1}^n (a+b+2q-i)\lambda_i + \sum_{i=1}^n \lambda_i^2$$

$$\geq \sum_{i=1}^n (a+b+2q-n)\lambda_i + \sum_{i=1}^n \lambda_i = (a+b+2q-n+1)|\lambda|.$$

Hence

$$(4.10) \quad \tau_\lambda^2 \geq (a+b+2q-n+1)^2|\lambda|^2 \geq (a+b+2q-n+1)^2$$

for all partitions λ such that $|\lambda| \geq 1$. Inserting this lower bound for τ_λ in (4.9), we obtain (4.7).

It also is evident that equality in (4.7) implies

$$[(a+b+2q-n+1)^{-2}\tau_\lambda^2 - 1]c_\tau^2 = 0, \quad |\lambda| \geq 1.$$

If $|\lambda| > 1$ then, by (4.10), $(a+b+2q-n+1)^{-2}\tau_\lambda^2 > 1$, so $c_\lambda = 0$ for $|\lambda| > 1$. For the case in which $|\lambda| = 1$, i.e. $\lambda = (1,0,\ldots,0)$, we have

$$(a+b+2q-n+1)^{-2}\tau_{(1)}^2 - 1 = (a+b+2q-n+1)^{-2}(a+b+2q)^2 - 1$$

$$= \begin{cases} > 0, & n \geq 2 \\ 0, & n = 1. \end{cases}$$

Therefore if $n = 1$ then equality holds if and only if $g = c_{(0)} + c_{(1)}P_{(1)}^{(a,b)}$, a polynomial of degree at most 1. If $n > 1$ then equality holds if and only if g is a constant. $\square$

We note that an explicit formula for the norm, $\|P_\lambda^{(a,b)}\|$, may be obtained as a special case of a result of Heckman (1987), Theorem 8.5, combined with the solution to Macdonald's constant term conjecture (Opdam, 1989, Theorem 4.1).

Sharper bounds on $\text{Var}\, g(X_1,\ldots,X_n)$ can be obtained by the same method of proof of Theorem 4.1. The following result illustrates the general procedure for deriving these bounds.

COROLLARY 4.2. *Under the hypotheses of* Theorem 4.1, *we have*

$$\text{Var}\, g(X_1,\ldots,X_n) \leq (a+b+2q-n+1)^{-2} E[\Delta g(X_1,\ldots,X_n)]^2$$

$$- 3 \sum_{|\lambda|=2} \frac{[E\, P_\lambda^{(a,b)}(X_1,\ldots,X_n)\Delta g(X_1,\ldots,X_n)]^2}{\tau_\lambda^2 \|P_\lambda^{(a,b)}\|^2}.$$

PROOF. By (4.8) and (4.9),

$$(a+b+2q-n+1)^{-2}E[\Delta g]^2 - \operatorname{Var} g \geq \sum_{|\lambda|\geq 1}(|\lambda|^2-1)c_\lambda^2\|P_\lambda^{(a,b)}\|^2$$
$$\geq \sum_{|\lambda|=2}(|\lambda|^2-1)c_\lambda^2\|P_\lambda^{(a,b)}\|^2$$
$$= 3\sum_{|\lambda|=2}c_\lambda^2\|P_\lambda^{(a,b)}\|^2.$$

By orthogonality of the $P_\lambda^{(a,b)}$, we have $Eg(X_1,\ldots,X_n)P_\lambda^{(a,b)}(X_1,\ldots,X_n) = c_\lambda\|P_\lambda^{(a,b)}\|^2$; hence, by (4.5),

$$c_\lambda = \|P_\lambda^{(a,b)}\|^{-2}Eg(X_1,\ldots,X_n)P_\lambda^{(a,b)}(X_1,\ldots,X_n)$$
$$= \|P_\lambda^{(a,b)}\|^{-2}\tau_\lambda^{-1}Eg(X_1,\ldots,X_n)\Delta P_\lambda^{(a,b)}(X_1,\ldots,X_n)$$
$$= \|P_\lambda^{(a,b)}\|^{-2}\tau_\lambda^{-1}EP_\lambda^{(a,b)(X_1,\ldots,X_n)}\Delta g(X_1,\ldots,X_n),$$

where the last equality holds since the Laplace-Beltrami operator is self-adjoint (cf. Macdonald, 1987). Therefore

$$c_\lambda^2\|P_\lambda^{(a,b)}\|^2 = \|P_\lambda^{(a,b)}\|^{-2}\tau_\lambda^{-2}[E\,P_\lambda^{(a,b)(X_1,\ldots,X_n)}\Delta g(X_1,\ldots,X_n)]^2,$$

and the proof is complete. □

4.2. The Wishart distribution. Consider an $n\times n$ symmetric positive-definite random matrix X which has a *Wishart distribution* with p.d.f. proportional to

$$(\det X)^{a-\frac{1}{2}(n+1)}\exp(-\operatorname{tr} X), \quad X > 0,$$

where $a > (n-1)/2$. Then the p.d.f. of $X_1,\ldots,X_n$, the eigenvalues of X, is proportional to

$$\prod_{i<j}^n |x_i-x_j|\prod_{i=1}^n x_i^{a-\frac{1}{2}(n+1)}\exp(-x_i),$$

where $x_1,\ldots,x_n > 0$.

Similar to the case of the multivariate beta distribution, we consider the more general context in which random variables $X_1,\ldots,X_n$ have a p.d.f. which is proportional to

(4.11) $$\prod_{i<j}^n |x_i-x_j|^{2/\alpha}\prod_{i=1}^n x_i^a \exp(-x_i),$$

where $x_1,\ldots,x_n > 0$, $a > -1$. In this case, the corresponding orthogonal polynomials are the *generalized Laguerre polynomials*, indexed by partitions λ, and which are defined by

(4.12) $$L_\lambda^{(a)}(x_1,\ldots,x_m) = \sum_{\mu\subseteq\lambda}(-1)^\mu \frac{\binom{\lambda}{\mu}}{(a+q)_\mu}\frac{J_\mu^{(\alpha)}(x_1,\ldots,x_n)}{J_\mu^{(\alpha)}(1,\ldots,1)},$$

where the generalized binomial coefficient $\binom{\lambda}{\mu}$ is defined in (4.4) and, for any partition $\lambda = (\lambda_1,\ldots,\lambda_m)$,

$$(a)_\lambda := \prod_{i=1}^m \left(a - \frac{1}{\alpha}(i-1)\right)_{\lambda_i},$$

and, for any nonnegative integer k, $(a)_k := a(a+1)\cdots(a+k-1)$.

For the case in which $\alpha = 2$ the generalized Laguerre polynomials were defined by Constantine (1966); and for general $\alpha > 0$, the polynomials were defined by Lassalle (1991b).

Proceeding as in the multivariate beta case, we can derive variance inequalities for symmetric functions $g(X_1, \ldots, X_n)$, where g is square-integrable with respect to the p.d.f. (4.11), by orthogonal expansions based on the generalized Laguerre polynomials; this will require the eigenfunction property (Lassalle, 1991b)

$$\Delta_1 L_\lambda^{(a)}(x_1, \ldots, x_n) = -|\lambda| L_\lambda^{(a)}(x_1, \ldots, x_n)$$

where

$$\Delta_1 = \sum_{i=1}^n x_i \frac{\partial^2}{\partial x_i^2} + \frac{2}{\alpha} \sum_{\substack{i,j=1 \\ i \neq j}}^n \frac{x_i}{x_i - x_j} \frac{\partial}{\partial x_i} - \sum_{i=1}^n x_i \frac{\partial}{\partial x_i} + (a+1) \sum_{i=1}^n \frac{\partial}{\partial x_i}.$$

However, a more efficient method is to develop these variance inequalities as a limiting case of Theorem 4.1. In the classical case it is well-known that if a scalar random variable $X \sim B(a, b)$, a beta distribution on $(0, 1)$ with parameters a, b then, as $b \to \infty$, X/b converges in distribution to a gamma distribution on $(0, \infty)$ with index a.

A similar result holds for the densities (4.3) and (4.11); that is, if the random vector $(X_1, \ldots, X_n)$ has the p.d.f. proportional to (4.3) then, as $b \to \infty$, $b^{-1}(X_1, \ldots, X_n)$ converges in distribution to a random vector which has the p.d.f. proportional to (4.11). Consequently,

$$\lim_{b \to \infty} P_\lambda^{(a,b)}\left(b^{-1}(x_1, \ldots, x_n)\right) = L_\lambda^{(a)}(x_1, \ldots, x_n),$$

a result established by Lassalle (1991b).

As a limiting case of Theorem 4.1 we obtain the following result.

THEOREM 4.3. *Let $(X_1, \ldots, X_n)$ be a random vector with p.d.f. proportional to (4.11), and let $g : \mathbf{R}_+^n \to \mathbf{R}$ be a symmetric function with $g \in C^2(\mathbf{R}_+^n)$ and $Var g(X_1, \ldots, X_n) < \infty$. Then*

$$Var g(X_1, \ldots, X_n) \leq E[\Delta_1 g(X_1, \ldots, X_n)]^2.$$

Equality holds if and only if g is a polynomial of degree at most 1.

It should be noted that, for the case in which $\alpha = 2$, the differential operator Δ_1 is the radial part of the Laplace-Beltrami operator on the cone of positive-definite real symmetric matrices.

5. Multivariate normal distributions

In the case of the normal distributions, Chen (1982) proved a result of which the following is a special case.

THEOREM 5.1 (Chen, 1982). *Let $X_1, \ldots, X_n$ be i.i.d. standard normal random variables, and let $g \in C^1(\mathbf{R}^n)$ with $Var g(X_1, \ldots, X_n) < \infty$. Then*

(5.1) $$Var g(X_1, \ldots, X_n) \leq \sum_{i=1}^n E[\partial_i g(X_1, \ldots, X_n)]^2,$$

where $\partial_i g \equiv \partial g / \partial x_i$.

By application of the method of orthogonal expansions we obtain the following result which has been derived earlier by different methods by Houdré, et al., (1998).

THEOREM 5.2 (Houdré, et al., 1998). *Let $X_1, \ldots, X_n$ be i.i.d. standard normal random variables; l be a fixed positive integer; and let $g \in C^{2l+1}(\mathbf{R}^n)$ with $\operatorname{Var} g(X_1, \ldots, X_n) < \infty$. Then*

(5.2)
$$\max_{1 \leq i \leq n} \sum_{j=1}^{2l} \frac{(-1)^{j+1}}{j!} E[\partial_i^j g(X_1, \ldots, X_n)]^2 \leq \operatorname{Var} g(X_1, \ldots, X_n)$$

$$\leq \min_{1 \leq i \leq n} \sum_{j=1}^{2l-1} \frac{(-1)^{j+1}}{j!} E[\partial_i^j g(X_1, \ldots, X_n)]^2,$$

where $\partial_i^j g \equiv \partial^j g / \partial x_i^j$.

PROOF. Let H_k denote the classical Hermite polynomial of degree k; the sequence of polynomials $\mathcal{H}_k(x) := H_k(2^{-1/2}x)$, $x \in \mathbf{R}$, $k = 0, 1, 2, \ldots$, is a complete orthogonal basis for the Hilbert space of functions which are square-integrable with respect to the standard normal distribution. It is well-known (and it follows also from Theorem 2.4) that the multivariate Hermite polynomials

(5.3)
$$\mathcal{H}_\kappa(x_1, \ldots, x_n) := \prod_{i=1}^{n} \mathcal{H}_{k_i}(x_i),$$

where $\kappa = (k_1, \ldots, k_n)$ and the k_i are nonnegative integers for all i, form a complete orthogonal basis for the class of functions $g : \mathbf{R}^n \to \mathbf{R}$ such that $\operatorname{Var} g(X_1, \ldots, X_n) < \infty$. Defining $|\kappa| := k_1 + \cdots + k_n$ and $\kappa! = k_1! \cdots k_n!$ then it is straightforward to verify that

(5.4)
$$E[\mathcal{H}_\kappa(X_1, \ldots, X_n)]^2 = \kappa! 2^{|\kappa|}.$$

Further, it follows from the classical formula,

(5.5)
$$\mathcal{H}_k'(x) = 2^{1/2} k \mathcal{H}_{k-1}(x), \qquad k \geq 1,$$

that for $1 \leq i \leq n$ and $j = 0, 1, 2, \ldots$,

$$\partial_i^j \mathcal{H}_\kappa(x_1, \ldots, x_n) = 2^{j/2} \frac{k_i!}{(k_i - j)!} \mathcal{H}_{\kappa_j}(x_1, \ldots, x_n),$$

where $\kappa_j := (k_1, \ldots, k_{i-1}, k_i - j, k_{i+1}, \ldots, k_n)$.

Expanding g in the basis of multivariate Hermite polynomials,

$$g(x_1, \ldots, x_n) = \sum_\kappa c_\kappa \mathcal{H}_\kappa(x_1, \ldots, x_n),$$

then, by orthogonality of the Hermite polynomials, we have

$$\operatorname{Var} g(X_1, \ldots, X_n) = \sum_{\kappa \neq 0} \kappa! 2^{|\kappa|} c_\kappa^2.$$

For any positive integer m, we now have

$$\sum_{j=1}^{m} \frac{(-1)^{j+1}}{j!} E[\partial_i^j g(X_1,\ldots,X_n)]^2$$

$$= \sum_{j=1}^{m} \frac{(-1)^{j+1}}{j!} E[\sum_{\kappa \neq 0} 2^{j/2} \frac{k_i!}{(k_i-j)!} c_\kappa \mathcal{H}_{\kappa_j}(X_1,\ldots,X_n)]^2$$

$$= \sum_{j=1}^{m} \frac{(-1)^{j+1}}{j!} \sum_{\kappa \neq 0} c_\kappa^2 2^j \left(\frac{k_i!}{(k_i-j)!}\right)^2 E[\mathcal{H}_{\kappa_j}(X_1,\ldots,X_n)]^2.$$

Applying (5.4) we obtain

$$\sum_{j=1}^{m} \frac{(-1)^{j+1}}{j!} E[\partial_i^j g(X_1,\ldots,X_n)]^2 = \sum_{j=1}^{m} \frac{(-1)^{j+1}}{j!} \sum_{\kappa \neq 0} \kappa! 2^{|\kappa|} c_\kappa^2 \frac{k_i!}{(k_i-j)!}$$

$$= \sum_{\kappa \neq 0} \kappa! 2^{|\kappa|} c_\kappa^2 \sum_{j=1}^{m} \frac{(-1)^{j+1}}{j!} \frac{k_i!}{(k_i-j)!}$$

$$= \sum_{\kappa \neq 0} \kappa! 2^{|\kappa|} c_\kappa^2 \left[1 + (-1)^{m+1} \binom{k_i-1}{m}\right],$$

where the last equality follows from a well-known combinatorial formula.

If m is odd, $m = 2l-1$, say, then we have

$$\sum_{j=1}^{2l-1} \frac{(-1)^{j+1}}{j!} E[\partial_i^j g(X_1,\ldots,X_n)]^2 = \sum_{\kappa \neq 0} \kappa! 2^{|\kappa|} c_\kappa^2 \left[1 + \binom{k_i-1}{2l-1}\right]$$

$$\geq \sum_{\kappa \neq 0} \kappa! 2^{|\kappa|} c_\kappa^2$$

$$= \operatorname{Var} g(X_1,\ldots,X_n).$$

Since i was chosen arbitrarily then the upper bound in (5.2) has been established.

Similarly, if m is even, $m = 2l$, say, then

$$\sum_{j=1}^{2l} \frac{(-1)^{j+1}}{j!} E[\partial_i^j g(X_1,\ldots,X_n)]^2 = \sum_{\kappa \neq 0} \kappa! 2^{|\kappa|} c_\kappa^2 \left[1 - \binom{k_i-1}{2l}\right]$$

$$\leq \sum_{\kappa \neq 0} \kappa! 2^{|\kappa|} c_\kappa^2$$

$$= \operatorname{Var} g(X_1,\ldots,X_n).$$

Since i was chosen arbitrarily, this proves the lower bound in (5.2). $\square$

REMARK 5.3. For the case in which $n = 1$, Theorem 5.2 reduces to an inequality of Houdré and Kagan (1995):

$$(5.6) \qquad \sum_{j=1}^{2l} \frac{(-1)^{j+1}}{j!} E[g^{(j)}(X)]^2 \leq \operatorname{Var} g(X) \leq \sum_{j=1}^{2l-1} \frac{(-1)^{j+1}}{j!} E[g^{(j)}(X)]^2,$$

where X is standard normal and $g \in C^{2l}(\mathbf{R})$. If $l = 1$ the upper bound in Theorem 5.2 gives a sharper result than that given in Theorem 5.1.

In closing, we make an application to the theory of Edgeworth expansions. Let X be a standard normal random variable, and let $g \in C^{(k)}\mathbf{R}$ with $Eg^{(j)}(X) < \infty$ for all $j = 0, 1, \ldots, k$. Retaining the notation $\mathcal{H}_j(x)$ for the Hermite polynomials orthogonal with respect to the distribution of X then

$$g(x) = \sum_{k=0}^{\infty} c_k \mathcal{H}_k(x),$$

and we obtain

(5.7) $$E\mathcal{H}_j(X)g(X) = j! 2^j c_j = 2^{j/2} Eg^{(j)}(X).$$

For $j = 1$, (5.7) is known as Stein's identity (Stein, 1974).

The formula (5.7) also has the following interesting consequence. Denoting the standard normal density function by $\phi(x)$, suppose that

$$f_k(x) = \left[\sum_{j=0}^{k} c_j \mathcal{H}_j(x) \right] \phi(x),$$

$c_0 = 1$, is a k-term Edgeworth approximation to a (possibly unknown) density function $f(x)$. Suppose we wish to evaluate $E_{f_k} g(X)$, the expectation of $g(X)$ with respect to the (improper) density function f_k. Then (5.7) gives the result:

$$E_{f_k} g(X) = \int f_k(x) g(x) \, dx = \sum_{j=0}^{k} c_j \int g(x) \mathcal{H}_j(x) \phi(x) \, dx = \sum_{j=0}^{k} 2^{j/2} c_j Eg^{(j)}(X).$$

As an example, if $g(x) = e^{tx}$, $t \in \mathbf{R}$, then we obtain the elegant result for the moment-generating function of the improper density f_k:

(5.8) $$E_{f_k} e^{tX} = e^{t^2/2} \sum_{j=0}^{k} 2^{j/2} c_j t^j.$$

It is also clear that analogs of (5.8) can be obtained by the use of orthogonal expansions for other probability distributions.

Acknowledgements. We are grateful to T. Wake Epps for pointing out the connection between our results and the computation of expectations with respect to Edgeworth approximations; to Michel Lassalle for comments on the evaluation of the norm of the generalized Jacobi polynomials in Section 4; and also to the referees for suggestions which helped us improve our exposition.

References

J. Aitchison, *The Statistical Analysis of Compositional Data*, Chapman & Hall, London (1986).
G. Alexits, *Convergence Problems of Orthogonal Series*, Pergamon Press, New York (1961).
H. J. Brascamp and E. H. Lieb, *On extensions of the Brünn-Minkowski and Prékopa-Leindler theorems, including inequalities of log-concave functions and with an application to the diffusion equation*, J. Funct. Anal. **22** (1976) 366–389.
T. Cacoullos, *On upper and lower bounds for the variance of a function of a random variable*, Ann. Probab. **10** (1982) 799–809.
L. H. Y. Chen, *An inequality for the multivariate normal distribution*, J. Multivar. Anal. **12** (1982) 306–312.
H. Chernoff, *A note on an inequality involving the normal distribution*, Ann. Probab. **9** (1981) 533–535.
R. J. Connor and J. E. Mosimann, *Concepts of independence for proportions with a generalization of the Dirichlet distribution*, J. Amer. Statist. Assoc. **64** (1969) 194–206.

A. G. Constantine, *The distribution of Hotelling's generalized T_0^2*, Ann. Math. Statist. **37** (1966) 215–225.

A. G. Constantine and A. T. James, *Generalized Jacobi polynomials as spherical functions of the Grassmann manifold*, Proc. London Math. Soc. **29(3)** (1974) 174–192.

G. A. Fomin, *Convergence of termwise derived Fourier-Jacobi series and expansions in the Faber-Schauder system*, Matematicheskie Zametki **53** (1993) 114–120.

G. Heckman, *Root systems and hypergeometric functions II*, Compositio Math. **64** (1987) 353–373.

J. R. Higgins, *Completeness and Basis Properties of Sets of Special Functions*, Cambridge Univ. Press, New York (1977).

C. Houdré and A. Kagan, *Variance inequalities for functions of Gaussian variables*, J. Theoret. Probab. **8** (1995) 23–30.

C. Houdré and V. Perez-Abreu, *Covariance identities and inequalities for functions on Wiener and Poisson spaces*, Ann. Probab. **23** (1995) 400–419.

C. Houdré, *Some applications of covariance identities and inequalities to functions of multivariate normal variables*, J. Amer. Statist. Assoc. **90** (1995) 965–968.

C. Houdré, V. Perez-Abreu and D. Surgailis, *Interpolation, correlation identities and inequalities for infinitely divisible variables*, J. Fourier Anal. Appl. (1998), to appear.

C. A. J. Klaassen *On an inequality of Chernoff*, Ann. Probab. **13** (1985) 966–974.

T. H. Koornwinder and A. L. Schwartz, *Product formulas and associated hypergroups for orthogonal polynomials on the simplex and on a parabolic biangle*, Constr. Approx. **13** (1997) 537–567.

M. Lassalle, *Une formule du binôme généralisée pour les polynômes de Jack*, C. R. Acad. Sci. Paris, Sér. I **310** (1990) 253–256.

M. Lassalle, *Polynômes de Jacobi généralisés*, C. R. Acad. Sci. Paris, Sér. I **312** (1991a) 425–428.

M. Lassalle, *Polynômes de Laguerre généralisés*, C. R. Acad. Sci. Paris, Sér. I **312** (1991b) 725–728.

R. H. Lochner, *A generalized Dirichlet distribution in Bayesian life-testing*, J. Roy. Statist. Soc., Ser. B **37** (1975) 103–113.

I. G. Macdonald, *Commuting differential operators and zonal spherical functions*, Lecture Notes in Math. **1271** (1987) 189–200.

I. G. Macdonald, *Symmetric Functions and Hall Polynomials*, 2nd. ed. Oxford Univ. Press, New York (1995).

R. J. Muirhead, *Aspect of Multivariate Statistical Theory*, Wiley, New York (1982).

J. Nash, *Continuity of solutions of parabolic and elliptic equations*, Amer. J. Math. **80** (1958) 931–954.

E. Opdam, *Some applications of hypergeometric shift operators*, Invent. Math. **98** (1989) 1–18.

J. Proriol, *Sur une famille de polynômes à deux variables orthogonaux dans un triangle*, C. R. Acad. Sci. Paris **245** (1957) 2459–2461.

D. St. P. Richards, *Analogs and extensions of Selberg's integral*, In: *q-Series and Combinatorics* (D. Stanton, ed.), IMA Vol. Math. Appl., **18**, 109–137, Springer, New York, 1989

C. Stein, *Estimation of the mean of a multivariate normal distribution*, Ann. Statist. **9** (1981) 1135–1151.

G. Szegö, *Orthogonal Polynomials*, 4th. ed., Amer. Math. Soc., Providence, RI. (1975).

R. A. Vitale, *A differential version of the Efron-Stein inequality: Bounding the variance of a function of an infinitely divisible variable*, Statist. Probab. Lett. **7** (1989) 105–112.

H. Weyl, *The method of orthogonal projection in potential theory*, Duke Math. J. **7** (1940) 411–444.

Division of Statistics, University of Virginia, Charlottesville, Virginia 22903; and Southeastern Rural Mental Health Research Center, University of Virginia, Charlottesville, Virginia 22908

Division of Statistics, University of Virginia, Charlottesville, Virginia 22903

A note on sums of independent random variables

Paweł Hitczenko and Stephen Montgomery-Smith

ABSTRACT. In this note a two sided bound on the tail probability of sums of independent, and either symmetric or nonnegative, random variables is obtained. We utilize a recent result by Latała on bounds on moments of such sums. We also give a new proof of Latała's result for nonnegative random variables, and improve one of the constants in his inequality.

1. Introduction

Recently Latała (1997) obtained the following remarkable result: for a sequence of random variables (X_n) and $1 \leq p < \infty$ define the following Orlicz norm

$$(1.1) \qquad |||(X_k)|||_p = \inf\{\lambda > 0 : \prod_n \mathbb{E}|1 + X_n/\lambda|^p \leq e^p\}.$$

Latała proved that

$$(1.2) \qquad \frac{e-1}{2e^2}|||(X_k)|||_p \leq \left(\mathbb{E}|\sum X_k|^p\right)^{1/p} \leq e|||(X_k)|||_p,$$

provided (X_n) are either symmetric or positive, and in the first case $p \geq 2$, and in the second case $p \geq 1$. The main novelty here is the fact that, contrary to the classical inequalities, the constants here are independent of p. Certain particular cases of Latała's result had been known earlier (see e.g., Hitczenko (1993), Gluskin and Kwapień (1995) or Hitczenko, Montgomery-Smith and Oleszkiewicz (1997)), but they can be easily deduced from Latała's inequality.

Of course, the ultimate goal is to obtain bounds on the tail probabilities for sums of random variables. Latała's result prompted us to investigate that problem. This program has been completed; our methods, which are based on estimates for the decreasing rearrangement of a random variable, work in a rather general setting. As a result we were able to obtain extensions of Latała's result in various directions. The details of that approach will be presented elsewhere. The goal of this note is quite different; we will present a very simple argument that allows one to deduce tail bounds from Latała's result. As a matter of fact, this approach

---

1991 *Mathematics Subject Classification.* Primary 60G50, 60E15; Secondary 46E30.
Key words and phrases. sums of independent random variables, tail distributions.
The first author was partially supported by NSF grant DMS 9401345.
The second author was partially supported by NSF grant DMS 9424396, and by the University of Missouri Research Board.

© 1999 American Mathematical Society

formally does not really depend on Latała's result, but it requires a knowledge of his bounds on moments in order to be employed successfully. We will also present a short proof (based on decoupling techniques) of Latała's result for non-negative random variables. Our proof gives a slightly better constant on the left-hand side of (1.2).

Our notation is standard; for a sequence (z_k) we let $z_n^* = \max_{1 \leq k \leq n} |z_k|$. The letters c and C denote absolute constants whose values may change from one use to the next. We will write $S = \sum_{k=1}^{\infty} X_k$, and $S_n = \sum_{k=1}^{n} X_k$, and $\|S\|_p = (\mathbb{E}|S|^p)^{1/p}$.

2. Tail estimates via moment estimates

In this section we will to obtain two-sided estimates for tails of sums of independent random variables. For the sake of brevity we will concentrate on symmetric random variables, although it will be clear that our arguments work for nonnegative random variables as well. In certain special cases tail inequalities have been obtained from moment inequalities (see Gluskin and Kwapień (1995), Hitczenko and Kwapień (1994) or Hitczenko, Montgomery-Smith and Oleszkiewicz (1997)). Also, in the case of multiples of Rademacher random variables, two-sided estimates have been obtained by Montgomery and Odlyzko (1988), and Montgomery-Smith (1990).

THEOREM 2.1. *There exist positive constants c, C, α and δ such that for all sequences of independent symmetric random variables (X_n), and for all t such that*

$$t \geq \frac{1}{2} \|\sum X_i I(|X_i| \leq t)\|_2 = \frac{1}{2} \left(\sum_{i=1}^{n} \|X_i I(|X_i| \leq t)\|_2^2 \right)^{1/2},$$

the following holds: Let p_t be the least p such that

$$\|\sum X_i I(|X_i| \leq t)\|_p \geq 2t.$$

Then we have the inequalities

(2.1) $$\mathbb{P}(|S_n| > t) \geq c\{\mathbb{P}(X_n^* > t) + \exp(-\alpha p_t)\},$$

and

(2.2) $$\mathbb{P}(|S_n| > 4t) \leq C\{\mathbb{P}(X_n^* > t) + \exp(-\delta p_t)\}.$$

If $t \leq \frac{1}{2} \|\sum X_i I(|X_i| \leq t)\|_2$, then

$$\mathbb{P}(|S_n| > t) \geq c.$$

PROOF. For a given t, let $Y_i = X_i I(|X_i| \leq t)$, and let $s_n = \sum_{j=1}^{n} Y_j$. Notice that $\|s_n\|_p$ is a continuous, increasing function of p, and that $\|s_n\|_2 \leq 2t$. Hence either $2 \leq p_t < \infty$ and $\|s_n\|_{p_t} = 2t$, or $p_t = \infty$ and $\|s_n\|_\infty \leq 2t$.

Let us start by proving (2.1). It follows from Levy's inequality and contraction principle (see e.g., Kwapień Woyczyński (1992, Propositions 1.1.2 and 1.2.1)) that

$$\mathbb{P}(X_n^* > t) \leq 2\mathbb{P}(|S_n| > t),$$

and,

$$\mathbb{P}(|s_n| > t) \leq 2\mathbb{P}(|S_n| > t).$$

Hence

(2.3) $$\mathbb{P}(|S_n| > t) \geq \frac{1}{4}\Big(\mathbb{P}(X_n^* > t) + \mathbb{P}(|s_n| > t)\Big).$$

Now we can see that if $p_t = \infty$, then the inequality is established. In the case that $p_t < \infty$, we need to obtain a lower estimate for the tail probability of a maximum of partial sums of uniformly bounded symmetric random variables (Y_i). But for such random variables, the following inequality is true (cf. Hitczenko (1994)): for all $q \geq p \geq 1$, we have

$$(2.4) \qquad \|s_n\|_q \leq C\frac{q}{p}\big\{\|s_n\|_p + \|Y_n^*\|_q\big\} \leq C\frac{q}{p}\big\{\|s_n\|_p + t\big\}.$$

We also use the Paley-Zygmund inequality that states that for any non-negative random variable Z, and $0 < \lambda < 1$,

$$\mathbb{P}(Z > \lambda EZ) \geq (1-\lambda)^2 \frac{(EZ)^2}{EZ^2}.$$

Since $t = \frac{1}{2}\|s_n\|_{p_t}$, we have that

$$\mathbb{P}(|s_n| > t) \geq \mathbb{P}(|s_n|^{p_t} > 2^{-p_t}\|s_n\|_{p_t}^{p_t}) \geq (1 - 2^{-p_t})^2 \frac{\|s_n\|_{p_t}^{2p_t}}{\|s_n\|_{2p_t}^{2p_t}}.$$

It follows from (2.4) that the denominator is no more than

$$C^{2p_t}\{\|s_n\|_{p_t} + t\}^{2p_t} \leq (\tfrac{3}{2}C)^{2p_t}\|s_n\|_{p_t}^{2p_t}.$$

Therefore, we get the estimate

$$\mathbb{P}(|s_n| \geq t) \geq (1 - 2^{-p_t})^2 (\tfrac{3}{2}C)^{-p_t} \geq \exp(-\alpha p_t),$$

which, together with (2.3) gives (2.1).

Inequality (2.2) is an easy consequence of Chebyshev's inequality. If $p_t < \infty$, then

$$\mathbb{P}(|S_n| > 4t) \leq \mathbb{P}(X_n^* > t) + \mathbb{P}(|s_n| > 4t) \leq \mathbb{P}(X_n^* > t) + \frac{E|s_n|^{p_t}}{(4t)^{p_t}}$$

$$\leq \mathbb{P}(X_n^* > t) + 2^{-p_t} = \mathbb{P}(X_n^* > t) + \exp(-\delta p_t).$$

If $p_t = \infty$, we use the same ideas, noticing that $\mathbb{P}(|s_n| > 2t) = 0$.

Finally, if $t \leq \frac{1}{2}\|s_n\|_2$, we apply the contraction principle, the Paley-Zygmund inequality, and (2.4), to get

$$2\mathbb{P}(|S_n| > t) \geq \mathbb{P}(|s_n|^2 > \tfrac{1}{4}E|s_n|^2) \geq \frac{9\|s_n\|_2^4}{16\|s_n\|_4^4} \geq \frac{9\|s_n\|_2^4}{16C^4(\|s_n\|_2+t)^4}$$

which is bounded below by a universal constant. $\square$

REMARK 1. The above theorem allows us to approximate tails of the sums of independent random variables in terms of tails of the individual summands. This follows from the fact that in view of Latała's result p_t can be approximated using only information about marginal distributions, and from the well known inequality

$$(2.5) \qquad \frac{\sum \mathbb{P}(|X_i| > u)}{1 + \sum \mathbb{P}(|X_i| > u)} \leq \mathbb{P}(X_n^* > u) \leq 2\frac{\sum \mathbb{P}(|X_i| > u)}{1 + \sum \mathbb{P}(|X_i| > u)},$$

which gives tails of X_n^* in terms of tails of individual summands.

3. Another proof of Latała's result for nonnegative rv's

Here we intend to give another proof of Latała's formula concerning $\|S\|_p$ for nonnegative random variables.

THEOREM 3.1. *Let (X_n) be a sequence of positive independent random variables. Then for all $p \geq 1$ we have that*

$$\kappa \||(X_n)\||_p \leq \|S\|_p \leq (e^p - 1)^{1/p} \||(X_n)\||_p, \tag{3.1}$$

where $\||(X_n)\||_p$ is given by (1.1), and κ is the positive number for which $f(\kappa) = e$, where

$$f(x) = \sum_{k=0}^{\infty} \frac{(2k+1)^k}{k!} x^k.$$

PROOF. First note that if $\||(X_n)\||_p \leq 1$, then since $1 + \sum_n X_n \leq \prod_n (1 + X_n)$, we have that $\|S\|_p^p \leq e^p - 1$. This proves the second inequality in (3.1) To prove the first, we use certain results concerning decoupling. These ideas appear often in the literature (usually in the context of mean-zero or symmetric random variables, see e.g., Kwapień and Woyczyński (1992)). However, since we will need control of constants, we cite the following, which is a special case of de la Peña, Montgomery-Smith and Szulga (1994, Theorem 2.1). □

LEMMA 3.2. *Let (X_n) be a sequence of real valued independent random variables. Let $(X_n^{(l)})$ be independent copies of (X_n) for $1 \leq l \leq k$. Furthermore, let $f_{i_1,\ldots,i_k}$ be elements of a Banach space such that $f_{i_1,\ldots,i_k} = 0$ unless the $i_1,\ldots,i_k$ are distinct. Then for any $1 \leq p \leq \infty$, we have that*

$$\left\| \sum_{i_1,\ldots,i_k} f_{i_1,\ldots,i_k} X_{i_1} \cdots X_{i_k} \right\|_p \leq (2k+1)^k \left\| \sum_{i_1,\ldots,i_k} f_{i_1,\ldots,i_k} X_{i_1}^{(1)} \cdots X_{i_k}^{(k)} \right\|_p.$$

Now let us finish the proof of Theorem 3.1. Note that

$$\prod_n \mathbb{E}|1 + X_n|^p = \left\| \prod_n (1 + X_n) \right\|_p^p,$$

and so by Minkowski's inequality we have that

$$\left\| \prod_n (1 + X_n) \right\|_p \leq 1 + \sum_{k=1}^{\infty} \left\| \sum_{i_1 < \cdots < i_k} X_{i_1} \cdots X_{i_k} \right\|_p.$$

But if $k \geq 1$

$$\left\| \sum_{i_1 < \cdots < i_k} X_{i_1} \cdots X_{i_k} \right\|_p = \frac{1}{k!} \left\| \sum_{\substack{i_1,\ldots,i_k \\ \text{distinct}}} X_{i_1} \cdots X_{i_k} \right\|_p$$

$$\leq \frac{(2k+1)^k}{k!} \left\| \sum_{\substack{i_1,\ldots,i_k \\ \text{distinct}}} X_{i_1}^{(1)} \cdots X_{i_k}^{(k)} \right\|_p$$

(where $(X_n^{(l)})$ are independent copies of (X_n) for $1 \leq l < \infty$)

$$\leq \frac{(2k+1)^k}{k!} \left\| \sum_{i_1,\ldots,i_k} X_{i_1}^{(1)} \cdots X_{i_k}^{(k)} \right\|_p$$

$$= \frac{(2k+1)^k}{k!} \|S\|_p^k.$$

Hence
$$\left\| \prod_n (1+X_n) \right\|_p \leq f(\|S\|_p),$$

So, if $\|S\|_p \leq \kappa$, then
$$\left\| \prod_n (1+X_n) \right\|_p \leq e,$$

that is,
$$\||(X_n)|\|_p \leq 1.$$

REMARK 2. Our constant in the second inequality of (3.1) is essentially the same as Latała's constant. But in the first inequality our constant, which may numerically be shown to be about 0.1549, is slightly better than Latała's constant, which is about 0.1162.

References

[1] V. de la Peña, S.J. Montgomery-Smith and J. Szulga *Contraction and decoupling inequalities for multilinear forms and U-statistics*, Ann. of Probab. **22** (1994), 1745–1765.

[2] E.D. Gluskin and S. Kwapień *Tail and moment estimates for sums of independent random variables with logarithmically concave tails*, Studia Math. **114** (1995), 303 - 309.

[3] P. Hitczenko *Domination inequality for martingale transforms of Rademacher sequence*, Israel J. Math. **84** (1993), 161–178.

[4] P. Hitczenko *On a domination of sums of random variables by sums of conditionally independent ones*, Ann. Probab. **22** (1994), 453–468.

[5] P. Hitczenko and S. Kwapień *On the Rademacher series*, Probability in Banach Spaces, Nine, Sandbjerg, Denmark, (J. Hoffmann-Jørgensen, J. Kuelbs, M.B. Marcus, ed.) Birkhäuser, Boston (1994), 31–36.

[6] P. Hitczenko, S.J. Montgomery-Smith and K. Oleszkiewicz *Moment inequalities for sums of certain independent symmetric random variables*, Studia Math **123** (1997), 15–42.

[7] S. Kwapień and W.A. Woyczyński *Random Series and Stochastic Integrals. Single and Multiple*, Birkhäuser, Boston (1992).

[8] R. Latała *Estimation of moments of sums of independent random variables*, Ann. Probab. **25** (1997), 1502–1513.

[9] H.L. Montgomery and A.M. Odlyzko *Large deviations of sums of independent random variables*, Acta Arithmetica **49** (1988), 427–434.

[10] S.J. Montgomery-Smith *The distribution of Rademacher sum*, Proc. Amer. Math. Soc. **109** (1990), 517–522.

DEPARTMENT OF MATHEMATICS, NORTH CAROLINA STATE UNIVERSITY, RALEIGH, NC 27695-8205
E-mail address: pawelmath.ncsu.edu

DEPARTMENT OF MATHEMATICS, UNIVERSITY OF MISSOURI–COLUMBIA, COLUMBIA, MO 65211
E-mail address: stephenmath.missouri.edu

Exponential integrability of diffusion processes

Yaozhong Hu

ABSTRACT. Let x_t be the solution of a stochastic differential equation
$$dx_t = b(t, x_t)dt + \sum_{i=1}^{m} \sigma_i(t, x_t)dw_t^i, \quad 0 \le t \le T, \quad x_0 = x.$$
We give conditions on the coefficients b, $\sigma_1, \cdots, \sigma_m$ such that the solution x_t is exponentially integrable, i.e. $\mathbb{E} e^{\alpha |x_T|^\beta} < \infty$ for some positive β and α. We also give conditions on the coefficients such that the solution x_t is uniformly exponentially integrable, i.e. $\mathbb{E} e^{\alpha \sup_{0 \le t \le T} |x_t|^\beta} < \infty$ for some positive β and α.

1. Introduction

The exponential integrability of a random variable is an important problem: when we use the Norvikov condition or the Kazamaki condition to show that a super-martingale is a martingale, we need to prove the exponential integrability of a certain random variable, see [6] and the references therein.

In [7], we use the exponential integrability to investigate the convergence rate of some numerical schemes for the Zakai equations.

In addition to the interest of the problem itself, we also know [1], [9] that the exponential integrability has some application to Sobolev inequality of diffusions.

For a (Banach space-valued) Gaussian variable X, it is known [11] that for any $\alpha > 0$, $\mathbb{E} e^{\alpha \|X\|} < \infty$. It is also known [4], [8] that there is a $\alpha_0 > 0$, $\mathbb{E} e^{\alpha_0 \|X\|^2} < \infty$. Recently, there are some studies on the exponential integrability of non-Gaussian random variables, [1], [2], [5], [9], [10], [14].

In this paper we are concerned with the exponential integrability of certain diffusion processes. This result provides the exponential integrability of a large family of (not necessary) random variables.

Let $\sigma_i(\cdot, \cdot)$, $i = 1, 2, \cdots, m$, $b(\cdot, \cdot)$ be continuous mappings from $\mathbb{R}_+ \times \mathbb{R}^d$ to $\mathbb{R}^d$ and let w_t^i, $i = 1, 2, \cdots, m$ be m independent Brownian motions on probability space (Ω, F, P) with filtration $(F_t)_{0 \le t < \infty}$. Consider the following stochastic

---

1991 *Mathematics Subject Classification.* Primary 60H10, 60J45; Secondary 46J12, 46E30.

The author was supported in part by the New Faculty Fund and the General Research Fund of University of Kansas.

© 1999 American Mathematical Society

differential equation

$$dx_t = b(t, x_t)dt + \sum_{i=1}^{m} \sigma_i(t, x_t)dw_t^i, \quad 0 \leq t \leq T, \quad x_0 = x, \tag{1.1}$$

where dw_t^i denotes Itô differential. The formal generator of this equation is

$$L_t f(x) = \frac{1}{2} \sum_{i,j=1}^{d} a_{ij}(t,x) \frac{\partial^2 f}{\partial x_i \partial x_j}(x) + \sum_{i=1}^{d} b_i(t,x) \frac{\partial f}{\partial x_i}(x) \tag{1.2}$$

where

$$(a_{ij}(t,x))_{1 \leq i,j \leq d} = A = \sum_{k=1}^{m} \sigma_k(t,x) \sigma_k^T(t,x).$$

If A is bounded and uniformly elliptic, $i.e.$ there are two constants $0 < \lambda < \mu < \infty$ such that

$$\lambda I \leq A \leq \mu I,$$

and if L is of divergence form, $i.e.$ $Lf(x) = \frac{1}{2} \sum_{i,j=1}^{d} \frac{\partial}{\partial x_i} \left(a_{ij}(t,x) \frac{\partial f}{\partial x_j}(x) \right)$, then from [12], we know that there is a constant M, dependent on λ and μ such that

$$P\left(\sup_{0 \leq s \leq t} |x_s - x| \geq r \right) \leq M e^{-\frac{r^2}{Mt}} \tag{1.3}$$

for all $(t,x) \in \mathbb{R}^d$ and $r > 0$.

Let us first assume that A is bounded and uniformly elliptic. Then we have

$$\mathbb{E} e^{\beta \sup_{0 \leq s \leq t} |x_s|^2} \leq e^{2\beta |x|^2} \mathbb{E} e^{2\beta \sup_{0 \leq s \leq t} |x_s - x|^2}.$$

It is easy to see that the right hand side is finite by (1.3). This means that if L is of divergence form and if A is uniformly bounded and uniformly elliptic, then x_t is uniformly integrable.

In what follows we deal with large class of stochastic differential equations whose coefficients are not necessarily bounded or uniformly elliptic. We shall not assume that the coefficients are globally Lipschitz continuous. But we do assume more conditions than those that guarantee the existence of the solution without explosion. More precisely, we assume that the coefficients satisfy the following assumptions:

(I) There are $0 < \lambda_1, \gamma, \beta < \infty$ and $\phi(x)$ such that ϕ is of class C^2, $\phi(x) \geq 0$, $\phi(x) \geq \gamma |x|^\beta$ (when $x \to \infty$) and

$$L_t \phi(x) \leq \lambda_1 \phi^{1-\gamma}(x), \quad \forall \ x \in \mathbb{R}^d.$$

(II) There is $0 < \lambda_2 < \infty$ such that

$$|\Gamma(\phi)(x)| \leq \lambda_2 \phi^{2-\gamma}(x), \quad \forall \ (t,x) \in \mathbb{R}_+ \times \mathbb{R}^d,$$

where $\Gamma(\phi)$ is the "carré du champ" operator:

$$\Gamma(\phi)(x) = \sum_{i,j=1}^{d} a_{ij}(t,x) \frac{\partial \phi(x)}{\partial x_i} \frac{\partial \phi(x)}{\partial x_j}.$$

We omit the dependence of Γ on t for simplicity. From the assumption (I), we know by [13] that the solution of (1.1) exists and is unique.

We say that x_T is exponentially integrable if there is a $\beta > 0$ such that

$$\mathbb{E} e^{\alpha |X_T|^\beta} < \infty$$

for some $\alpha > 0$. We say that the solution $x_\cdot$ of (1.1) is uniformly exponentially integrable if there is a $\beta > 0$ such that

$$\mathbb{E} e^{\alpha \sup_{0 \leq t \leq T} |X_t|^\beta} < \infty$$

for some $\alpha > 0$. It is easy to see that uniformly exponential integrability implies exponential integrability.

Now we state the main results of this paper.

The first theorem concerns with the uniformly exponential integrability of the solution

THEOREM 1.1. *Let $x_t, 0 \leq t \leq T$ be the solution of (1.1) and let the coefficients satisfy* (I) *and* (II).
a) *If $\gamma > \frac{3}{2}$, then*

$$\mathbb{E} e^{\alpha \sup_{0 \leq t \leq T} \phi(x_t)} < \infty$$

for any real value α.
a) *If $\gamma = \frac{3}{2}$, then*

$$\mathbb{E} e^{\alpha \sup_{0 \leq t \leq T} \phi(x_t)} < \infty$$

for all real values $\alpha \leq \alpha_0$, where $\alpha_0 > 0$ is positive number.

The second theorem will be concerned with the exponential integrability of x_T. Since we are not concerned with the sup norm, we expect to have "more" exponential integrability. The condition (I) can be weakened as
(I)′ There are $0 < \lambda_1, \beta < \infty$ and $\phi(x)$ such that ϕ is of class C^2, $\phi(x) \geq 0$, $\phi(x) \geq \gamma |x|^\beta$ (when $x \to \infty$) and

$$L_t \phi(x) \leq \lambda_1 \phi(x).$$

THEOREM 1.2. *Let $x_t, 0 \leq t \leq T$ be the solution of (1.1) and let the coefficients satisfy* (I)′ *and* (II).
a) *If $\gamma > 1$, then*

$$\mathbb{E} e^{\alpha \phi(x_T)} < \infty$$

for any real value α.
a) *If $\gamma = 1$, then*

$$\mathbb{E} e^{\alpha \phi(x_T)} < \infty$$

for all real values $\alpha \leq \alpha_0$, where $\alpha_0 > 0$ is positive number.

In Section 2, we shall give proof for Theorem 1.1 and in Section 3 we shall give proof of Theorem 1.2.

To explain the results obtained, we consider its application to the solution of the following simple equations:

$$dx_t = x_t^\alpha dw_t, \quad 0 \leq t \leq T, \quad x_0 = x.$$

It is easy to see that when $\alpha = 1$ (*i.e.* this is the case we get the so-called geometric Brownian motion), the solution is *not* exponentially integrable. But we can show that when $\alpha < 1$, the solution is uniformly exponentially integrable.

2. Proof of Theorem 1.1

We start by proving a lemma:

LEMMA 2.1. *Let the coefficients satisfy* (II). *Then*
$$\sum_{k=1}^{m} |\langle \nabla \phi(x), \sigma_k(x) \rangle|^2 \leq \lambda_2 \phi(x)^{2-\gamma}.$$

PROOF. Let Tr denote the trace of a matrix and $\nabla \phi \times \nabla \phi$ denote the tensor product of $\nabla \phi$, *i.e.* a matrix with entries $\left(\dfrac{\partial \phi}{\partial x_i} \cdot \dfrac{\partial \phi}{\partial x_j} \right)$. Then

$$\begin{aligned}
\sum_{k=1}^{m} |\langle \nabla \phi(x), \sigma_k(x) \rangle|^2 &= \sum_{k=1}^{m} \text{Tr}\left[\sigma_k(x) \sigma_k(x)^T \nabla \phi \times \nabla \phi \right] \\
&= \sum_{i,j=1}^{d} a_{ij}(t,x) \frac{\partial \phi}{\partial x_i} \frac{\partial \phi}{\partial x_j} \\
&\leq \lambda_2 \phi^{2-\gamma}(x)
\end{aligned}$$

This completes the proof. $\square$

We turn to the proof of Theorem 1.1
It is easy to check that
$$L_t f^n = n f^{n-1} L_t f + n(n-1) f^{n-2} \sum_{i,j=1}^{d} a_{ij} \frac{\partial f}{\partial x_i} \frac{\partial f}{\partial x_j}.$$

Applying Itô formula
$$f(x_t) = f(x_0) + \int_0^t L_s f(x_s) ds + \sum_{k=1}^{m} \int_0^t \langle \nabla f(x_s), \sigma_k(s, x_s) \rangle dw_s^k$$

to $f(x) = \phi^n(x)$, we have
$$\begin{aligned}
\phi^n(x_t) =\ & \phi^n(x_0) + n \int_0^t \phi^{n-1}(x_s) L_s \phi(x_s) ds \\
& + n(n-1) \int_0^t \phi^{n-2}(x_s) \sum_{i,j=1}^{d} a_{ij}(s, x_s) \frac{\partial \phi}{\partial x_i}(x_s) \frac{\partial \phi}{\partial x_j}(x_s) ds \\
& + n \sum_{k=1}^{m} \int_0^t \phi^{n-1}(x_s) \langle \nabla \phi(x_s), \sigma_k(s, x_s) \rangle dw_s^k.
\end{aligned}$$

Thus
$$\begin{aligned}
\mathbb{E} \sup_{0 \leq s \leq t} \phi^n(x_s) \leq\ & \phi^n(x_0) + n \mathbb{E} \sup_{0 \leq s \leq t} \int_0^s \phi^{n-1}(x_\rho) L_\rho \phi(x_\rho) d\rho \\
& + n(n-1) \mathbb{E} \sup_{0 \leq s \leq t} \int_0^s \phi^{n-2}(x_\rho) \sum_{i,j=1}^{d} a_{ij}(\rho, x_\rho) \frac{\partial \phi}{\partial x_i}(x_\rho) \frac{\partial \phi}{\partial x_j}(x_\rho) d\rho \\
& + n \mathbb{E} \sup_{0 \leq s \leq t} \sum_{k=1}^{m} \int_0^s \phi^{n-1}(x_\rho) \langle \nabla \phi(x_\rho), \sigma_k(\rho, x_\rho) \rangle dw_\rho^k
\end{aligned}$$

$$\leq \phi^n(x_0) + \lambda_1 n \mathbb{E} \int_0^t \phi^{n-\gamma}(x_\rho) d\rho + \lambda_2 n^2 \int_0^t \mathbb{E}\, \phi^{n-\gamma}(x_\rho) d\rho$$
$$+ n \left\{ \mathbb{E} \sup_{0 \leq s \leq t} \left| \sum_{k=1}^m \int_0^s \phi^{n-1}(x_\rho) \langle \nabla \phi(x_\rho), \sigma_k(\rho, x_\rho) \rangle dw_\rho^k \right|^2 \right\}^{1/2}$$

The last term inside $\{\,\cdot\,\}$ is estimated as follows: By Doob's inequality, we have for any $1 \leq k \leq m$,

$$\mathbb{E} \sup_{0 \leq s \leq t} \left| \int_0^s \phi^{n-1}(x_s) \langle \nabla \phi(x_s), \sigma_k(s, x_s) \rangle dw_s^k \right|^2$$
$$\leq 4 \mathbb{E} \left| \int_0^t \phi^{n-1}(x_s) \langle \nabla \phi(x_s), \sigma_k(s, x_s) \rangle dw_s^k \right|^2$$
$$\leq 4 \mathbb{E} \int_0^t \left| \phi^{n-1}(x_s) \langle \nabla \phi(x_s), \sigma_k(s, x_s) \rangle \right|^2 ds$$
$$\leq 4 \mathbb{E} \int_0^t \phi^{2n-2}(x_s) \sum_{k=1}^m \langle \nabla \phi(x_s), \sigma_k(s, x_s) \rangle^2 ds$$
$$\leq 4 \lambda_2 \mathbb{E} \int_0^t \phi^{2n-\gamma}(x_s) ds$$

Therefore

$$\mathbb{E} \sup_{0 \leq s \leq t} \phi^n(x_s) \leq \phi^n(x_0) + \lambda_1 n \mathbb{E} \int_0^t \phi^{n-\gamma}(x_s) ds + \lambda_2 n^2 \mathbb{E} \int_0^t \phi^{n-\gamma}(x_s) ds$$
$$+ 2n \sqrt{\lambda_2} \left\{ \mathbb{E} \int_0^t \phi^{2n-\gamma}(x_s) ds \right\}^{1/2}$$
$$= \phi^n(x_0) + C n^2 \mathbb{E} \int_0^t \phi^{n-\gamma}(x_s) ds$$
$$+ 2n \sqrt{\lambda_2} \left\{ \mathbb{E} \int_0^t \phi^{2n-\gamma}(x_s) ds \right\}^{1/2}$$
$$\leq \phi^n(x_0) + Cn + Cn^2 \mathbb{E} \int_0^t \phi^{n-\gamma}(x_s) ds$$
$$+ 2n \sqrt{\lambda_2} \left\{ \mathbb{E} \int_0^t \phi^{2n-\gamma}(x_s) ds \right\}^{1/2}$$

Let $Z_t = \mathbb{E} \sup_{0 \leq s \leq t} \phi^n(x_s)$. Then

$$Z_t \leq \phi^n(x_0) + Cn + Cn^2 \int_0^t Z_s^{\frac{n-\gamma}{n}}(x_s) ds$$
$$+ Cn \left\{ \int_0^t Z_s^{\frac{2n-\gamma}{n}} ds \right\}^{1/2},$$

where and in what follows C is a generic constant independent of n, whose value may be different in different appearance. Let $Y_t = Z_t^2$. Then

$$\begin{aligned} Y_t &\leq 9\phi(x_0)^{2n} + Cn + Cn^2 \int_0^t Y_s^{\frac{n-\gamma}{n}} ds + Cn^2 \int_0^t Y_s^{\frac{n-\gamma}{n}} ds \\ &\leq 9\phi(x_0)^{2n} + Cn + Cn^2 \int_0^t Y_s^{\frac{n-\gamma}{n}} ds \end{aligned}$$

By an inequality from [3] (p. 135, Theorem 3),

$$\begin{aligned} Y_T &\leq \left\{ \left(9\phi(x_0)^{2n} + Cn\right)^{\gamma/n} + \frac{\gamma}{n}(Cn^2)^2 \right\}^{n/\gamma} \\ &\leq \{C + C\, n^3\}^{n/\gamma} \\ &\leq C\, n^{3n/\gamma}, \end{aligned}$$

By the definition of Y_T and Jensen's inequality, we have

$$\mathbb{E} \sup_{0 \leq t \leq T} \phi(x_t)^n \leq C\, n^{3n/2\gamma}.$$

By Stirling's formula $n! \sim n^n e^{-n}\sqrt{2\pi n}$, it is straightforward to check that if $\gamma > \frac{3}{2}$, then

$$\mathbb{E}\, e^{\alpha \sup_{0 \leq t \leq T} \phi(x_t)} < \infty$$

for any real value α and if $\gamma = \frac{3}{2}$, then

$$\mathbb{E}\, e^{\alpha \sup_{0 \leq t \leq T} \phi(x_t)} < \infty$$

for all real values $\alpha \leq \alpha_0$, where $\alpha_0 > 0$ is positive number, proving the theorem.

3. Proof of Theorem 1.2

Using Itô's formula, we have

$$\begin{aligned} \mathbb{E}\,\phi(x_t)^n &= \phi(x_0)^n + n\mathbb{E} \int_0^t \phi^{n-1}(x_s) L_s \phi(x_s) ds \\ &\quad + n(n-1) \int_0^t \mathbb{E}\, \phi^{n-2}(x_s) \sum_{i,j=1}^d a_{ij}(s, x_s) \frac{\partial \phi}{\partial x_i}(x_s) \frac{\partial \phi}{\partial x_j}(x_s) ds \\ &\leq \phi(x_0)^n + \lambda_1 n \mathbb{E} \int_0^t \phi^n(x_s) ds \\ &\quad + \lambda_2 n^2 \int_0^t \mathbb{E}\, \phi^{n-\gamma}(x_s) ds \end{aligned}$$

Let $Z_s = \mathbb{E}\,\phi^n(x_s)$. Then

$$Z_t \leq \phi^n(x_0) + \lambda_1 n \int_0^t Z_s ds + \lambda_2 n^2 \int_0^t Z_s^{1-\frac{\gamma}{n}} ds.$$

Denote

$$g(t) = \lambda_1 nt + \lambda_2 n^2 t^{1-\frac{\gamma}{n}}.$$

The above inequality can be written as

$$Z_t = \phi(x_0)^n + \int_0^t g(Z_s) ds.$$

By an inequality of Bihari and Langenhop [3] (p. 135, Theorem 3), we have
$$Z_t \leq G^{-1}\left(G\left(\phi(x_0)^n\right) + t\right),$$
where
$$G(u) = \int_1^u \frac{1}{g(t)} dt.$$
We are going to bound $G(u)$. When $u \geq 1$,
$$\begin{aligned} G(u) &= \int_1^u \frac{1}{\lambda_1 n t + \lambda_2 n^2 t^{1-\frac{\gamma}{n}}} dt \\ &\leq \int_1^u \frac{1}{\lambda_1 n t} dt \\ &\leq \frac{1}{\lambda_1 n} \log u. \end{aligned}$$
Thus
$$G\left(\phi(x_0)^n\right) \leq \frac{\log \phi(x_0)}{\lambda_1}.$$
We are going to bound $G(u)$ from below by a nondecreasing function. When $u \geq n^{\frac{n}{\gamma}}$,
$$\begin{aligned} G(u) &= \int_1^u \frac{dt}{\lambda_1 n t + \lambda_2 n^2 t^{1-\frac{\gamma}{n}}} dt \\ &\geq \int_{n^{\frac{n}{\gamma}}}^u \frac{dt}{\lambda_1 n t + \lambda_2 n^2 t t^{-\frac{\gamma}{n}}} dt \\ &\geq \int_{n^{\frac{n}{\gamma}}}^u \frac{dt}{\lambda_1 n t + \lambda_2 n^2 t \left(n^{\frac{n}{\gamma}}\right)^{-\frac{\gamma}{n}}} dt \\ &= \int_{n^{\frac{n}{\gamma}}}^u \frac{dt}{Cnt} \\ &\geq \frac{1}{Cn} \left[\log u - \log n^{\frac{n}{\gamma}}\right] \\ &= \frac{1}{Cn} \log \frac{u}{n^{\frac{n}{\gamma}}}. \end{aligned}$$
Since $G(u) \geq \frac{1}{Cn} \log \frac{u}{n^{\frac{n}{\gamma}}}$
$$G^{-1}\left(\frac{\log \phi(x_0)}{\lambda_1} + T\right) \leq n^{\frac{n}{\gamma}} e^{Cn}.$$
We obtain that
$$\mathbb{E}\, \phi^n(x_T) \leq e^{Cn} n^{\frac{n}{\gamma}}.$$
By Stirling's formula $n! \sim n^n e^{-n} \sqrt{2\pi n}$, it is straightforward to check that if $\gamma > 1$, then
$$\mathbb{E}\, e^{\alpha \phi(x_T)} < \infty$$
for any real value α and if $\gamma = 1$, then
$$\mathbb{E}\, e^{\alpha \phi(x_T)} < \infty$$
for all real values $\alpha \leq \alpha_0$, where $\alpha_0 > 0$ is positive number, proving the theorem.

4. Examples

Let us consider the following stochastic differential equation

(4.1) $$dx_t = x_t^\alpha dw_t, \quad 0 \leq t \leq T, \quad x_0 = x.$$

The corresponding formal generator is

$$Lf(x) = \frac{1}{2}x^{2\alpha}f''(x).$$

$$\Gamma(f)(x) = x^{2\alpha}(f'(x))^2$$

Let $\phi : \mathbb{R} \to \mathbb{R}_+$ be a C^3 function and when $|x| > 1$, $\phi(x) = |x|^\beta$, where $\beta > 0$. By the choice of ϕ we know that when $|x| > 1$

$$\phi'(x) = \beta \text{sign}(x)|x|^{\beta-1} \quad \text{and} \quad \phi''(x) = \beta(\beta-1)|x|^{\beta-2}.$$

It is easy to see that when $|x| > 1$

$$L\phi(x) = \beta(\beta-1)\frac{1}{2}|x|^{2\alpha+\beta-2}$$

On the other hand $\phi(x)^{1-\gamma} = |x|^{\beta-\gamma\beta}$ when $|x| > 1$. The assumption (I) is satisfied if $2\alpha + \beta - 2 \leq \beta(1-\gamma)$ which is implied by the following choice

$$\beta = \frac{2-2\alpha}{\gamma} \quad \text{when} \quad \alpha < 1.$$

$\Gamma(\phi)$ is computed as

$$\Gamma(\phi) = \beta^2 x^{2\alpha+2\beta-2}$$

Thus the assumption (II) is satisfied if

$$2\alpha + 2\beta - 2 \leq (2-\gamma)\beta$$

which is implied by also choosing

$$\beta = \frac{2-2\alpha}{\gamma} \quad \text{when} \quad \alpha < 1.$$

Therefore, for any $\alpha < 1$, and $\gamma > 0$, if we take $\beta = \frac{2-2\alpha}{\gamma}$, then the function ϕ constructed as above satisfies (I) and (II) with β being

$$\frac{4-4\alpha}{3} \quad \text{when} \quad \gamma = \frac{3}{2}$$

or

$$< \frac{4-4\alpha}{3} \quad \text{when} \quad \gamma > \frac{3}{2}.$$

Thus for any $\alpha < 1$, the solution x_t of (4.1) is exponentially integrable. More precisely,

COROLLARY 1. *When $\alpha < 1$, the equation (4.1) has a unique solution x_t.*
(I) *For any $C > 0$, $T > 0$, and $\beta < \frac{4-4\alpha}{3}$, we have*

$$\mathbb{E} e^{C \sup_{0 \leq t \leq T} |x_t|^\beta} < \infty$$

(II) *There is a $C_0 > 0$ (dependent on T, the data of the equation (4.1)), such that*

$$\mathbb{E} e^{C_0 \sup_{0 \leq t \leq T} |x_t|^\beta} < \infty, \quad \text{where} \quad \beta = \frac{4-4\alpha}{3}.$$

In a similar way we have that

COROLLARY 2. *When $\alpha < 1$, the equation (4.1) has a unique solution x_t.*
(I) *For any $C > 0$, $T > 0$, and $\beta < 2 - 2\alpha$, we have*

$$\mathbb{E}\, e^{C|x_t|^\beta} < \infty$$

(II) *There is a $C_0 > 0$ (dependent on T, the data of the equation (4.1)), such that*

$$\mathbb{E}\, e^{C_0 \sup_{0 \leq t \leq T} |x_t|^\beta} < \infty, \quad \text{where} \quad \beta = \beta < 2 - 2\alpha.$$

The above results are sharp in the sense that when α is not strictly smaller than 1, then the solution x_T is *not* exponentially integrable. Let us consider the following equation

(4.2) $$dx_t = x_t dw_t, \quad 0 \leq t \leq T, \quad x_0 = x.$$

(In this case $\alpha = 1$). The solution of this equation is called geometric Brownian motion and can be represented explicitly as

$$x_t = x e^{w_t - \frac{1}{2}t}.$$

It is easy to see that for any positive $\beta > 0$,

$$\mathbb{E}\left\{ e^{|x_T|^\beta} \right\} = \infty.$$

When $\alpha = 0$, the solution x_t of (4.1) is the Brownian motion, *i.e.* $x_t = w_t$. Theorem 1.1 implies then

$$\mathbb{E}\, e^{C \sup_{0 \leq t \leq T} |w_t|^{\frac{4}{3}}} < \infty$$

for some $C > 0$ which is known to be true by the Theorem of Fernique mentioned in the introduction.

And in fact we know that

$$\mathbb{E}\, e^{C \sup_{0 \leq t \leq T} |w_t|^2} < \infty$$

for some $C > 0$. Thus our Theorem 1.1 is not as strong as Fernique's result when applied to Brownian motions. But it is stronger than the Skorohod's result even in this case.

Motivated by this we conjecture that

Let $x_t, 0 \leq t \leq T$ be the solution of (1.1) and let the coefficients satisfy (I) and (II).
a) *If $\gamma > 1$, then*

$$\mathbb{E}\, e^{\alpha \sup_{0 \leq t \leq T} \phi(x_t)} < \infty$$

for any real value α.
a) *If $\gamma = 1$, then*

$$\mathbb{E}\, e^{\alpha \sup_{0 \leq t \leq T} \phi(x_t)} < \infty$$

for all real values $\alpha \leq \alpha_0$, where $\alpha_0 > 0$ is positive number.

Acknowledgment. The author would like to thank Jan Rosiński and Adam Jakubowski for helpful discussions.

References

[1] Aida S. Masuda T. and Shigekawa I. *Logarithmic Sobolev inequalities and exponential integrability.* J. Funct. Anal. 126 (1994), 83-101.

[2] de Acosta A. *Strong exponential integrability of sums of independent B-valued random vectors.* Prob. Math. Statist. 1 (1980), 133-150.

[3] Beckenbach E.F. and Bellman R. *Inequalities.* Springer, 1965.

[4] Fernique X. *Intégrabilité des vecteurs gausiens.* Compté Rendus Acad. Sci. Paris 270 (1974), 1698-1699.

[5] Fukuda R. *Exponential integrability of sub-Gaussian vectors.* Probab. Theory Related Fields 85 (1990), no. 4, 505–521.

[6] Hu Y.Z. and Kallianpur G. *Exponential Integrability and Singular Infinite Dimensional Stochastic Differential Equations.* J. Appl. Math. Optim. 37 (1998), 295-353.

[7] Hu Y.Z. *Numerical approximations of Zakai equations, in preparation.*

[8] Kuo H. H. *Gaussian measures in Banach spaces.* Lecture Notes in Mathematics, Vol. 463. Springer-Verlag, Berlin-New York, 1975.

[9] Ledoux M. *Remarks on logarithmic Sobolev constants, exponential integrability and bounds on the diameter.* J. Math. Kyoto Univ. 35 (1995), 211-220.

[10] Rosiński J. *Remarks on strong exponential integrability of vector-valued random series and triangular arrays.* Ann. Probab. 23 (1995), 464-473.

[11] Skorohod A. V. *Integration in Hilbert space.* Translated from the Russian by Kenneth Wickwire. Ergebnisse der Mathematik und ihrer Grenzgebiete, Band 79. Springer-Verlag, New York-Heidelberg, 1974.

[12] Stroock D. W. *Diffusion semigroups corresponding to uniformly elliptic divergence form operators.* Séminaire de Probabilités, XXII, 316–347, Lecture Notes in Math., 1321, Springer, Berlin-New York, 1988.

[13] Stroock D. W. and Varadhan S.R.S. *Multidimensional diffusion processes.* Springer, 1979.

[14] Üstünel A. S. *Intégrabilité exponentielle de fonctionnelles de Wiener.* C. R. Acad. Paris Sér I Math 315 (1992), 997-1000.

DEPARTMENT OF MATHEMATICS, UNIVERSITY OF KANSAS, LAWRENCE, KS 66045-2142
E-mail address: hu@math.ukans.edu

Local dependencies in random fields via a Bonferroni-type inequality

Adam Jakubowski and Jan Rosiński

ABSTRACT. We provide an inequality which is a useful tool in studying both large deviation results and limit theorems for sums of random fields with "negligible" small values. In particular, the inequality covers cases of stable limits for random variables with heavy tails and compound Poisson limits of $0-1$ random variables.

1. Bonferroni-type inequalities in limit theorems for sums of stationary sequences

The simplest Bonferroni-type inequality can be formulated in the following way (see inequality I.17, p. 16, [GS96]):

$$(1.1) \qquad 0 \leq \sum_{i=1}^{n} P(A_i) - P(\bigcup_{i=1}^{n} A_i) \leq \sum_{1 \leq i < j \leq n} P(A_i \cap A_j),$$

where $A_1, A_2, \ldots, A_n$ are events in some probability space.

In general this inequality gives very bad estimate for the difference $\sum_{i=1}^{n} P(A_i) - P(\bigcup_{i=1}^{n} A_i)$ (see p. 19, [GS96] for discussion of typical examples). However, when properly used, it brings essential simplification in many areas. Perhaps the most known (and the simplest) is the limit theory for order statistics of stationary sequences, as presented in [LLR83] or [G78]. It may be instructive to provide the reader with a brief outline of the reasoning leading to the basic result of this theory (Theorem 3.4.1, Chapter 3, [LLR83]).

Let $X_1, X_2, \ldots$, be a stationary sequence and let $M_n = \max_{1 \leq i \leq n} X_i$ be partial maxima for this sequence. Given a sequence $\{u_n\}$ of numbers we want to calculate the limit for $P(M_n \leq u_n)$. For a large class of stationary sequences (satisfying so called condition $D(u_n)$), we can asymptotically replace $P(M_n \leq u_n)$ with

$$P(M_{[n/k_n]} \leq u_n)^{k_n},$$

1991 *Mathematics Subject Classification.* Primary 60E15, 60F10; Secondary 60F05, 60E07.

Research of the first author was done during visits to Université de Lille I and University of Tennessee, Knoxville.

Research of the second author was supported in part by the NSF Grant DMS-97-04744.

© 1999 American Mathematical Society

with some $k_n \to \infty$. This in turn is asymptotically the same as
$$\exp(-k_n P(M_{[n/k_n]} > u_n)).$$
For fixed n, set $A_i = \{X_i > u_n\}$ and observe that by (1.1)
$$k_n |P(M_{[n/k_n]} > u_n) - [n/k_n] P(X_1 > u_n)|$$
$$\leq k_n \sum_{1 \leq i < j \leq [n/k_n]} P(X_i > u_n, X_j > u_n).$$
If so called condition $D'(u_n)$ is also satisfied, then the last expression above tends to zero as $n \to \infty$ and we can calculate the limit for $P(M_n \leq u_n)$ as if the random variables X_i were independent, i.e.
$$\lim_{n \to \infty} P(M_n \leq u_n) = \exp(- \lim_{n \to \infty} n P(X_1 > u_n)).$$
Condition $D(u_n)$ represents here "mixing" or "weak dependence" properties of the sequence in the form proper for maxima, while condition $D'(u_n)$ asserts that in the sequence $\{X_i\}$ there are no local (within intervals of length $[n/k_n]$) clusters of values exceeding levels u_n. Since independent random variables satisfy condition $D'(u_n)$ for sequences $\{u_n\}$ of interest, one can also say that the sequence essentially has no "local dependencies" between random variables. The latter terminology is even more convincing when one realizes that condition $D'(u_n)$ cannot hold for 1-dependent random variables $X_i = Y_{i-1} \vee Y_i$, where Y_i is a sequence of independent and identically distributed random variables and u_n is such that $\liminf_n nP(Y_1 > u_n) > 0$. Clearly, such X_i's exhibit "local dependencies" and admit "local clusters" of values exceeding levels u_n.

It was R.A. Davis who first observed that similar results hold also for sums of stationary sequences with heavy tails. Using the technique of extreme value theory as well as the series representation for stable laws due to LePage, Woodroofe and Zinn, Davis [**D83**] proved that asymptotics of sums of "weakly dependent" stationary random variables *with* heavy tails and *without* local dependencies is essentially the same as if they were independent. Subsequent papers [**JK89**], [**DH95**], [**K95**] showed that the essence of Davis' method was representing sums as integrals with respect to point processes on $\mathbb{R}^1 \setminus \{0\}$ built upon the sequence X_i. If one defines $N_n(A) = \sum_{i=1}^n I(X_i/B_n \in A)$, then
$$\frac{X_1 + X_2 + \ldots + X_n}{B_n} = \int_{\mathbb{R}^1 \setminus \{0\}} x N_n(dx),$$
and weak convergence of N_n's implies weak convergence of S_n/B_n. In particular, results for sums of dependent sequences with heavy tails can be obtained in a similar way as results for sums of independent sequences were derived in [**R86**] (this analogy is not applicable for functional convergence).

The difference between weakly dependent and independent case is that in the absence of conditions excluding clusters of "big" values (like $D'(u_n)$ in the theory for extremes), the parameters of the limiting stable law are determined by local dependence properties. Davis and Hsing [**DH95**] provide a probabilistic representation for these parameters. In some cases (e.g. for m-dependent random variables) another, much simpler representation is available [**JK89**], which is valid also for generalized Poisson limits [**K95**]. Comparing to stable limit theorems for m-dependent random vectors obtained by purely analytical methods by L. Heinrich in [**H82**], [**H85**], probabilistic reasoning gave both deeper insight into the structure of the limiting stable laws and allowed avoiding many of technicalities in formulation of results. When specialized to sums of m-dependent $0-1$ random variables, the point processes method provides sufficient and *necessary* conditions for convergence to compound Poisson distribution [**K95**], contrary to the earlier methods based on Poisson approximations via the Chen-Stein method (see e.g. [**AGG90**]), where only sufficient conditions are given.

It is interesting that most of the above results can be obtained without employing point processes techniques and using the following Bonferroni-type inequality.

THEOREM 1.1 (Lemma 3.2, [**J97**]). *Let $Z_1, Z_2, \ldots$ be stationary random vectors taking values in a linear space $(E, \mathcal{B}_E)$. Set $S_0 = 0$, $S_k = \sum_{j=1}^{k} Z_j$, $k \in \mathbb{N}$.*

If $U \in \mathcal{B}_E$ is such that $0 \notin U$, then for every $n \in \mathbb{N}$ and every m, $0 \leq m \leq n$, the following inequality holds:

$$|P(S_n \in U) - n(P(S_{m+1} \in U) - P(S_m \in U))|$$

(1.2)
$$\leq 2mP(Z_1 \neq 0) + 2 \sum_{\substack{1 \leq i < j \leq n \\ j-i > m}} P(Z_i \neq 0, Z_j \neq 0).$$

Although inequality (1.2) does not fit the formal definition of the Bonferroni-type inequality given on p. 10 in [**GS96**], we call it Bonferroni-type for the following reasons.

1. When $m = 0$ we obtain from (1.2)

$$|P(S_n \in U) - nP(Z_1 \in U)| \leq 2 \sum_{1 \leq i < j \leq n} P(Z_i \neq 0, Z_j \neq 0),$$

what is formally similar to (1.1). Notice that the constant 2 above is sharp.

2. The inequality becomes interesting only if we deal with at least "weak dependence", that is under purely probabilistic assumption.

3. The inequality is proved by integrating its pointwise version and in this sense its proof is similar to proofs of the Bonferroni-type inequalities obtained by the "indicator method" (see [**GS96**]).

4. In Section 3 we provide a unifying framework for both inequalities (1.1) and (1.2).

The inequality looks very restrictive and may seem applicable only to 0-1 stationary random variables $Z_j = I_{A_j}$, in which case it reads as follows.

$$\left|P(\sum_{j=1}^{n} I_{A_j} = k) - n\left(P(\sum_{j=1}^{m+1} I_{A_j} = k) - P(\sum_{j=1}^{m} I_{A_j} = k)\right)\right|$$
$$\leq 2mP(A_1) + 2 \sum_{\substack{1 \leq i < j \leq n \\ j-i > m}} P(A_i \cap A_j).$$

The above inequality can be directly applied to give an alternative (and much simpler!) proof of results due to Kobus [**K95**] for m-dependent $0-1$ random variables.

Originally however inequality (1.2) was designed to manipulate with probabilities of large deviation for sums of random variables with heavy tails. An extensive discussion of such results as well as their meaning for stable limit theorems (essential part of necessary and sufficient conditions) can be found in [**J93**], [**J97**] and

[JNZ97] (for necessary results on stable laws we refer to [JW94] and [ST94]). Here let us sketch basic ideas only.

Let $X_1, X_2, \ldots$ be a stationary sequence, $S_n = X_1 + X_2 + \ldots + X_n$, $B_n \to \infty$ be a $1/p$-regularly varying sequence, where $0 < p < 2$ and let $x_n \to \infty$. We are interested in asymptotic behavior of large deviation probabilities $P(S_n/B_n > x_n)$. More precisely, we want to prove that under relatively mild assumptions

(1.3) $$x_n^p P(S_n/B_n > x_n) \to c_+,$$

where the constant $0 < c_+ < \infty$ can be identified. The first step consists in proving that as $n \to \infty$

$$x_n^p \left(P(S_n/B_n > x_n) - P\left(\sum_{j=1}^n Z_{n,j}^{\delta_n} > x_n\right) \right) \to 0,$$

where

$$Z_{n,j}^{\delta_n} = \begin{cases} 0 & \text{if } |X_j| < B_n \cdot x_n \cdot \delta_n, \\ X_j/B_n & \text{otherwise.} \end{cases}$$

This requires some polynomial domination condition on tail probabilities of X_j's and, if $1 \leq p < 2$, some assumptions on the size of variances of random variables $T_n^{\delta_n} = S_n/B_n - \sum_{j=1}^n Z_{n,j}^{\delta_n}$.

In the next, essential step, we apply inequality (1.2) to random variables $Z_{n,j}^{\delta_n}$, $j = 1, 2, \ldots, n$, and $U = (x_n, \infty)$. Careful control of the size of x_n and δ_n *plus* information on dependence (e.g. m-dependence) *plus* return to original random variables allow reducing (1.3) to

$$x_n^p n \big(P(X_1 + X_2 + \ldots + X_{m+1} > x_n B_n) - P(X_1 + X_2 \ldots + X_m > x_n B_n) \big) \to c_+.$$

This shows that the limiting parameter c_+ can be calculated using only finite dimensional (of size $m + 1$) distributions of the sequence $X_1, X_2, \ldots$, and that its value depends on local dependence structures, as desired.

Clearly, variants of the above reasoning with m varying are also workable.

In the present paper we are going to prove an analog of (1.2) for random fields and in nonstationary case. Following the line of [J97] it allows deriving results for m-dependent random fields, similar to stable limit theorems of Heinrich [H86], [H87] or results on convergence to compound Poisson distributions [AGG90]. We leave their extensive discussion to other place.

2. A Bonferroni-type inequality for random fields

In what follows we choose and fix two integer numbers:
d - the dimension of the lattice $\mathbb{Z}^d$ indexing random fields $\{Z_t\}_{t \in \mathbb{Z}^d}$;
m - the admissible size of local clusters, $m \geq 0$.

If $\{Z_t\}$ is a random field and $\Lambda \subset \mathbb{Z}^d$ is a finite set, we define

(2.1) $$S_\Lambda = \sum_{t \in \Lambda} Z_t, \quad S_\emptyset = 0.$$

Let

$$B = \{0, 1, \ldots, m\}^d,$$

and $B_t = B + t$, $t \in \mathbb{Z}^d$. Further, let

$$\mathcal{E} = \{0, 1\}^d = \{\varepsilon = (\varepsilon_1, \varepsilon_2, \ldots, \varepsilon_d) : \varepsilon_j = 0 \text{ or } 1\}$$

and let

(2.2) $$B_t^\varepsilon = B_t \cap B_{t+\varepsilon}, \quad \varepsilon \in \mathcal{E}.$$

Define, for $U \in \mathcal{B}_E$ and $\boldsymbol{t} \in \mathbb{Z}^d$,

$$\Delta_t(U) = \sum_{\varepsilon \in \mathcal{E}} (-1)^{|\varepsilon|} P(S_{B_t^\varepsilon} \in U)$$

where

$$|\varepsilon| = \varepsilon_1 + \varepsilon_2 + \cdots + \varepsilon_d.$$

Put $\mathbf{1} = (1,\ldots,1) \in \mathcal{E}$. Define the *"boundary"* of a set $\Lambda \subset \mathbb{Z}^d$ by

(2.3) $$\partial \Lambda = \{\boldsymbol{s} \notin \Lambda : \exists_{t \in \Lambda}\ \boldsymbol{s} \in B_t\} \cup \{\boldsymbol{t} \in \Lambda : \exists_{s \in \Lambda^c}\ \boldsymbol{t} \in B_s \setminus B_{s+1}\}$$

Notice that the second part of $\partial \Lambda$, consisting of points from Λ, is empty when $d=1$.

THEOREM 2.1. *Let Z_t, $\boldsymbol{t} \in \mathbb{Z}^d$ be a random field with values in a linear space $(E, \mathcal{B}_E)$. If $U \in \mathcal{B}_E$ and $0 \notin U$ then*

(2.4) $$|P(S_\Lambda \in U) - \sum_{t \in \Lambda} \Delta_t(U)| \leq c_1(d,m) \sum_{s \in \partial \Lambda} P(Z_s \neq 0)$$
$$+ c_2(d,m) \sum_{\substack{s,t \in \Lambda \\ \|t-s\|_\infty > m}} P(Z_s \neq 0, Z_t \neq 0),$$

where $c_1(d,m) = 2^d((m+1)^d - 1)$ and $c_2(d,m) = 2^{-1}(1 + 2^d(2m+1)^d)$.

PROOF. Let

(2.5) $$\delta_t(U) = \sum_{\varepsilon \in \mathcal{E}} (-1)^{|\varepsilon|} I(S_{B_t^\varepsilon} \in U).$$

Since $\Delta_t(U) = E\delta_t(U)$, it is enough to establish a "pointwise" version of (2.4), i.e.

(2.6) $$|I(S_\Lambda \in U) - \sum_{t \in \Lambda} \delta_t(U)| \leq c_1(d,m) \sum_{s \in \partial \Lambda} I(Z_s \neq 0)$$
$$+ c_2(d,m) \sum_{\substack{s,t \in \Lambda \\ \|t-s\|_\infty > m}} I(Z_s \neq 0, Z_t \neq 0),$$

We shall deal with a modification of Z_t which vanishes outside our set Λ:

$$Z_t' = \begin{cases} Z_t & \text{if } \boldsymbol{t} \in \Lambda, \\ 0 & \text{if } \boldsymbol{t} \notin \Lambda. \end{cases}$$

Let S_Λ' and $\delta_t'(U)$ denote quantities defined by replacement of Z_t with Z_t' in formulas (2.1) and (2.5), respectively. Then by the very definition we have

$$I(S_\Lambda \in U) = I(S_\Lambda' \in U).$$

Further, $\delta_t(U) \neq \delta'_t(U)$ implies that there exists $s \in B_t \cap \Lambda^c$ such that $Z_s \neq 0$. Hence we can estimate

$$|(I(S_\Lambda \in U) - \sum_{t \in \Lambda} \delta_t(U)) - (I(S'_\Lambda \in U) - \sum_{t \in \Lambda} \delta'_t(U))|$$

(2.7)
$$\leq \sum_{\substack{t \in \Lambda \\ B_t \cap \Lambda^c \neq \emptyset}} 2^d I(\exists_{s \in B_t \cap \Lambda^c} Z_s \neq 0)$$

$$\leq 2^d \sum_{\substack{t \in \Lambda \\ B_t \cap \Lambda^c \neq \emptyset}} \sum_{s \in B_t \cap \Lambda^c} I(Z_s \neq 0) =: R_1,$$

where the factor 2^d comes from the cardinality of $\mathcal{E}$. Furthermore, if $\partial_1 \Lambda$ denotes the first part of the boundary (2.3) consisting of points from Λ^c, then

(2.8)
$$R_1 = 2^d \sum_{t \in \Lambda} \sum_{s \in \Lambda^c} I_{B_t \cap \Lambda^c}(s) I(Z_s \neq 0)$$

$$= 2^d \sum_{\substack{s \in \Lambda^c \\ \exists_{t \in \Lambda} s \in B_t}} (\sum_{t \in \Lambda} I_{B_t \cap \Lambda^c}(s)) I(Z_s \neq 0)$$

$$\leq 2^d ((m+1)^d - 1) \sum_{s \in \partial_1 \Lambda} I(Z_s \neq 0).$$

Hence it suffices to prove (2.6) under the assumption that

(2.9)
$$Z_t = 0 \text{ for } t \notin \Lambda.$$

In this case the first sum on the right hand side of (2.6) will be over the second part of the boundary (2.3) consisting of points from Λ. Now define a random set

$$\Lambda_0 = \{s \in \mathbb{Z}^d : Z_t \neq 0\}$$

and let

$$\mathrm{diam}(\Lambda_0) = \sup\{\|s - u\|_\infty : s, u \in \Lambda_0\}.$$

Notice that (2.9) gives

(2.10)
$$\Lambda_0 \subset \Lambda$$

so that $\mathrm{diam}(\Lambda_0)$ is a bounded random variable. For a <u>fixed</u> point ω in the probability space we will consider three particular cases of $\mathrm{diam}(\Lambda_0(\omega))$:

Case 1. $\mathrm{diam}(\Lambda_0) \leq m$.

This assumption implies that $\Lambda_0 \subset B_{t_0}$ for some $t_0 \in \mathbb{Z}^d$. Hence

(2.11)
$$I(S_\Lambda \in U) = I(S_{\Lambda_0} \in U) = I(S_{B_{t_0}} \in U).$$

The first observation is that if $t \notin B_{t_0}$, then $\delta_t(U) = 0$ and consequently

(2.12)
$$\sum_{t \in \Lambda} \delta_t(U) = \sum_{t \in B_{t_0} \cap \Lambda} \delta_t(U).$$

Indeed, for $t = (t_1, t_2, \ldots, t_d) \notin B_{t_0}$ we have

$$\delta_t(U) = \sum_{\varepsilon \in \mathcal{E}} (-1)^{|\varepsilon|} I(S_{B_t^\varepsilon} \in U)$$

$$= \sum_{\varepsilon \in \mathcal{E}} (-1)^{|\varepsilon|} I(S_{B_t \cap B_{t+\varepsilon} \cap B_{t_0}} \in U).$$

If $t_k \geq t_k^0$, $k = 1, 2, \ldots, d$, where $\boldsymbol{t_0} = (t_1^0, t_2^0, \ldots, t_d^0)$, then either $\boldsymbol{t} \in B_{\boldsymbol{t_0}}$ or $B_{\boldsymbol{t}} \cap B_{\boldsymbol{t_0}} = \emptyset$. The former possibility has been excluded and by the latter $\delta_{\boldsymbol{t}}(U) = 0$. So assume that $t_k < t_k^0$ for some k, $1 \leq k \leq d$. Let $\boldsymbol{\varepsilon'} = (\varepsilon_1, \ldots, \varepsilon_{k-1}, 0, \varepsilon_{k+1}, \ldots, \varepsilon_d)$ and $\boldsymbol{\varepsilon''} = (\varepsilon_1, \ldots, \varepsilon_{k-1}, 1, \varepsilon_{k+1}, \ldots, \varepsilon_d)$. Then

$$S_{B_{\boldsymbol{t}} \cap B_{\boldsymbol{t}+\boldsymbol{\varepsilon'}} \cap B_{\boldsymbol{t_0}}} = S_{B_{\boldsymbol{t}} \cap B_{\boldsymbol{t}+\boldsymbol{\varepsilon''}} \cap B_{\boldsymbol{t_0}}},$$

hence

$$(-1)^{|\boldsymbol{\varepsilon'}|} I(S_{B_{\boldsymbol{t}} \cap B_{\boldsymbol{t}+\boldsymbol{\varepsilon'}} \cap B_{\boldsymbol{t_0}}} \in U) + (-1)^{|\boldsymbol{\varepsilon''}|} I(S_{B_{\boldsymbol{t}} \cap B_{\boldsymbol{t}+\boldsymbol{\varepsilon''}} \cap B_{\boldsymbol{t_0}}} \in U) = 0,$$

and so $\delta_{\boldsymbol{t}}(U) = 0$. Thus (2.12) follows and now we will prove that

$$(2.13) \qquad I(S_{B_{\boldsymbol{t_0}}} \in U) = \sum_{\boldsymbol{t} \in B_{\boldsymbol{t_0}}} \delta_{\boldsymbol{t}}(U).$$

Define

$$q_{\boldsymbol{t}} = I(S_{B_{\boldsymbol{t}} \cap B_{\boldsymbol{t_0}}} \in U)$$

and for $A \subset \mathbb{Z}^d$

$$Q(A) = \sum_{\boldsymbol{t} \in A \cap B_{\boldsymbol{t_0}}} q_{\boldsymbol{t}}.$$

Notice that for some $\boldsymbol{t} \in B_{\boldsymbol{t_0}}$ the points $\boldsymbol{t} + \boldsymbol{\varepsilon}$ may lie outside of $B_{\boldsymbol{t_0}}$, but then $q_{\boldsymbol{t}+\boldsymbol{\varepsilon}} = 0$ and so

$$Q(B_{\boldsymbol{\varepsilon}}) = \sum_{\boldsymbol{t} \in B_{\boldsymbol{t_0}}} q_{\boldsymbol{t}+\boldsymbol{\varepsilon}}.$$

By the assumption currently in force

$$(2.14) \qquad \begin{aligned} \sum_{\boldsymbol{t} \in B_{\boldsymbol{t_0}}} \delta_{\boldsymbol{t}}(U) &= \sum_{\boldsymbol{t} \in B_{\boldsymbol{t_0}}} \sum_{\boldsymbol{\varepsilon} \in \mathcal{E}} (-1)^{|\boldsymbol{\varepsilon}|} I(S_{B_{\boldsymbol{t}} \cap B_{\boldsymbol{t}+\boldsymbol{\varepsilon}} \cap B_{\boldsymbol{t_0}}} \in U) \\ &= \sum_{\boldsymbol{t} \in B_{\boldsymbol{t_0}}} \sum_{\boldsymbol{\varepsilon} \in \mathcal{E}} (-1)^{|\boldsymbol{\varepsilon}|} q_{\boldsymbol{t}+\boldsymbol{\varepsilon}} \\ &= \sum_{\boldsymbol{\varepsilon} \in \mathcal{E}} (-1)^{|\boldsymbol{\varepsilon}|} \sum_{\boldsymbol{t} \in B_{\boldsymbol{t_0}}} q_{\boldsymbol{t}+\boldsymbol{\varepsilon}} \\ &= \sum_{\boldsymbol{\varepsilon} \in \mathcal{E}} (-1)^{|\boldsymbol{\varepsilon}|} Q(B_{\boldsymbol{\varepsilon}}) \end{aligned}$$

The function Q is additive, hence by the inclusion-exclusion formula

$$(2.15) \qquad Q\left(\bigcup_{k=1}^{d} B_{\boldsymbol{e}_k}\right) = \sum_{\substack{\boldsymbol{\varepsilon} \in \mathcal{E} \\ \boldsymbol{\varepsilon} \neq 0}} (-1)^{|\boldsymbol{\varepsilon}|-1} Q\left(\bigcap_{\{k: \varepsilon_k = 1\}} B_{\boldsymbol{e}_k}\right) = \sum_{\substack{\boldsymbol{\varepsilon} \in \mathcal{E} \\ \boldsymbol{\varepsilon} \neq 0}} (-1)^{|\boldsymbol{\varepsilon}|-1} Q(B_{\boldsymbol{\varepsilon}}),$$

where $\boldsymbol{e}_k$ are the standard unit vectors in $\mathbb{Z}^d$. Combining (2.14) and (2.15) we obtain

$$\sum_{\boldsymbol{t} \in B_{\boldsymbol{t_0}}} \delta_{\boldsymbol{t}}(U) = Q(B_{\boldsymbol{0}}) - Q\left(\bigcup_{k=1}^{d} B_{\boldsymbol{e}_k}\right) = q_{\boldsymbol{0}} = I(S_{B_{\boldsymbol{t_0}}} \in U).$$

Hence, in view of (2.11)–(2.12),

$$I(S_\Lambda \in U) - \sum_{\boldsymbol{t} \in \Lambda} \delta_{\boldsymbol{t}}(U) = \sum_{\boldsymbol{s} \in B_{\boldsymbol{t_0}} \cap \Lambda^c} \delta_{\boldsymbol{s}}(U).$$

Observe that $\delta_{\boldsymbol{s}}(U) \neq 0$ implies that there is $\boldsymbol{t} \in B_{\boldsymbol{s}} \setminus B_{\boldsymbol{s}+1}$ such that $Z_{\boldsymbol{t}} \neq 0$. Indeed, if this is not the case then $S_{B_{\boldsymbol{s}}^{\boldsymbol{\varepsilon}}} = S_{B_{\boldsymbol{s}}^{\boldsymbol{1}}}$ for every $\boldsymbol{\varepsilon} \in \mathcal{E}$, and

$$(2.16) \qquad \delta_{\boldsymbol{s}}(U) = (\sum_{\boldsymbol{\varepsilon} \in \mathcal{E}} (-1)^{|\boldsymbol{\varepsilon}|}) I(S_{B_{\boldsymbol{s}}^{\boldsymbol{1}}} \in U) = 0.$$

Consequently,

$$|I(S_\Lambda \in U) - \sum_{\boldsymbol{t} \in \Lambda} \delta_{\boldsymbol{t}}(U)| \leq 2^d \sum_{\boldsymbol{s} \in \Lambda^c} \sum_{\boldsymbol{t} \in \Lambda} I(\boldsymbol{t} \in B_{\boldsymbol{s}} \setminus B_{\boldsymbol{s}+1}) I(Z_{\boldsymbol{t}} \neq 0)$$
$$\leq 2^d ((m+1)^d - m^d - 1) \sum_{\boldsymbol{t} \in \partial \Lambda} I(Z_{\boldsymbol{t}} \neq 0)$$

which is clearly dominated by the right hand side of (2.6).

Case 2. $m < \mathrm{diam}(\Lambda_0) \leq 2m$.

In this case there exist $\boldsymbol{t}_0, \boldsymbol{s}_0 \in \Lambda$ such that $Z_{\boldsymbol{t}_0} \neq 0$, $Z_{\boldsymbol{s}_0} \neq 0$, and $\|\boldsymbol{t}_0 - \boldsymbol{s}_0\|_\infty > m$. Hence

$$(2.17) \qquad 2^{-1} \sum_{\substack{\boldsymbol{s},\boldsymbol{t} \in \Lambda \\ \|\boldsymbol{t}-\boldsymbol{s}\|_\infty > m}} I(Z_{\boldsymbol{s}} \neq 0, Z_{\boldsymbol{t}} \neq 0) \geq 1$$

and trivially,

$$(2.18) \qquad I(S_\Lambda \in U) \leq 2^{-1} \sum_{\substack{\boldsymbol{s},\boldsymbol{t} \in \Lambda \\ \|\boldsymbol{t}-\boldsymbol{s}\|_\infty > m}} I(Z_{\boldsymbol{s}} \neq 0, Z_{\boldsymbol{t}} \neq 0).$$

By the present assumption there exists $\boldsymbol{t}_0$ such that

$$\Lambda_0 \subset \{0,\ldots,2m\}^d + \boldsymbol{t}_0 := K_{\boldsymbol{t}_0}.$$

Similarly as in Case 1 we argue that $\delta_{\boldsymbol{t}}(U) = 0$ for $\boldsymbol{t} \notin K_{\boldsymbol{t}_0}$. Since each such term is bounded by 2^{d-1}, we get

$$(2.19) \qquad \begin{aligned} \sum_{\boldsymbol{t} \in \Lambda} |\delta_{\boldsymbol{t}}(U)| &= \sum_{\boldsymbol{t} \in \Lambda \cap K_{\boldsymbol{t}_0}} |\delta_{\boldsymbol{t}}(U)| \\ &\leq 2^{d-1}(2m+1)^d \, 2^{-1} \sum_{\substack{\boldsymbol{s},\boldsymbol{t} \in \Lambda \\ \|\boldsymbol{t}-\boldsymbol{s}\|_\infty > m}} I(Z_{\boldsymbol{s}} \neq 0, Z_{\boldsymbol{t}} \neq 0). \end{aligned}$$

Now (2.18) and (2.19) together imply (2.6).

Case 3. $\mathrm{diam}(\Lambda_0) > 2m$.

The assumption implies (2.17), hence (2.18). Moreover, for every $\boldsymbol{u} \in \mathbb{Z}^d$ there exists $\boldsymbol{s} \in \Lambda_0$ such that $\|\boldsymbol{u} - \boldsymbol{s}\|_\infty > m$. From the proof of Case 1 (see (2.16)) we know that $\delta_{\boldsymbol{t}}(U) \neq 0$ implies that $Z_{\boldsymbol{u}} \neq 0$ for some $\boldsymbol{u} \in B_{\boldsymbol{t}} \setminus B_{\boldsymbol{t}+1}$ and, under the present assumption, for such an $\boldsymbol{u}$ there is $\boldsymbol{s}$ such that $Z_{\boldsymbol{s}} \neq 0$ and $\|\boldsymbol{u} - \boldsymbol{s}\|_\infty > m$.

Hence we have the following estimates

$$\sum_{t\in\Lambda}|\delta_t(U)| \leq 2^{d-1} \sum_{t\in\Lambda} \sum_{\substack{u\in(B_t\setminus B_{t+1})\cap\Lambda}} \sum_{\substack{s\in\Lambda \\ \|s-u\|_\infty>m}} I(Z_s\neq 0, Z_u\neq 0)$$

$$= 2^{d-1} \sum_{\substack{s,u\in\Lambda \\ \|s-u\|_\infty>m}} \Big[\sum_{t\in\Lambda} I(u\in B_t\setminus B_{t+1})\Big] I(Z_s\neq 0, Z_u\neq 0)$$

$$\leq 2^{d-1}((m+1)^d - m^d) \sum_{\substack{s,u\in\Lambda \\ \|s-u\|_\infty>m}} I(Z_s\neq 0, Z_u\neq 0)$$

$$\leq 2^d(2m+1)^d\, 2^{-1} \sum_{\substack{s,u\in\Lambda \\ \|s-u\|_\infty>m}} I(Z_s\neq 0, Z_u\neq 0).$$

The proof of Theorem 2.1 is complete. □

3. An abstract form of the Bonferroni–type inequality

Consider a family of events $\mathcal{A} = \{A_T\}$ indexed by finite subsets T of $\mathbb{Z}^d$. A family of events $\mathcal{C} = \{C_t\}$ indexed by points $t \in \mathbb{Z}^d$ is said to be a *complete cover* of $\mathcal{A}$ if for every finite sets $T, T_1, T_2 \subset \mathbb{Z}^d$, $A_T \subset \bigcup_{t\in T} C_t$ and

(3.1) $$A_{T_1} \triangle A_{T_2} \subset \bigcup_{t \in T_1 \triangle T_2} C_t.$$

Define

$$\Delta_t = \sum_{\varepsilon\in\mathcal{E}} (-1)^{|\varepsilon|} P(A_{B_t^\varepsilon}),$$

where B_t^ε is given by (2.2). Following the steps of the proof of Theorem 2.1 we can prove the following "abstract form" of the Bonferroni–type inequality.

THEOREM 3.1. *Let $\mathcal{C} = \{C_t\}$ be a complete cover of $\mathcal{A} = \{A_T\}$ as defined above. Then for every finite set $\Lambda \subset \mathbb{Z}^d$*

(3.2)
$$|P(A_\Lambda) - \sum_{t\in\Lambda} \Delta_t| \leq c_1(d,m) \sum_{s\in\partial\Lambda} P(C_s)$$
$$+ c_2(d,m) \sum_{\substack{s,t\in\Lambda \\ \|s-t\|_\infty>m}} P(C_s\cap C_t)$$

where constants c_1 and c_2 are the same as in Theorem 2.1.

PROOF. We will only indicate the main steps of the proof. Since $\Delta_t = E\delta_t$, where

(3.3) $$\delta_t = \sum_{\varepsilon\in\mathcal{E}} (-1)^{|\varepsilon|} I(A_{B_t^\varepsilon}),$$

it is enough to prove that

(3.4)
$$|I(A_\Lambda) - \sum_{t \in \Lambda} \delta_t| \le c_1(d,m) \sum_{s \in \partial \Lambda} I(C_s)$$
$$+ c_2(d,m) \sum_{\substack{s,t \in \Lambda \\ \|t-s\|_\infty > m}} I(C_s \cap C_t)$$

holds everywhere on the probability space. First we will show that it suffices to prove (3.4) for the modifications $\mathcal{A}' = \{A'_T\}$ and $\mathcal{C}' = \{C'_t\}$ defined as follows

$$A'_T = A_{T \cap \Lambda}$$

and

$$C'_t = \begin{cases} C_t & \text{if } t \in \Lambda, \\ \emptyset & \text{if } t \notin \Lambda. \end{cases}$$

Note that $\mathcal{C}'$ is a complete cover of $\mathcal{A}'$. Define δ'_t by replacing $A_{B^\varepsilon_t}$ with $A'_{B^\varepsilon_t}$ in (3.3). If $\delta_t \ne \delta'_t$, then (3.1) yields

$$|\delta_t - \delta'_t| \le \sum_{\varepsilon \in \mathcal{E}} |I(A_{B^\varepsilon_t}) - I(A'_{B^\varepsilon_t})|$$
$$\le \sum_{\varepsilon \in \mathcal{E}} \sum_{s \in B^\varepsilon_t \cap \Lambda^c} I(C_s)$$
$$\le 2^d \sum_{s \in B_t \cap \Lambda^c} I(C_s)$$

which makes the reduction from $\mathcal{A}, \mathcal{C}$ to $\mathcal{A}', \mathcal{C}'$ possible, analogously to the first part of the proof of Theorem 2.1, (2.7)–(2.8). Now we can assume that $C_t = \emptyset$ for $t \notin \Lambda$. Define a random set

$$\Lambda_0 = \{s \in \mathbb{Z}^d : I(C_s) = 1\}.$$

(3.4) can now be established by considering the three cases of $\text{diam}(\Lambda_0)$, exactly as in the proof of Theorem 2.1. □

Theorem 3.1 gives Bonferroni-type inequalities in a variety of important cases. We mention below some of them.

Examples.

(i) $\mathcal{C}$ is an arbitrary family of events and

$$A_T = \bigcup_{t \in T} C_t.$$

In this case Theorem 3.1 generalizes the classical Bonferroni inequality (1.1).

(ii) $A_T = \{\max_{t \in T} Z_t > \lambda\}$ and $C_t = \{Z_t > \lambda\}$, where $\{Z_t\}$ is a real-valued random field. This is a special important case of (i).

(iii) $A_T = \{\sum_{t \in T} Z_t \in U\}$, where Z_t and U are as in Section 2, and $C_t = \{Z_t \ne 0\}$. This shows that Theorem 2.1 is a special case of Theorem 3.1.

(iv) $A_T = \{\prod_{t \in T} Z_t \in U\}$ and $C_t = \{Z_t \ne 1\}$, where $\{Z_t\}$ is a complex-valued random field and U is a Borel subset of the complex plane such that $1 \notin U$.

(v) Let $\{Z_t\}$ be a random field taking values in a measurable semigroup G with the neutral element I. Fix a linear order in $\mathbb{Z}^d$ to avoid the ambiguity in the definition of $\Pi_T = \prod_{t \in T} Z_t$ in the case when G is non Abelian (for instance, the lexicographical order). Then $A_T = \{\Pi_T \in U\}$ and $C_t = \{Z_t \neq I\}$ satisfy the assumptions of Theorem 3.1, provided $I \notin U$. In particular, Theorem 3.1 gives the Bonferroni-type inequality for products of random matrices.

References

[AGG90] R. Arratia, L. Goldstein and L. Gordon, *Poisson approximation and the Chen-Stein method*, Statist. Sci. **5** (1990), 403–434.

[D83] R.A. Davis, *Stable limits for partial sums of dependent random variables*, Ann. Probab. **11** (1983), 262–269.

[DH95] R.A. Davis and T. Hsing, *Point processes and partial sum convergence for weakly dependent random variables with infinite variance*, Ann. Probab. **23** (1995), 879–917.

[DJ89] M. Denker and A. Jakubowski, *Stable limit distributions for strongly mixing sequences*, Stat. Probab. Lett. **8** (1989), 477–483.

[G78] J. Galambos, *The Asymptotic Theory of Extreme Order Statistics*, Wiley, New York 1978.

[GS96] J. Galambos and I. Simonelli, *Bonferroni-type Inequalities with Applications*, Springer, New York 1996.

[H82] L. Heinrich, *A method of derivation of limit theorems for sums of m-dependent random variables*, Z. Wahrscheinlichkeitstheorie verw. Gebiete **64** (1982), 501–515.

[H85] L. Heinrich, *Stable limits for sums of m-dependent random variables*, Serdica, **11** (1985), 189–199.

[H86] L. Heinrich, *Stable limit theorems for sums of multiply indexed m-dependent random variables*, Math. Nachr. **127** (1986), 193–210.

[H87] L. Heinrich, *On the central limit problem for sequences and fields of m-dependent random variables*, in: Limit Theorems in Probability Theory and Related Fields (collection), Wiss. Theorie Prax., Tech. Univ. Dresden, Dresden 1987, pp. 25–48.

[J93] A. Jakubowski, *Minimal conditions in p-stable limit theorems*, Stoch. Proc. Appl. **44** (1993), 291–327.

[J97] A. Jakubowski, *Minimal conditions in p-stable limit theorems* II, Stoch. Proc. Appl. **68** (1997), 1–20.

[JK89] A. Jakubowski and M. Kobus, *α-stable limit theorems for sums of dependent random vectors*, J. Multivariate Anal. **29** (1989), 219–251.

[JNZ97] A. Jakubowski, A.V. Nagaev and A. Zaigraev *Large deviation probabilities for sums of heavy-tailed random vectors*, Stochastic Models, **13** (1997), 647–660.

[JW94] A. Janicki and A. Weron, *Simulation and Chaotic Behavior of α-stable Stochastic Processes*, Marcel Dekker, New York 1994.

[K95] M. Kobus, *Generalized Poisson distributions as limits of sums for arrays of dependent random vectors*, J. Multivariate Anal. **52** (1995), 199–244.

[LLR83] M.R. Leadbetter, G. Lindgren and H. Rootzén, *Extremes and Related Properties of Random Sequences and Processes*, Springer, Berlin 1983.

[R86] S.I. Resnick, *Point processes, regular variation and weak convergence*, Adv. in Appl. Probab. **18** (1986), 66–138.

[ST94] G. Samorodnitsky and M.S. Taqqu, *Stable Non-Gaussian Random Processes. Stochastic Models with Infinite Variance*, Chapman and Hall, London 1994.

FACULTY OF MATHEMATICS AND COMPUTER SCIENCE, NICHOLAS COPERNICUS UNIVERSITY, CHOPINA 12/18, 87-100 TORUŃ, POLAND
E-mail address: adjakubo@mat.uni.torun.pl

DEPARTMENT OF MATHEMATICS, UNIVERSITY OF TENNESSEE, KNOXVILLE, TN 37996, USA
E-mail address: rosinski@math.utk.edu

Pricing-differentials and bounds for lookback options, and prophet problems in probability

Robert P. Kertz

ABSTRACT. Both lookback call options and standard call options can be used to hedge against similar types of risks and to take advantage of similar market movements. Over a broad collection of markets, to what extent can the prices of these two types of options differ? This paper answers this question for discrete-time market models, when the options are of European type and when they are of perpetual American type. The analysis uses results from the mathematics literature on 'prophet' inequalities and on optimal stopping, and emphasizes comparisons in 'extremal' markets.

1. Introduction

A call option for a stock is a contract between two parties which gives the holder of the option the right (but not an obligation) to buy a share of the stock from the writer of the option at some future date (the option's exercise or expiration date) at some predetermined price (the strike price). In buying the stock, the holder is said to exercise the option. For a standard European call option, the holder exercising the option sells the purchased stock immediately for the stock's market price at the expiration date; but for the lookback call option with fixed strike price, the holder exercising the option sells the purchased stock immediately back to the writer for the maximum stock price over the time interval between the option's start-up date and expiration date. The price of each of these options is the amount paid from the buyer to the writer at the option's start-up date. Both of these options can be used to take advantage of rises in the stock's price or to hedge in similar ways against other market behavior. However, the notable difference in the two types of options allows the lookback option holder, and not the standard option holder to take advantage of upward movements that reverse direction before the option's expiration date. Clearly, for the removal of the regret after the 'up-down swing', the price of this lookback option should be greater than that of the standard call option. But how great a difference in pricing is this? This difference can be determined

---

1991 *Mathematics Subject Classification*. Primary 60G40, 90A09; Secondary 60E15.
Key words and phrases. Sharp inequalities for stochastic processes, prophet inequalities, option pricing, lookback options.

© 1999 American Mathematical Society

exactly for specific market models when all model parameters are known. But what can be said about this difference without this perfect knowledge?

In this paper, the extent to which these option prices can differ is determined, over certain collections of discrete-time market models. In particular, exact differential-pricing extremes are found and examples of markets leading to these extremes are given. This is done for European-style options (with exercise of the option only possible at the expiration date) in Section 4. And this is done for perpetual American-style options (with no expiration date, and exercise at any random time depending only on past information) in Section 5.

Some of these results are direct consequences of a number of inequalities for stochastic processes called 'prophet' inequalities. These prophet inequalities are discussed in Section 2. Most of the other mathematical techniques are rather elementary, and involve recursive procedures, use of limit and optimization concepts in the calculus, results on specific collections of stochastic processes (such as submartingales, independent random variables, and sums of independent random variables), etc.

Another way in which prophet problems are used in asset pricing theory is referenced in Section 6. This involves pricing of lookback options in markets with standard call options as the underlying financial instruments, work done by Hobson [**Hob**].

Market formulation is the content of most of Section 3 (for finite horizon models), and part of Section 5 (for infinite horizon models).

2. Prophet inequalities

The term 'prophet' has been used to describe concepts in many mathematical contexts, but here it is used in a rather specific sense. The terminology 'prophet' inequality is used here to describe an inequality which compares the quantity $E(\max_{n=0,\ldots,N} Z_n)$, the expected reward for a 'prophet' with foresight of the future, and the quantity $V(Z_0,\ldots,Z_N) := \sup\{EZ_\tau : \tau \in \mathcal{T}_N\}$, the optimal expected reward for a gambler who must use non-anticipating stopping times to stop at or before time N, which holds for all sequences $(Z_n)_{n\geq 0}$ of integrable random variables in some class. In the value $V(Z_0,\ldots,Z_N)$, the supremum is taken over the set $\mathcal{T}_N$ of stopping times τ, $0 \leq \tau \leq N$, relative to the natural filtration $(\mathcal{F}_n)_{n\geq 0}$ of the sequence $(Z_n)_{n\geq 0}$. For surveys on prophet inequalities, see Hill and Kertz [**H-K92**], and Kertz [**K**].

In this Section, a number of prophet inequalities are recalled which give the 'flavor' of the area and which are pertinent to the objective in future sections of determining pricing-differentials between 'lookback' and 'terminal payoff' options in standard markets.

The first of these is a ratio prophet inequality that was developed in Dubins and Pitman [**D-P**] and Hill and Kertz [**H-K83**].

(2.1) For integrable, nonnegative r.v.'s $Z_0,\ldots,Z_N$,

$$E\left(\max_{0\leq n\leq N} Z_n\right) \leq (N+1)V(Z_0,\ldots,Z_N).$$

The inequality is sharp (within martingales). The sharpness of the inequality is seen by the following martingale sequence: let $\hat{Z}_0 = 1$; and for $n \geq 0$, with $L > 1$, let $\hat{Z}_{n+1}$ satisfy the following:

if $\hat{Z}_n = L^n$, then
$P(\hat{Z}_{n+1} = L^{n+1} \mid \hat{Z}_0, \ldots, \hat{Z}_n) = L^{-1} = 1 - P(\hat{Z}_{n+1} = 0 \mid \hat{Z}_0, \ldots, \hat{Z}_n)$,
and if $\hat{Z}_n = 0$, then
$P(\hat{Z}_{n+1} = 0 \mid \hat{Z}_0, \ldots, \hat{Z}_n) = 1$. Thus, $V(\hat{Z}_0, \ldots, \hat{Z}_N) = E\hat{Z}_0 = 1$ and
$E\left(\max_{0 \leq n \leq N} \hat{Z}_n\right) = N(1 - L^{-1}) + 1$, so that
$E\left(\max_{0 \leq n \leq N} \hat{Z}_n\right) / V(\hat{Z}_0, \ldots, \hat{Z}_N) \nearrow N + 1$ as $L \nearrow \infty$.

The sequence $(\hat{Z}_0, \ldots, \hat{Z}_N)$ of (2.1) are random variables of 'long shot' type for L large.

For comparison and application purposes, two types of difference prophet inequalities are given. For the first comparison, see [**H-K83**].

(2.2) For r.v.'s $Z_0, \ldots, Z_N$ taking values in finite interval $[a, b]$,

$$E\left(\max_{0 \leq n \leq N} Z_n\right) - V(Z_0, \ldots, Z_N) \leq \left(\frac{N}{N+1}\right)^{N+1}(b-a).$$

The inequality is sharp (attained with martingales). For attainment of the inequality use $Z_n^* = a + (b-a)\hat{Z}_n$, $n = 0, \ldots, N$, and let $\hat{Z}_0 = \left(\frac{N}{N+1}\right)^N$; and for $n = 0, \ldots, N-1$, let $\hat{Z}_{n+1}$ satisfy the following:
if $\hat{Z}_n = \left(\frac{N}{N+1}\right)^{N-n}$, then
$P\left(\hat{Z}_{n+1} = \left(\frac{N}{N+1}\right)^{N-(n+1)} \mid \hat{Z}_0, \ldots, \hat{Z}_n\right) =$
$= \frac{N}{N+1} = 1 - P\left(\hat{Z}_{n+1} = 0 \mid \hat{Z}_0, \ldots, \hat{Z}_n\right)$
and if $\hat{Z}_n = 0$, then $P\left(\hat{Z}_{n+1} = 0 \mid \hat{Z}_0, \ldots, \hat{Z}_n\right) = 1$. Thus,
$V(Z_0^*, \ldots, Z_N^*) = a + (b-a)V(\hat{Z}_0, \ldots, \hat{Z}_N) = a + (b-a)\left(\frac{N}{N+1}\right)^N$ and
$E\left(\max_{0 \leq n \leq N} Z_n^*\right) = a + (b-a)E\left(\max_{0 \leq n \leq N} \hat{Z}_n\right)$
$= a + (b-a)\left\{(N+1)\left(\frac{N}{N+1}\right)^N - N\left(\frac{N}{N+1}\right)^{N+1}\right\}$, and so
$E\left(\max_{0 \leq n \leq N} Z_n^*\right) - V(Z_0^*, \ldots, Z_N^*) = (b-a)\left(\frac{N}{N+1}\right)^{N+1}$.

Here again there is a 'maximal spreading of values' for the extremal random variables, within constraints imposed by the range restriction and the subtraction of the value function. If one replaces the range restriction by a variance constraint, a difference prophet inequality of quite a different nature is obtained; for its proof and further discussion, see Dubins and Schwarz [**D-S**] and Kennedy and Kertz [**K-K**].

(2.3) For $N \geq 1$, there exists a minimal universal constant k_N such that for any martingale $Z_0, Z_1, \ldots$ with $\mathrm{Var}(Z_n) \leq \sigma^2 < \infty$ for all $n \geq 0$,

$$E\left(\max_{0 \leq n \leq N} Z_n\right) - V(Z_0, \ldots, Z_N) \leq k_N \sigma.$$

The constants $(k_N)_{N \geq 1}$ satisfy $k_N \leq 1$, $k_N \nearrow$, and $\lim_{N \to \infty} k_N = 1$. The following martingale sequences $(\hat{Z}_0, \ldots, \hat{Z}_N)$, $N \geq 1$, satisfy the conditions $\mathrm{Var}\,\hat{Z}_n = E\hat{Z}_n^2 \leq E\hat{Z}_N^2 = \mathrm{Var}\,\hat{Z}_N = 1$ for $0 \leq n \leq N$ and $\lim_{N \to \infty} E\left(\max_{0 \leq n \leq N} \hat{Z}_n\right) - V(\hat{Z}_0, \ldots, \hat{Z}_N) = 1$. For $n = 0, \ldots, N$, r.v.

$\hat{Z}_n$ takes on $n+2$ values; for $n = 0, \ldots, N-1$, define parameters $a_n = (N-n)\ln\left(\frac{N-n}{N+1}\right) - (N+1-n)\ln\left(\frac{N+1-n}{N+1}\right)$ and $b_n = -\ln\left(\frac{N-n}{N+1}\right)$. First, r.v. $\hat{Z}_0$ takes values a_0 and b_0 with probabilities $\frac{1}{N+1}$ and $\frac{N}{N+1}$ respectively; second, for $n = 0, \ldots, N-2$, let $\hat{Z}_{n+1}$ satisfy the following:
if $\hat{Z}_n = b_n$, then
$P(\hat{Z}_{n+1} = a_{n+1} | \hat{Z}_0, \ldots, \hat{Z}_n) = \frac{1}{N-n} = 1 - P(\hat{Z}_{n+1} = b_{n+1} | \hat{Z}_0, \ldots, \hat{Z}_n)$,
and if $\hat{Z}_n = a_i$, then
$P(\hat{Z}_{n+1} = a_i | \hat{Z}_0, \ldots, \hat{Z}_n) = 1$, for $i = 0, \ldots, n$;
and third, r.v. $\hat{Z}_N$ satisfies the following: if $\hat{Z}_{N-1} = b_{N-1}$, then
$P(\hat{Z}_N = b_{N-1} + \frac{1}{p} | \hat{Z}_0, \ldots, \hat{Z}_{N-1}) = \frac{p}{e} =$
$= 1 - P(\hat{Z}_N = b_{N-1} - \frac{1}{e-p} | \hat{Z}_0, \ldots, \hat{Z}_{N-1})$, and
if $\hat{Z}_{N-1} = a_i$, then $P(\hat{Z}_N = a_i | \hat{Z}_0, \ldots, \hat{Z}_{N-1}) = 1$, for $i = 0, \ldots, N-1$,
where p is the unique number in $(0,1)$ for which $E(\hat{Z}_N^2) = 1$.

3. Market, portfolio and pricing background

The mathematical models used here to capture the financial market and agent's investment include the following features and requirements.

(3.1)
- The small investor's (agent's) actions do not affect the prices in the financial market.
- The agent has positive initial capital and no other income streams or consumption.
- The agent's investment strategy consists of purchasing shares in the money market and shares of stock. During this process, there are no 'frictions' – there are no transaction costs, no taxes, and no liquidity problems.
- The agent is not allowed to use 'insider' information in choosing a portfolio.
- Markets are competitive. Market participants prefer more wealth than less.
- There are no counterparty risks.
- There are no arbitrage opportunities.
- There are no borrowing and no short selling constraints on investors, except for a nonnegative wealth constraint.
- Money market rates for borrowing and lending are the same.

The stock-and-bond market is now defined. This model is the standard discrete-time market model described, for example, in Harrison and Kreps [**Ha-K**], Harrison and Pliska [**Ha-P**], Shiryaev, et al. [**SKKM**], Pliska [**Pl**], and Duffie [**D**].

The following processes are defined on the underlying probability space $(\Omega, \mathcal{F}, P)$, where $|\Omega| < \infty$ and each element of Ω has positive probability under $P(\cdot)$ (*'positive' a means $a > 0$ in this paper*). Fix positive integer N, and let $(\mathcal{F}_n)_{0 \leq n \leq N}$ denote the 'information' filtration, with $\mathcal{F}_0 = \{\phi, \Omega\} \subset \mathcal{F}_1 \subset \cdots \subset \mathcal{F}_N = \mathcal{F}$.

The *stock and bond processes* $(S_n)_{0 \leq n \leq N}$ and $(B_n)_{0 \leq n \leq N}$ describe

(3.2) $S_n =$ the price of one share of the stock throughout time interval $(n, n+1)$, and

$B_n =$ the price of one share of the money market throughout $(n, n+1)$,

for $n = 0, \ldots, N$. The money market is a type of security, such as a bond or bank account, that is riskless to the investor. It is assumed that the B_n and S_n are $\mathcal{F}_n$-measurable and positive for each $n = 0, \ldots, N$. For simplicity of expressions and without loss of generality, assume $B_0 = 1$. It is also assumed that $B_n \leq B_{n+1}$ for all $n = 0, \ldots, N$. If we denote

(3.3) $r_n = \Delta B_n / B_{n-1}$ = the interest rate from the money market during $(n-1, n)$, and

$\rho_n = \Delta S_n / S_{n-1}$ = the proportional return from the stock during $(n-1, n)$,

for $n = 1, \ldots, N$, (with $\Delta a_n = a_n - a_{n-1}$), then the money market and stock price processes have respective representations

(3.4) $$B_n = \prod_{j=1}^{n}(1 + r_j) \quad \text{and} \quad S_n = S_0 \prod_{j=1}^{n}(1 + \rho_j)$$

for $n = 0, \ldots, N$ (with $\prod_{j=1}^{0} a_j \equiv 1$). Note that under these assumptions, $r_n \geq 0$ for all $n = 0, \ldots, N$, that is, the interest rates are nonnegative r.v.'s.

The agent's choice mechanism is described through the agent's *portfolio* $\pi = (\pi_n = (\beta_n, \gamma_n))_{0 \leq n \leq N}$. The agent's initial capital is $x = \beta_0 B_0 + \gamma_0 S_0$, where β_0 = the number of shares in the money market held by the agent initially and γ_0 = the number of shares of stock held by the agent initially. For $n = 1, \ldots, N$,

(3.5) β_n = the number of shares in the money market held by the agent during the time interval $(n - 1, n)$, and

γ_n = the number of shares of stock held by the agent during $(n-1, n)$.

It is assumed that the portfolio $\pi = (\pi_n)$ is a predictable process, that is, (β_n, γ_n) is $\mathcal{F}_{n-1}$-measurable for $n = 1, \ldots, N$. Borrowing and short-selling (negative β_n's and γ_n's) are allowed.

Let $X^\pi = (X_n^\pi)_{0 \leq n \leq N}$ be the agent's *wealth process* associated with portfolio π, defined for $0 \leq n \leq N$ by

(3.6) X_n^π = agent's wealth under portfolio π at time n
$= \beta_n B_n + \gamma_n S_n$

The agent's portfolio $\pi = (\beta_n, \gamma_n)_{0 \leq n \leq N}$ is assumed to be *admissible* in that π is *self-financing*, that is, $B_{n-1}\Delta\beta_n + S_{n-1}\Delta\gamma_n = 0$ for $n = 1, \ldots, N$. (Equivalently, $X_n^\pi = \beta_{n+1} B_n + \gamma_{n+1} S_n$ for $n = 0, \ldots, N-1$; and so,

(3.7) $$X_n^\pi = X_0^\pi + \sum_{j=1}^{n}(\beta_j \Delta B_j + \gamma_j \Delta S_j)$$

for $n = 1, \ldots, N$.)

Define the *discount factors* $\alpha_n = 1/B_n$ for $n = 0, \ldots, N$; the *discounted stock price process* $(\alpha_n S_n)_{0 \leq n \leq N}$ and the *discounted wealth process* $(\alpha_n X_n^\pi)_{0 \leq n \leq N}$ associated with admissible portfolio π. Observe that under the given assumptions, the discount factors $(\alpha_n)_{0 \leq n \leq N}$ are non-increasing r.v.'s.

(3.8) **Assumption.** There is a probability measure P_0 on $(\Omega, \mathcal{F})$, equivalent to P, under which the discounted price process $(\alpha_n S_n, \mathcal{F}_n)_{0 \leq n \leq N}$ is a martingale.

The stock-and-bond markets satisfying Assumption (3.8) are called *arbitrage-free markets*. This terminology is justified since it is known from [**Ha-P**] that Assumption (3.8) is equivalent to the lack of arbitrage opportunities for the investor. (An *arbitrage opportunity* exists for the investor if there is an admissible portfolio π for which $X_0^\pi = 0$ and $X_N^\pi > 0$ with positive probability under $P(\cdot)$.) Any such probability P_0 is called an *equivalent martingale measure* for the model. Under such a p.m. P_0, and for any admissible portfolio π under which $(X_n^\pi)_{0 \leq n \leq N}$

are P_0-integrable r.v.'s, the discounted wealth process $(\alpha_n X_n^\pi, \mathcal{F}_n)_{0 \leq n \leq N}$ is also a martingale (see Harrison and Pliska [**Ha-P**], Shiryaev, et al. [**SKKM**], and Williams [**W**]).

A *European Contingent Claim* (ECC) is any financial instrument consisting of a payment f_N at the expiration date N, where f_N is a nonnegative, integrable, $\mathcal{F}_N$-measurable function on $(\Omega, \mathcal{F})$.

A [*minimal*] *hedging strategy* against the ECC with initial capital $x > 0$ is any admissible portfolio π for which $X_0^\pi = x$ and $X_N^\pi \geq f_N$ [$X_N^\pi = f_N$]. Let $\mathcal{H}(x)$ denote the collection of hedging strategies against the ECC with initial capital x.

The *fair price* of the ECC is defined to be the number $v_N := \inf\{x > 0 : \mathcal{H}(x) \neq \emptyset\}$. The following result is proved for example in Pliska [**Pl**].

(3.9) For any arbitrage-free stock-and-bond market and any ECC with payment function f_N and expiration date N, for which a minimal hedging strategy exists, there is a unique fair price for the ECC given by $v_N = E_0(\alpha_N f_N)$ for any and every equivalent martingale measure P_0.

This result (3.9) leaves open the question of whether or not a minimal hedging strategy exists for the ECC. Thus it is standard to introduce the following Assumption.

(3.10) **Assumption.** The stock-and-bond market is *complete*, that is, for any nonnegative random variable Z on $(\Omega, \mathcal{F})$, there exists a self-financing portfolio π for which $Z = X_N^\pi$.

For characterizations of complete markets in the finite sample space case see, for example, Proposition 2.12 of Harrison and Pliska [**Ha-P**] or results (1.22) and (4.17) in Pliska [**Pl**]. From Harrison and Kreps [**Ha-K**] for any arbitrage-free stock-and-bond market, completeness of the market is equivalent to the uniqueness of equivalent martingale measure P_0 of the model, and equivalent to the statement that each martingale $(M_n, \mathcal{F}_n)_{0 \leq n \leq N}$ under P_0 has representation $M_n = M_0 + \sum_{k=1}^n c_k \Delta m_k$ with c_k being an $\mathcal{F}_{k-1}$-measurable r.v. and $\Delta m_k = \alpha_k S_k - \alpha_{k-1} S_{k-1}$ for $1 \leq k \leq N$. See also Harrison and Pliska [**Ha-P**] or Shiryaev, et al. [**SKKM**]. This leads to the following pricing result.

(3.11) For any arbitrage-free, complete stock-and-bond market with equivalent martingale measure P_0, an ECC for this market with payment function f_N and expiration date N has fair price $v_N = E_0(\alpha_N f_N)$.

EXAMPLE 3.1. (Multiplicative) Binomial Market (see, for example, Cox and Rubenstein [**C-R**], Williams [**W**], Price [**Pr**], Shiryaev, et al. [**SKKM**], and Pliska [**Pl**]) Let $\Omega = \{-1, +1\}^N = \{\omega = (\omega_1, \ldots, \omega_N) : \omega_i = \pm 1\}$ and $\mathcal{F} = 2^\Omega$. Let $(\epsilon_k)_{1 \leq k \leq N}$ denote the coordinate mappings and $\mathcal{F}_k = \sigma(\epsilon_1, \ldots, \epsilon_k)$ for $1 \leq k \leq N$. For $0 < p < 1$, let P denote the product probability measure under which $\epsilon_1, \ldots, \epsilon_N$ are independent, identically distributed r.v.'s with $P(\epsilon_i = 1) = p = 1 - P(\epsilon_i = -1)$. For $r \geq 0$ and $S_0 > 0$, and $-1 < d < r < u$, let $B_n := (1+r)^n$ and $S_n := S_0 \prod_{i=1}^n (1 + \rho_i)$ for $n = 1, \ldots, N$, where the ρ_i's are the i.i.d. r.v.'s satisfying $(\rho_i = u) = (\epsilon_i = 1)$ and $(\rho_i = d) = (\epsilon_i = -1)$. These (Multiplicative) Binomial Markets are arbitrage-free, complete markets; the unique equivalent martingale measure P_0 is the product p.m. P_0 with $p = \frac{r-d}{u-d}$. Under the normalization $B_0 = 1$, an ECC with payment function f_N and expiration date N has fair price $v_N = E_0((1+r)^{-N} f_N)$.

As defined in Shiryaev, et al. [**SKKM**], the '*symmetric case*' is that subcase of the Binomial Market in which $u = \lambda - 1$ and $d = \lambda^{-1} - 1$ for some $\lambda > 0$. (So in this case $\lambda^{-1} < r + 1 < \lambda$ and $p = (r + 1 - \lambda^{-1})/(\lambda - \lambda^{-1})$).

An *American Contingent Claim* (ACC) is any financial instrument consisting of the following: (i) an expiration date $N \in (0, \infty)$; (ii) a stopping time τ (with respect to $(\mathcal{F}_n)_{0 \leq n \leq N}$) which is called the exercise date; and (iii) a payment f_τ at exercise date τ, where the payment sequence $(f_n)_{0 \leq n \leq N}$ is a sequence of nonnegative, P_0-integrable r.v.'s adapted to $(\mathcal{F}_n)_{0 \leq n \leq N}$.

A *hedging strategy* against the ACC with initial capital x is any admissible portfolio π for which $X_0^\pi = x$ and $X_n^\pi \geq f_n$ for all $0 \leq n \leq N$. Let $\tilde{\mathcal{H}}(x)$ denote the collection of all hedging strategies against the ACC with initial capital x.

The *fair price* of the ACC is defined to be the number $\tilde{v}_N := \inf\{x > 0 : \tilde{\mathcal{H}}(x) \neq \emptyset\}$. See Duffie [**D**], Lamberton and Lapeyre [**L-L**] and Shiryaev, et al. [**SKKM**] for the following.

(3.12) For any arbitrage-free, complete stock-and-bond market, with equivalent martingale measure P_0, an ACC for this market with payment functions $(f_n)_{0 \leq n \leq N}$ and expiration date N has fair price $\tilde{v}_N = V_0(f_0, \ldots, f_N)$ where

$$\begin{aligned} V_0(f_0, \ldots, f_N) &= \sup\{E_0(\alpha_\tau f_\tau) : \\ &\quad \tau \text{ is a stopping time with respect to } (\mathcal{F}_n)_{0 \leq n \leq N}\} \\ &= E_0(\alpha_{\tilde{\tau}} f_{\tilde{\tau}}) \end{aligned}$$

and $\tilde{\tau} = \min\{0 \leq n \leq N : \alpha_n f_n = W_n\}$. The sequence $(W_n)_{0 \leq n \leq N}$ is called the Snell envelope for this optimal stopping problem, and is defined by $W_n := \operatorname{ess\,sup}\{E_0(\alpha_\sigma f_\sigma \mid \mathcal{F}_n) : \sigma \text{ is a stopping time with values in } \{n, \ldots, N\}\}$.

EXAMPLE 3.2. (Duffie [**D**]) For the arbitrage-free, complete stock-and-bond market with $B_0 \leq B_1 \leq \cdots \leq B_N$, the ACC with expiration date N and payment functions $f_n = (S_n - q)^+$, $0 \leq n \leq N$ (with $q > 0$ constant) has fair price $\tilde{v}_N = E_0(\alpha_N f_N)$ with exercise time $\tilde{\tau} \equiv N$. This will be clear from the proof of Theorem (4.1), where it will be shown that $(\alpha_n(S_n - q)^+, \mathcal{F}_n)_{0 \leq n \leq N}$ is a submartingale under P_0, and $V_0(f_0, \ldots, f_n) = E_0(\alpha_N f_N)$ for any submartingale.

EXAMPLE 3.3. (Shiryaev, et al. [**SKKM**]) For the symmetric Binomial Market, the ACC with expiration date N and payment functions $f_n = \beta^n(S_n - 1)^+$, $0 \leq n \leq N$ (with $0 < \beta < 1$ constant), has fair price $\tilde{v}_N = E_0((\alpha\beta)^{\tilde{\tau}}(S_{\tilde{\tau}} - 1)^+)$ with $\alpha = (1 + r)^{-1}$ and exercise time $\tilde{\tau}$ of 'threshold' form $\tilde{\tau} = \min\{0 \leq n \leq N : S_n \geq \lambda^{k_{N-n}}\} = \min\{0 \leq n \leq N : x + \sum_{i=1}^n \epsilon_i \geq k_{N-n}\}$ where $S_0 \lambda^x$ and threshold constants $k_1 \leq k_2 \leq \cdots \leq k_N$ are chosen appropriately, depending on the model parameters.

4. Pricing-differentials and bounds, with finite horizon

Several ECC's associated with the stock-and-bond market are given by identification of the payment function; each of these ECC's have expiration date N.

(4.1)
European call option with strike price q $\quad f_N = (S_N - q)^+$
Fixed-strike lookback call option $\quad\quad f_N = (\max_{0 \leq n \leq N} S_n - q)^+$
with strike price q $\quad\quad\quad\quad\quad\quad = \max_{0 \leq n \leq N}(S_n - q)^+$
Floating-strike lookback put option $\quad f_N = (\max_{0 \leq n \leq N} S_n - S_N)^+$

For standard known results on lookback options, see, for example, Conze and Viswanathan [**C-V**], Zhang [**Z**], Duffie [**D**], and Pliska [**Pl**].

In this Section, price comparisons between European call options and fixed-strike lookback options, over the collection of arbitrage-free, complete, stock-and-bond markets, are given (Theorem 4.1 and Corollary 4.5), and bounds are given for the prices of floating-strike lookback put options over this collection (Theorem 4.4).

THEOREM 4.1. *Over the collection of arbitrage-free, complete, stock-and-bond markets, the fixed-strike lookback option price is no more than $N + 1$ times the European call option price, if both options have the same expiration date N and same strike price. This bound of '$N + 1$' is sharp, even within the collection of symmetric Binomial Markets.*

PROOF. Fix the horizon N and the strike price of q and consider no-arbitrage discrete stock-and-bond market with $\alpha_0 \geq \cdots \geq \alpha_N$. Let P_0 be an associated equivalent martingale measure. It is shown that

$$(4.2) \quad E_0\left(\alpha_N \left(\max_{0 \leq n \leq N} S_n - q\right)^+\right) \leq (N+1) E_0(\alpha_N(S_N - q)^+)$$

and that this inequality is sharp within the collection of symmetric Binomial Markets. Observe first that $(\alpha_n(S_n - q)^+, \mathcal{F}_n)_{0 \leq n \leq N}$ is a submartingale since

$$\begin{aligned} E_0(\alpha_{n+1}(S_{n+1} - q)^+ \mid \mathcal{F}_n) &\geq (E_0(\alpha_{n+1} S_{n+1} \mid \mathcal{F}_n) - E_0(\alpha_{n+1} \mid \mathcal{F}_n) q)^+ \\ &\geq (\alpha_n S_n - E_0(\alpha_{n+1} \mid \mathcal{F}_n) q)^+ \\ &\geq \alpha_n(S_n - q)^+; \end{aligned}$$

and it follows that

$$\begin{aligned} E_0(\alpha_N(\max_{0 \leq n \leq N} S_n - q)^+) &\leq E_0(\max_{0 \leq n \leq N} \alpha_n(S_n - q)^+) \\ &\leq \sum_{n=0}^{N} E_0(\alpha_n(S_n - q)^+) \\ &\leq (N+1) E_0(\alpha_N(S_N - q)^+). \end{aligned}$$

Alternatively, to prove this inequality, one could use the ratio prophet inequality (2.1), upon noting that $V_0(\alpha_0(S_0 - q)^+, \ldots, \alpha_N(S_N - q)^+) = E_0(\alpha_N(S_N - q)^+)$. The first part of the conclusion now follows from the assumed completeness of the market, (3.11), and the definition of these options.

To obtain the sharpness claim, proceed as follows. Consider symmetric Binomial Markets with $B_0 = 1$, $S_0 = s > 0$ and $0 < r < \lambda - 1$. Without loss of generality, fix $q = 1$. For $m = 0, 1, \ldots$, define the *proportionate pricing differential* of the fixed-strike lookback option to the European call option by
(4.3)

$$R_{m+1}(s; r, \lambda) := \frac{E_0\left(\max_{0 \leq k \leq m} S_k - 1\right)^+}{E_0(S_m - 1)^+} = \frac{E_0\left(\max_{0 \leq k \leq m} s\lambda^{\epsilon_0 + \cdots + \epsilon_k} - 1\right)^+}{E_0(s\lambda^{\epsilon_0 + \cdots + \epsilon_m} - 1)^+}$$

where $\epsilon_0 \equiv 0$ and $\frac{0}{0}$ is interpreted as one. Consider the following statement for $m \geq 0$:

(4.4) For $\lambda \to \infty$ and $s = s(\lambda) > \lambda^m$, $\lim_{\substack{r \to 0 \\ \lambda \to \infty}} R_{m+1}(s; r, \lambda) = m + 1$.

For $m = 0$, statement (4.4) holds. Assume statement (4.4) holds for $0 \leq m < n$. It is shown that (4.4) holds for $m = n$. For $s > \lambda^n$,

$$R_{n+1}(s; r, \lambda) = E_0 \left(\max_{0 \leq k \leq n} s\lambda^{\epsilon_0 + \cdots + \epsilon_k} - 1 \right)^+ / E_0(s\lambda^{\epsilon_0 + \cdots + \epsilon_n} - 1)^+$$

$$= E_0 \left(\max_{0 \leq k \leq n} s\lambda^{\epsilon_0 + \cdots + \epsilon_k} - 1 \right) / E_0(s\lambda^{\epsilon_0 + \cdots + \epsilon_n} - 1)$$

$$\geq E_0 \left(\max_{0 \leq k \leq n} \lambda^{\epsilon_0 + \cdots + \epsilon_k} \right) / E_0(\lambda^{\epsilon_0 + \cdots + \epsilon_n})$$

$$= (r+1)^{-n} E_0 \left(\max_{0 \leq k \leq n} \lambda^{\epsilon_0 + \cdots + \epsilon_k} \right)$$

so it suffices to show that $\lim_{\substack{r \to 0 \\ \lambda \to \infty}} E_0 \left(\max_{0 \leq k \leq n} \lambda^{\epsilon_0 + \cdots + \epsilon_k} \right) = n + 1$. But this follows easily by induction upon observing that $c_n := E_0 \left(\max_{0 \leq k \leq n} \lambda^{\epsilon_0 + \cdots + \epsilon_k} \right)$ satisfies

$$c_n = (1 - \lambda^{-1}) P_0(\epsilon_1 < 0, \epsilon_1 + \epsilon_2 < 0, \ldots, \epsilon_1 + \cdots + \epsilon_n < 0) + (r+1) c_{n-1}$$

for $n \geq 1$; $1 \geq P_0(\epsilon_1 < 0, \ldots, \epsilon_1 + \cdots + \epsilon_n < 0) \geq (1-p)^n$; and $\lim_{\substack{r \to 0 \\ \lambda \to \infty}} p = 0$. □

COROLLARY 4.2. *Over the collection of arbitrage-free, complete, stock-and-bond markets, the fixed-strike lookback option with strike price q has fair price which is no larger than $N + 1$ times the the fair price of the ACC with payment functions $\left(f_n = (S_n - q)^+ \right)_{0 \leq n \leq N}$, when the two options have common expiration date N. This bound of '$N + 1$' is sharp, even within the collection of symmetric Binomial Markets.*

PROOF. Use the discussion within Example 3.2 and the proof of Theorem 4.1. □

REMARKS 4.3. (i) The symmetric Binomial Market in the proof of Theorem 4.1 contains the 'long-shot' characteristic of extremal distributions for other ratio prophet inequalities such as (2.1). Indeed, for λ large, the period-by-period proportional returns from the stock are large (at $u = \lambda - 1$) with small probability ($p = \frac{r+1-\lambda^{-1}}{\lambda - \lambda^{-1}}$) and are diminishing (at $d = \lambda^{-1} - 1$) with large probability ($q = \frac{\lambda - (r+1)}{\lambda - \lambda^{-1}}$).

(ii) To obtain sharpness for the comparison of Theorem 4.1 by using symmetric Binomial Markets with $S_0 = s > 0$ and $0 < r < \lambda - 1$, some condition on the initial stock price s is needed, such as the one in (4.4). For example, in the case of $m = 1$ and s fixed, if $s \leq 1$, then $R_2(s; r, \lambda) \equiv 1$, and if $1 < s$, then $R_2(s; r, \lambda) \to 2 - \frac{1}{s}$ as $\lambda \to \infty$ and $r \to 0$.

(iii) In some market models, stock prices are allowed to be zero. In this case, the example within (2.1) could be used to form a market leading to the sharpness assertion for inequality (4.2). In this paper stock prices have been assumed to be (strictly) positive. Instead of using the symmetric Binomial market construction in the proof of Theorem 4.1 to produce sharpness of (4.2), one could exhibit appropriate stock-and-bond markets by 'perturbing' the example of (2.1) in a way that also produces arbitrage-free, complete markets. For example, for parameters s, δ and L with $s > 0$, $0 < \delta \ll 1 \ll L$, and $B_n = 1$ for $n = 0, \ldots, N$, and with background data similar to that of Example 3.1, one can define (discounted) stock price process $(S_n)_{0 \leq n \leq N}$ by $S_0 = s$, and the two-valued r.v.'s $S_1, \ldots, S_n$ through conditional expectations as follows: for $k = 0, \ldots, N-1$, on $\{S_k = sL^k\}$, $P_0\left(S_{k+1} = sL^{k+1} \mid \mathcal{F}_k\right) = p_{k+1} := \frac{L^k - \delta^{k+1}}{L^{k+1} - \delta^{k+1}} =: 1 - q_{k+1} = P_0\left(S_{k+1} = s\delta^{k+1} \mid \mathcal{F}_k\right)$, and for $k = 1, \ldots, N-1$, on $\{S_k = s\delta^k\}$, $P_0\left(S_{k+1} = sL^{k+1} \mid \mathcal{F}_k\right) = \bar{p}_{k+1} := \frac{\delta^k - \delta^{k+1}}{L^{k+1} - \delta^{k+1}} =: 1 - \bar{q}_{k+1} = P_0\left(S_{k+1} = s\delta^{k+1} \mid \mathcal{F}_k\right)$. Observe that $L^{-1}(1 - \delta) \leq p_{k+1} \leq L^{-1}$ for $k = 0, \ldots, N-1$, and $\bar{p}_{k+1} \leq \delta$ for $k = 1, \ldots, N-1$, and obtain for $1 < s < \delta^{-N}$ that

$$E_0\left(\max_{0 \leq n \leq N} S_n - 1\right)^+ / E_0(S_N - 1)^+$$
$$\geq \frac{(s-1)q_1\bar{q}_2 \cdots \bar{q}_N + (sL-1)p_1 q_2 \bar{q}_3 \cdots \bar{q}_N + \cdots + (sL^N - 1)p_1 \cdots p_N}{(sL^N - 1)(p_1 \cdots p_N + C\delta)}$$
$$\geq \frac{(s-1)(1-L^{-1})(1-\delta)^{N-1} + (sL-1)L^{-1}(1-\delta)(1-L^{-1})(1-\delta)^{N-2}}{(sL^N - 1)(L^{-N} + C\delta)}$$
$$+ \cdots + \frac{(sL^N - 1)\left(\frac{1-\delta}{L}\right)^N}{(sL^N - 1)(L^{-N} + C\delta)}$$
$$\to N + 1$$

as $\delta \downarrow 0$, $L \uparrow \infty$ and $s \uparrow \infty$ (here C is a constant not depending on δ, L or s).

THEOREM 4.4. *Consider the collection of arbitrage-free, complete, stock-and-bond markets. Let $\hat{v}_N$ denote the fair price of the floating-strike lookback put option with expiration date N.*

(i) *If the discounted price process satisfies $\alpha_n S_n \leq L$ for $0 \leq n \leq N$, for some positive constant L, then $\hat{v}_N \leq \left(\frac{N}{N+1}\right)^{N+1} L$. This inequality is sharp within this collection of markets.*

(ii) *If the discounted price process has variances satisfying $\mathrm{Var}_0(\alpha_n S_n) \leq \sigma^2 < \infty$ for $0 \leq n \leq N$, under the equivalent martingale measure P_0, then $\hat{v}_N \leq k_N \sigma$, where the constants $(k_N)_{N \geq 1}$ satisfying $k_N \leq 1$, $k_N \nearrow$ and $\lim_{N \to \infty} k_N = 1$.*

PROOF. Under these assumptions, the discounted price process $(\alpha_n S_n)_{0 \leq n \leq N}$ is a martingale under the equivalent martingale measure P_0, and $\alpha_0 = 1 \geq \cdots \geq \alpha_N$. From (3.11) it follows that

$$\hat{v}_N = E_0\left(\alpha_N \left(\max_{0 \leq n \leq N} S_n - S_N\right)\right)$$
$$\leq E_0\left(\max_{0 \leq n \leq N} \alpha_n S_n\right) - V(\alpha_0 S_0, \ldots, \alpha_N S_N)$$

and the conclusions are consequences of (2.2) and (2.3).

To obtain that the inequality $\hat{v}_N \leq \left(\frac{N}{N+1}\right)^{N+1} L$ is sharp under the hypotheses of (i) for arbitrage-free, complete stock-and-bond markets with (strictly) positive prices, fix $N \geq 1$, $L > 0$ and $r = 0$, and let $0 < \delta < \frac{1}{N+1}$. Under background data similar to that of Example 3.1, define stock price process $(S_n)_{0 \le n \le N}$ by $S_0 = \left(\frac{N}{N+1}\right)^N L$ and the two-valued r.v.'s $S_1, \ldots, S_N$ through conditional expectations as follows: for $k = 0, \ldots, N-1$, on $\left\{S_k = \left(\frac{N}{N+1}\right)^{N-k} L\right\}$,

$$P_0\left(S_{k+1} = \left(\frac{N}{N+1}\right)^{N-(k+1)} L \mid \mathcal{F}_k\right) = p_{k+1} := \frac{\frac{N}{N+1} - \delta^{k+1}}{1 - \delta^{k+1}} =: 1 - q_{k+1}$$

$$= P_0\left(S_{k+1} = \left(\frac{N}{N+1}\right)^{N-(k+1)} \delta^{k+1} L \mid \mathcal{F}_k\right)$$

and for $k = 1, \ldots, N-1$, on $\left\{S_k = \left(\frac{N}{N+1}\right)^{N-k} \delta^k L\right\}$,

$$P_0\left(S_{k+1} = \left(\frac{N}{N+1}\right)^{N-(k+1)} L \mid \mathcal{F}_k\right) = \bar{p}_{k+1} := \frac{\frac{N}{N+1}\delta^k - \delta^{k+1}}{1 - \delta^{k+1}} =: 1 - \bar{q}_{k+1}$$

$$= P_0\left(S_{k+1} = \left(\frac{N}{N+1}\right)^{N-(k+1)} \delta^{k+1} L \mid \mathcal{F}_k\right).$$

Then observe that $\frac{N}{N+1} - \delta < p_{k+1} < \frac{N}{N+1}$ for $k = 0, \ldots, N-1$, and $\bar{p}_{k+1} < \delta$ for $k = 1, \ldots, N-1$, and calculate as follows:

$$E_0\left(\max_{0 \le n \le N} S_n - S_N\right)$$

$$\ge \left(\frac{N}{N+1}\right)^N L q_1 \bar{q}_2 \cdots \bar{q}_N + \left(\frac{N}{N+1}\right)^{N-1} L p_1 q_2 \bar{q}_3 \cdots \bar{q}_N +$$

$$+ \cdots + \left(\frac{N}{N+1}\right) L p_1 \cdots p_{N-1} q_N + L p_1 \cdots p_N - \left(\frac{N}{N+1}\right)^N L$$

$$\ge \left(\frac{N}{N+1}\right)^N L \left(\frac{1}{N+1}\right)(1-\delta)^{N-1}$$

$$+ \left(\frac{N}{N+1}\right)^{N-1} L \left(\frac{N}{N+1} - \delta\right)\left(\frac{1}{N+1}\right)(1-\delta)^{N-2} +$$

$$+ \cdots + \left(\frac{N}{N+1}\right) L \left(\frac{N}{N+1} - \delta\right)^{N-1}\left(\frac{1}{N+1}\right)$$

$$+ L\left(\frac{N}{N+1} - \delta\right)^N - \left(\frac{N}{N+1}\right)^N L$$

$$\to \left(\frac{N}{N+1}\right)^{N+1} L$$

as $\delta \downarrow 0$. $\square$

COROLLARY 4.5. *For the Markets of Theorem 4.4, let Δ_N denote the difference of the fixed-strike lookback option price minus the European call option price, when both options have the same expiration date N and the same strike price. Under the hypotheses of Theorem 4.4, the constants Δ_N satisfy the same bounds as the constants $\hat{v}_N$ of Theorem 4.4.*

PROOF. Let q be the common strike price of the two options. Then, under the equivalent martingale measure P_0, it follows that

$$\Delta_N = E_0\left(\alpha_N \left(\max_{0 \le n \le N} S_n - q\right)^+\right) - E_0(\alpha_N(S_n - q)^+)$$

$$\le E_0\left(\alpha_N \left(\max_{0 \le n \le N} S_n - S_N\right)\right) = \hat{v}_N,$$

and the conclusion follows from Theorem 4.4. □

Note that under the conditions of Theorem 4.4(i), and if strike price q satisfies $q \le \left(\frac{N}{N+1}\right)^N L$, then the inequality $\Delta_N \le \left(\frac{N}{N+1}\right)^{N+1} L$ is sharp, with use of the same stock-and-bond markets as constructed in the proof of Theorem 4.4.

5. Pricing-differentials, American style, in infinite horizon

There is an interesting extension of these ideas within the infinite horizon setting. This is a comparison between the perpetual American call option and the perpetual American lookback call option. To define these financial instruments, the stock-and-bond markets of Section 3 must be extended from finite to infinite horizon.

For this purpose, introduce the background data $(\Omega, \mathcal{F}, (\mathcal{F}_n)_{n \ge 0}, P)$ as follows. Let Ω_i, $i \ge 1$, be spaces with $|\Omega_i| < \infty$ for all $i \ge 1$, and

$$\Omega := \Omega_1 \times \Omega_2 \times \cdots = \{\omega = (\omega_1, \omega_2, \ldots) : \omega_i \in \Omega_i \text{ for all } i \ge 1\}.$$

Let $(\epsilon_i)_{i \ge 1}$ be the coordinate mappings, so $\epsilon_i(\omega) = \omega_i$ for $i \ge 1$; and define the σ-algebras $\mathcal{F}_0 := \{\emptyset, \Omega\}$, $\mathcal{F}_n := \sigma(\epsilon_1, \ldots, \epsilon_n)$ for $n \ge 1$, and $\mathcal{F} = \mathcal{F}_\infty = \sigma(\epsilon_1, \epsilon_2, \ldots)$. Assume that $P(\cdot)$ is a probability measure on Ω satisfying $P(\{\omega_1\} \times \cdots \times \{\omega_n\} \times \Omega_{n+1} \times \cdots) > 0$ for all $\omega_1 \in \Omega_1, \ldots, \omega_n \in \Omega_n$, $n \ge 1$.

Directly analogous to the definitions of Section 3, define the stock price process $(S_n)_{n \ge 0}$, money market (bond) price process $(B_n)_{n \ge 0}$, interest rates $(r_n)_{n \ge 0}$, discount factors $(\alpha_n = 1/B_n)_{n \ge 0}$, and discounted stock price process $(\alpha_n S_n)_{n \ge 0}$. As in the finite horizon market model, assume again that $(B_n)_{n \ge 0}$ are non-decreasing r.v.'s, with $B_0 = 1$, so that the interest rates $(r_n)_{n \ge 0}$ are nonnegative r.v.'s, and the discount factors $(\alpha_n)_{n \ge 0}$ are non-increasing r.v.'s taking values in $[0, 1]$.

We impose the following Assumption on the model.

(5.1) **Assumption.** There is a unique p.m. P_0 on $(\Omega, \mathcal{F})$ such that
(i) for each $n \ge 1$, $P_0^{(n)}$ is equivalent to $P^{(n)}$, where $P_0^{(n)} = P_0|_{\mathcal{F}_n}$ and $P^{(n)} = P|_{\mathcal{F}_n}$, and
(ii) under P_0, the discounted stock price process $(\alpha_n S_n, \mathcal{F}_n)_{n \ge 0}$ is a martingale.

For the infinite-horizon setting, a restriction is needed on allowable portfolios to remove the presence of 'doubling strategies' which can accumulate unlimited debt in order to produce assured gains (see, for example, Chapter 7 of Pliska [**Pl**], or Duffie [**D**]). The agent's *portfolio* $\pi = (\pi_n = (\beta_n, \gamma_n))_{n \ge 0}$ has interpretation as in (3.4), is a predicatable process, and has associated wealth process $X^\pi = (X_n^\pi)_{n \ge 0}$ as defined in (3.5). The agent's portfolio is assumed to be *admissible* in that (i) π is self-financing, that is, $B_{n-1}\Delta\beta_n + S_{n-1}\Delta\gamma_n = 0$ for all $n \ge 1$ (and so (3.6) holds for all $n \ge 1$), and (ii) the sequence $(X_n^\pi)_{n \ge 0}$ of r.v.'s is bounded below by a constant, almost everywhere (a.e.) $[P_0]$.

Observe that under Assumption (5.1) each of the finite-horizon stock-and-bond markets associated with this model is arbitrage-free and complete. Note also that Assumption (5.1) is a natural infinite-horizon analogue of Assumptions (3.10) and (3.12). However, there is a problem with arbitrage opportunities 'in-the-limit' if one stays within the infinite horizon. Of course, application of the results would necessitate pulling the results back to finite horizon market models at some point, in some way. See the comment at the very end of this Section. For a discussion of these arbitrage issues for infinite-horizon, discrete-time market models, see Schachermayer [**S**].

EXAMPLE 5.1. Symmetric Binomial Market, Infinite Horizon. Let $\Omega_i = \{-1, +1\}$ for $i \geq 1$, so $\Omega = \{-1, +1\}^\infty$. For $0 < p < 1$, let P denote infinite dimensional product measure on $(\Omega, \mathcal{F})$, so that $(\epsilon_i)_{i \geq 1}$ are i.i.d. r.v.'s with $P(\epsilon_i = 1) = p = 1 - P(\epsilon_i = -1)$. Suppose $r \geq 0$ and $0 < \lambda^{-1} < r + 1 < \lambda$ and (using $r_n = r$, $\rho_n = \lambda^{\epsilon_n} - 1$, and $\epsilon_0 = 0$) $B_n = (1+r)^n$ and $S_n = S_0 \lambda^{\epsilon_0 + \cdots + \epsilon_n}$ for $n \geq 0$. The unique p.m. P_0 of Assumption (5.1) is that infinite dimensional p.m. with $p = p_0 := (r + 1 - \lambda^{-1})/(\lambda - \lambda^{-1})$. Since $\alpha_n = (1+r)^{-n} =: \alpha^n$ for $n \geq 0$, the discounted stock price process, with initial price $S_0 = s = \lambda^x$, is given by

$$\alpha_n S_n = \alpha^n S_n = \alpha^n s \lambda^{\epsilon_0 + \cdots + \epsilon_n} = \alpha^n \lambda^{x + \epsilon_0 + \cdots + \epsilon_n}, \qquad \text{for } n \geq 0.$$

In this Section, a *stopping time* τ is any random variable taking values in $\{0, 1, \ldots, \infty\}$ for which $\{\tau = n\}$ is in $\mathcal{F}_n$ for $n \geq 0$. A *finite stopping time* τ is any stopping time τ for which $P_0(\tau < \infty) = 1$.

When needed in the following, for a sequence $(y_n)_{n \geq 0}$, define $y_\infty := \limsup_{n \to \infty} y_n$. For sequences of functions, this limit is defined pointwise. In particular, identification of y_∞ is needed if y_τ is to be evaluated on a set $\{\tau = \infty\}$ for a stopping time τ.

In the following, a *payment sequence* $(f_n)_{n \geq 0}$ is defined to be any sequence of nonnegative, P_0-integrable r.v.'s adapted to $(\mathcal{F}_n)_{n \geq 0}$ for which $\limsup_{n \to \infty} f_n < \infty$ a.e. $[P_0]$.

Define the *Perpetual American Contingent Claim* (PACC) in the infinite-horizon stock-and-bond market as any financial instrument consisting of the following: (i) a stopping time τ, called the exercise date; and (ii) for a payment sequence $(f_n)_{n \geq 0}$, a payment f_τ at exercise date τ. For a PACC with exercise date τ, the event $\{\tau = \infty\}$ is interpreted as those outcomes at which the PACC is never exercised. Note that if $\lim_{n \to \infty} \alpha_n = 0$ and $(f_n)_{n \geq 0}$ is a payment sequence, then $\alpha_\infty f_\infty = \lim_{n \to \infty} \alpha_n f_n = 0$ and there is no discounted payment if there is no exercise of the associated PACC.

Define the *Perpetual American Contingent Claim with finite exercise date* to be a PACC for which the exercise date τ is a finite stopping time, that is, τ is finite a.e. $[P_0]$.

A *hedging strategy* against the PACC with initial capital x is any admissible portfolio π for which $X_0^\pi = x$ and $X_n^\pi \geq f_n$ for all $n \geq 0$; let $\mathcal{H}^*(x)$ denote the collection of all hedging strategies against the PACC with initial capital x.

The *fair price* of the PACC is defined to be the number $v^* := \inf\{x > 0 : \mathcal{H}^*(x) \neq \emptyset\}$.

This defining procedure for the infinite-horizon setting is analogous to the infinite-horizon treatment in Chapters 1 and 2 of Karatzas and Shreve [**Ka-S**]. See also Shiryaev, et al. [**SKKM**] and Kramkov and Shiryaev [**Kr-Sh**].

It is natural at this point to formulate the analogue of (3.12), which in this setting connects the fair price of the PACC with the value of the appropriate optimal stopping problem. Before we do this, we recall the following result from optimal stopping theory (see, e.g., Theorem 4.7 of [**CRS**]).

For payment sequence $(f_n)_{n\geq 0}$ and discount factors $(\alpha_n)_{n\geq 0}$, for each $n \geq 0$,

(5.2a) $\sup\{E_0(\alpha_\tau f_\tau) : \tau$ is a stopping time with $\tau \geq n\}$
$= \sup\{E_0(\alpha_\tau f_\tau) : \tau$ is a finite stopping time with $\tau \geq n\}$
and

(5.2b) ess $\sup\{E_0(\alpha_\sigma f_\sigma \mid \mathcal{F}_n) : \sigma$ is a stopping time with $\sigma \geq n\}$
$=$ ess $\sup\{E_0(\alpha_\sigma f_\sigma \mid \mathcal{F}_n) : \sigma$ is a finite stopping time with $\sigma \geq n\}$.

The common supremum of (5.2a) is denoted V_n, and the common essential supremum of (5.2b) is denoted W_n^* for $n \geq 0$. The sequence $(W_n^*)_{n\geq 0}$ is called the Snell envelope for the optimal stopping problem. Now, the analogue of (3.12) is stated and proved.

THEOREM 5.2. *Given an infinite-horizon stock-and-bond market satisfying (5.1), and assume that given payment sequence $(f_n)_{n\geq 0}$ and discount factors $(\alpha_n)_{n\geq 0}$ satisfy $E_0(\sup_{n\geq 0} \alpha_n f_n) < \infty$. Then any PACC for this market with payment sequence $(f_n)_{n\geq 0}$ has fair price given by*

$$v^* = V_0 = E_0(\alpha_{\tau_s^*} f_{\tau_s^*})$$

where $\tau_s^* := \min\{n \geq 0 : \alpha_n f_n = W_n^*\}$ if this set is nonempty, and $= \infty$ otherwise.

PROOF. The following is an infinite horizon version of a finite horizon argument which is standard, found, e.g., in [**SKKM**]. First, it is shown that $v^* \geq V_0$. Given $\pi \in \mathcal{H}^*(x)$ for some $x > 0$; thus $\pi = (\pi_n = (\beta_n, \gamma_n))_{n\geq 0}$ is an admissible portfolio with wealth $X^\pi = (X_n^\pi)_{n\geq 0}$ satisfying $X_0^\pi = x$ and $X_n^\pi \geq f_n$ for all $n \geq 0$. Then $(\alpha_n X_n^\pi, \mathcal{F}_n)_{n\geq 0}$ is a martingale with respect to $P_0(\cdot)$; and for any finite stopping time τ, and each fixed integer N,

$$x = E_0(\alpha_0 X_0^\pi) = E_0(\alpha_{\tau \wedge N} X_{\tau \wedge N}^\pi) \geq E_0(\alpha_{\tau \wedge N} f\tau \wedge N).$$

From Fatou's Lemma it follows that

$$x \geq \liminf_{N \to \infty} E_0(\alpha_{\tau \wedge N} f_{\tau \wedge N}) \geq E_0(\alpha_\tau f_\tau),$$

and hence $v^* \geq V_0$, from (5.2a) and the definition of v^*.

Next, it is shown that $v^* \leq V_0$. Under the assumption $E_0(\sup_{n\geq 0} \alpha_n f_n) < \infty$ (and $\alpha_n, f_n \geq 0$ for $n \geq 0$), it is known that the Snell envelope $(W_n^*)_{n\geq 0}$ of (5.2b) satisfies (i) $(W_n^*, \mathcal{F}_n)_{n\geq 0}$ is a supermartingale under P_0, (ii) $W_n^* \geq \alpha_n f_n$ for all $n \geq 0$, and (iii) $(W_n^*)_{n\geq 0}$ is minimal among supermartingales which dominate $(\alpha_n f_n)_{n\geq 0}$ (pointwise as in (ii)) (see, e.g., Chapter 4 of [**CRS**]), and is uniformly integrable. From the standard Doob decomposition for supermartingales, it follows that $(W_n^*)_{n\geq 0}$ has representation $W_n^* = M_n - A_n$ for all $n \geq 0$, where $(M_n, \mathcal{F}_n)_{n\geq 0}$ is a uniformly integrable martingale with $M_0 = W_0^* = V_0$, and $(A_n)_{n\geq 0}$ is a stochastic process which is increasing, that is, $A_0 = 0 \leq \cdots \leq A_n \leq A_{n+1} \leq \cdots$ with $E_0(A_n) < \infty$ for $n \geq 0$; $(A_n)_{n\geq 0}$ is predictable, that is, A_n is $\mathcal{F}_{n-1}$-measurable for $n \geq 1$; and $(A_n)_{n\geq 0}$ is uniformly integrable. Since $(M_n, \mathcal{F}_n)_{n\geq 0}$ is a martingale with respect to P_0, with $M_0 = V_0$, then for each $n \geq 1$, $(M_l, \mathcal{F}_l)_{0\leq l\leq n}$ is a martingale with respect to P_0. As mentioned in the remark prior to Example 5.1, each horizon-n stock-and-bond market model is complete under P_0. And as mentioned in the remarks prior to (3.11), (e.g., from [**Ha-P**] or Section 5 of [**SKKM**]), it follows that $(M_l, \mathcal{F}_l)_{0 \leq l \leq n}$ has representation

$$M_l = M_0 + \sum_{k=1}^{l} \gamma_k^{(n)} \Delta M_k, \qquad 0 \leq l \leq n$$

where $\gamma_k^{(n)}$ is an $\mathcal{F}_{k-1}$-measurable r.v. and $\Delta M_k = \alpha_k S_k - \alpha_{k-1} S_{k-1}$ for $k = 1, \ldots, n$.

Define $\pi = (\pi_n = (\gamma_n, \beta_n))_{n \geq 0}$ by γ_0 and β_0 are constants such that $X_0^\pi = V_0 = \beta_0 + \gamma_0 S_0 = \beta_0 + \gamma_0 s$, and $\gamma_n := \gamma_n^{(n)}$ and $\beta_n := \frac{X_{n-1}^\pi - \gamma_n^{(n)} S_{n-1}}{B_{n-1}}$, with $X_n^\pi := \beta_n B_n + \gamma_n S_n$, for $n \geq 1$. Note that for $n \geq 1$, γ_n and β_n are $\mathcal{F}_{n-1}$-measurable, and $\beta_{n-1} B_{n-1} + \gamma_{n-1} S_{n-1} = X_{n-1}^\pi = \beta_n B_{n-1} + \gamma_n S_{n-1}$, so that $B_{n-1} \Delta \beta_n + S_{n-1} \Delta \gamma_n = 0$. Note also that $X_0^\pi = V_0 = M_0$, and for $n \geq 1$, if $\alpha_j X_j^\pi = M_j$ for $j = 0, \ldots, n-1$, then

$$\begin{aligned}
\alpha_n X_n^\pi &= \frac{\beta_n B_n + \gamma_n^{(n)} S_n}{B_n} = \beta_n + \gamma_n^{(n)} S_n \\
&= \frac{X_{n-1}^\pi - \gamma_n^{(n)} S_{n-1}}{B_{n-1}} + \gamma_n^{(n)} S_n \\
&= \alpha_{n-1} X_{n-1}^\pi + \gamma_n^{(n)} (\alpha_n S_n - \alpha_{n-1} S_{n-1}) \\
&= M_{n-1} + \gamma_n^{(n)} \Delta M_n \\
&= M_n.
\end{aligned}$$

Thus
$$X_n^\pi = B_n M_n = B_n(W_n^* + A_n) \geq B_n W_n^* \geq \alpha_n f_n B_n = f_n$$

for $n \geq 1$ (and also $X_n^\pi \geq 0$ for $n \geq 1$). We have shown that π is a hedging portfolio for the PACC with initial capital V_0. By definition of v^*, we obtain $v^* \leq V_0$.

From Chapter 4 of [**CRS**] we have that $\tau_s^* := \min\{n \geq 0 : \alpha_n f_n = W_n^*\}$ is a stopping time that satisfies $V_0 = E_0(\alpha_{\tau_s^*} f_{\tau_s^*})$ under the assumption $E_0(\sup_{n \geq 0} \alpha_n f_n) < \infty$, so $v^* = E_0(\alpha_{\tau_s^*} f_{\tau_s^*})$. $\square$

For this Section we are interested in the following PACC's, identified with their payment sequences $(f_n)_{n \geq 0}$.

(5.3)
  Perpetual American call option $f_n = \beta^n (S_n - q)^+$ for $n \geq 0$
  with strike price q,
  under dividend factor β

  Perpetual American lookback $f_n = \beta^n (\max_{0 \leq i \leq n} S_i - q)^+$
  call option with strike price q,   for $n \geq 0$
  under dividend factor β

where the dividend factor β satisfies $0 < \beta < 1$, and is used to model a constant rate at which dividends are paid out each period. Another PACC which is directly related to the second PACC of (5.3) is the 'Russian option,' with payment sequence $f_n = \beta^n \max_{0 \leq i \leq n} S_i$, for $n \geq 0$. For results on these perpetual options in the discrete-time setting see Shiryaev, et al. [**SKKM**], Darling, et al. [**DLT**], and Kramkov and Shiryaev [**Kr-Sh**], and in the continuous-time setting, see Karatzas and Shreve [**Ka-S**], McKean [**M**], Shepp and Shiryaev [**S-S**], and Duffie and Harrison [**D-H**].

THEOREM 5.3. *For the symmetric Binomial market of Example 5.1 with $0 < \lambda^{-1} < r+1 < \lambda$, $r > 0$ and $0 < p_0 < \frac{1}{2}$, and with dividend factor $0 < \beta < 1$, the fair price of the perpetual American call option with strike price q is given by*

$$
\begin{aligned}
v_C &= v_C(s, r, \lambda, \beta) \\
&:= \sup\{E_0((\alpha\beta)^\tau (S_\tau - q)^+) : \tau \text{ is a finite stopping time}\}
\end{aligned}
\tag{5.4}
$$

and the fair price of the perpetual American lookback call option with strike price q is given by

$$
\begin{aligned}
v_L &= v_L(s, r, \lambda, \beta) \\
&:= \sup\{E_0((\alpha\beta)^\tau (\max_{0 \leq n \leq \tau} S_n - q)^+) : \tau \text{ is a finite stopping time.}\}
\end{aligned}
\tag{5.5}
$$

PROOF. Observe that for each $k \geq 0$,

$$\beta_0^k (S_k - q)^+ \leq \beta_0^k \left(\max_{0 \leq n \leq k} S_n - q \right)^+ \leq \sup_{n \geq 0} \beta_0^n (S_n - q)^+$$

where $\beta_0 := \alpha\beta$. After some elementary background concepts are introduced, it will be shown that $E_0(\sup_{n \geq 0} \beta_0^n (S_n - q)^+) < \infty$ within the proof of Proposition 5.4. Thus, Theorem 5.2 can be applied to obtain the conclusions. □

For the setting of Theorem 5.3, there is another quantity useful in comparisons with the fair prices v_C and v_L; this is the quantity

$$m_0 = m_0(s, r, \lambda, \beta) := E_0 \left(\sup_{n \geq 0} (\alpha\beta)^n (S_n - q)^+ \right). \tag{5.6}$$

This quantity was used in an essential way within the proof of Theorem 5.3 (and Theorem 5.2). One can think of this quantity as a limit of prices of a type of finite-horizon lookback option in markets with dividends paid each period, since

$$E_0 \left(\sup_{n \geq 0} (\alpha\beta)^n (S_n - q)^+ \right) = \lim_{N \to \infty} E_0 \left(\max_{0 \leq n \leq N} (\alpha\beta)^n (S_n - q)^+ \right).$$

In particular, it is immediate that

$$v_C \leq v_L \leq m_0. \tag{5.7}$$

To give a more precise comparison of the quantities v_C, v_L and m_0 over the collection of symmetric Binomial markets of Example 5.1, we give results on representations and bounds for each of these quantities. Throughout we assume without loss of generality that $q = 1$.

For the results on the quantity $E_0(\sup_{n \geq 0} (\alpha\beta)^n (S_n - 1)^+)$, recall from Example 5.1 that $(\epsilon_i)_{i \geq 1}$ and i.i.d. r.v.'s with ϵ_i taking the values $+1$ and -1 with respective probabilities $p_0 = (r+1-\lambda^{-1})/(\lambda-\lambda^{-1})$ and $1-p_0 = (\lambda-(r+1))/(\lambda-\lambda^{-1})$. Also write $\epsilon_0 := 0$. Define sequences of r.v.'s $(N_k)_{k \geq 0}$, $(\epsilon_i^{(1)})_{i \geq 1}$, $(\epsilon_i^{(2)})_{i \geq 1}, \ldots$ as follows: $N_0 := 0$ and

$$
\begin{aligned}
N_1 &:= \inf\{n \geq 1 : \epsilon_1 + \cdots + \epsilon_n = 1\} \text{ if this set is nonempty,} \\
&\text{and} = \infty \text{ otherwise,} \\
\epsilon_i^{(1)} &:= \epsilon_{N_1+i} \text{ for } i \geq 1;
\end{aligned}
\tag{5.8}
$$

and for $k \geq 2$, if $N_1, (\epsilon_i^{(1)})_{i \geq 1}, \ldots, N_{k-1}$, and $(\epsilon_i^{(k-1)})_{i \geq 1}$ have been defined,

(5.9)
$$N_k := \inf\{n \geq 1 : \epsilon_1^{(k-1)} + \cdots + \epsilon_n^{(k-1)} = 1\}$$
if this set is nonempty, and $= \infty$ otherwise,
$$\epsilon_i^{(k)} := \epsilon_{N_1 + \cdots + N_k + i} = \epsilon_{N_{k-1} + i}^{(k-1)} \text{ for } i \geq 1;$$

and define the lifetime

(5.10)
$$\begin{aligned} L &= 0 &&\text{if } N_1 = \infty \\ &= k &&\text{if } N_1 < \infty, \ldots, N_k < \infty, N_{k+1} = \infty, \text{ for } 1 \leq k < \infty, \\ &= \infty &&\text{if } N_i < \infty \text{ for all } i = 1, 2, \ldots.\end{aligned}$$

Recall, for example, from Feller Volume I [**F**] or Resnick [**Res**], that $N_1, N_2, \ldots$, are i.i.d. r.v.'s taking values in $\{1, 2, \ldots, \infty\}$ with probability generating function $\Phi_0(s) = E_0(S^{N_1})$ satisfying

(5.11)
  (i) $(1 - p_0)s\Phi^2(s) - \Phi(s) + p_0 s = 0$
  (ii) $\Phi(s) = \frac{1 - \sqrt{1 - 4p_0(1-p_0)s^2}}{2(1-p_0)s}$ for $s \neq 0$, and $= 0$ for $s = 0$.
  (iii) $P_0(N_1 < \infty) = \Phi(1) = \begin{cases} p_0/(1-p_0) & \text{if } 0 \leq p_0 \leq \frac{1}{2} \\ 1 & \text{if } \frac{1}{2} \leq p_0 \leq 1. \end{cases}$

Observe that $0 < p_0 < \frac{1}{2}$ if and only if $\lambda > r + 1 + \sqrt{(r+1)^2 - 1}$, and for these values of p_0 denote

(5.12)
$$\hat{p}_0 := P_0(N_1 = \infty) = \frac{1 - 2p_0}{1 - p_0} = \frac{\lambda + \lambda^{-1}}{\lambda - (r+1)},$$

and obtain that r.v. L is geometric distributed with parameter $\hat{p}_0$, that is, $P_0(L = k) = (1 - \hat{p}_0)^k \hat{p}_0$ for $k = 0, 1, \ldots$. To simplify further expressions, write

(5.13)
$$\beta_0 := \alpha\beta = \beta/(r+1).$$
$$\phi_0(\lambda) := \Phi(\beta_0) = E_0(\beta_0^{N_1}) = E_0(\beta_0^{N_1} I_{(N_1 < \infty)})$$

From straightforward (but tedious) calculus, for $0 < \lambda^{-1} < r + 1 < \lambda$, we have

(5.14)
  (i) as λ increases from $r + 1$ to ∞, $\phi_0(\lambda)$ decreases from β_0 to 0; and
  (ii) as λ increases from $r + 1$ to $r + 1 + \sqrt{(r+1)^2 - 1}$,
  $\lambda\phi_0(\lambda)$ decreases from β to $\beta\left(\frac{r+1+\sqrt{(r+1)^2-1}}{r+1+\sqrt{(r+1)^2-\beta^2}}\right)$,
  and as λ increases from $r + 1 + \sqrt{(r+1)^2 - 1}$ to ∞,
  $\lambda\phi_0(\lambda)$ increases from $\beta\left(\frac{r+1+\sqrt{(r+1)^2-1}}{r+1+\sqrt{(r+1)^2-\beta^2}}\right)$ to β.

Within the next Proposition and throughout the remainder of this Section, we use the convention that $\lfloor a \rfloor =$ greatest integer $\leq a$, and $\lceil a \rceil =$ least integer $\geq a$.

PROPOSITION 5.4. *In the setting of* Example 5.1, *assume also that* $0 < \beta < 1$ *and* $0 < p_0 < \frac{1}{2}$. *Then*
(5.15)
$$m_0(s, r, \lambda, \beta) = E_0\left(\sup_{n \geq 0} \beta_0^n (S_n - 1)^+\right) = E_0\left(\sup_{0 \leq k \leq L} \beta^{N_0 + \cdots + N_k}(s\lambda^k - 1)^+\right)$$

With $\hat{p}_0(\lambda) := \hat{p}_0$ as in (5.12) and $\phi_0(\lambda) = E_0(\beta_0^{N_1})$ as in (5.13), for $x < 0$,

$$(5.16) \quad \frac{\lambda^x(\phi_0(\lambda))^{\lceil -x \rceil}}{1 - \lambda\phi_0(\lambda)} \geq m_0(\lambda^x, r, \lambda, \beta)$$

$$\geq \frac{\lambda^x p_0(\lambda)(\lambda\phi_0(\lambda))^{\lceil -x \rceil}}{1 - \lambda\phi_0(\lambda)} - \frac{p_0(\lambda)(\phi_0(\lambda))^{\lceil -x \rceil}}{1 - \phi_0(\lambda)}$$

and for $x \geq 0$,

$$(5.17) \quad \frac{\lambda^x}{1 - \lambda\phi_0(\lambda)} \geq m_0(\lambda^x, r, \lambda, \beta) \geq \frac{\lambda^x p_0(\lambda)}{1 - \lambda\phi_0(\lambda)} - \frac{p_0(\lambda)}{1 - \phi_0(\lambda)}$$

PROOF. Representation (5.15) is immediate from the definitions since

$$\sup_{n \geq 0} \beta_0^n (s\lambda^{\epsilon_0 + \cdots + \epsilon_n} - 1)^+ = \sup_{k \geq 0} \beta_0^{N_0 + \cdots + N_k}(s\lambda^k - 1)^+$$

$$= \sup_{0 \leq k \leq L} \beta_0^{N_0 + \cdots + N_k}(s\lambda^k - 1)^+.$$

Inequalities (5.16) and (5.17) have similar verifications; we prove (5.17). For $x \geq 0$, we have (recall that $\lambda > 1$)

$$E_0\left(\sup_{0 \leq k \leq L} \beta_0^{N_0 + \cdots + N_k}(\lambda^{x+k} - 1)^+\right)$$

$$\leq E_0\left(\sum_{k=0}^L \beta_0^{N_0 + \cdots + N_k} \lambda^{x+k}\right) = E_0\left(\sum_{l=0}^\infty \sum_{k=0}^l \beta_0^{N_0 + \cdots + N_k} \lambda^{x+k} I_{(L=l)}\right)$$

$$= \lambda^x + \lambda^x E_0\left(\sum_{k=1}^\infty \beta_0^{N_1 + \cdots + N_k} \lambda^k I_{(N_1 < \infty, \ldots, N_k < \infty)}\right)$$

$$= \lambda^x \sum_{k=0}^\infty \left(\lambda E_0(\beta_0^{N_1})\right)^k = \lambda^x/(1 - \lambda E_0 \beta_0^{N_1}),$$

and

$$E_0\left(\sup_{0 \leq k \leq L} \beta_0^{N_0 + \cdots + N_k}(\lambda^{x+k} - 1)^+\right)$$

$$\geq E_0\left(\beta_0^{N_1 + \cdots + N_L}(\lambda^{x+L} - 1)\right)$$

$$= \lambda^x \left\{P_0(N_1 = \infty) + \sum_{k=1}^\infty (\lambda E_0(\beta_0^{N_1} I_{(N_1 < \infty)}))^k P_0(N_{k+1} = \infty)\right\}$$

$$- \left\{P_0(N_1 = \infty) + \sum_{k=1}^\infty (E_0(\beta_0^{N_1} I_{(N_1 < \infty)}))^k P_0(N_{k+1} = \infty)\right\}$$

$$= \frac{\lambda^x P_0(N_1 = \infty)}{1 - \lambda E_0(\beta_0^{N_1})} - \frac{P_0(N_1 = \infty)}{1 - E_0(\beta_0^{N_1})}.$$

$\square$

Next, results are given on the value $v_C(s, r, \lambda, \beta)$ of (5.4). For this purpose, certain parameters and random variables are introduced.

(5.18) The parameter $\gamma = \gamma(\lambda) = \gamma(r, \lambda, \beta)$ is defined through $\lambda^\gamma = \psi = \phi_0^{-1}$ with ϕ_0, defined in (5.13), having representation

$$\phi_0 = \frac{A_0}{2} - \sqrt{\frac{A_0^2}{4} - B_0}, \text{ where } A_0 = \frac{1}{\beta_0(1-p_0)}, \quad B_0 = \frac{p_0}{1-p_0}, \text{ and,}$$

$$\psi = \frac{A}{2} + \sqrt{\frac{A^2}{4} - B}, \text{ where } A = \frac{1}{\beta_0 p_0}, \quad B = \frac{1-p_0}{p_0}$$

so ϕ_0 satisfies $(1-p_0)\beta_0\phi_0^2 - \phi_0 + p_0\beta_0 = 0$ with $0 < \phi_0 < \lambda^{-1}$, and ψ satisfies $p_0\beta_0\psi^2 - \psi + (1-p_0)\beta_0 = 0$ with $\psi > \lambda$.

If we use the convention $a(\lambda) \sim b(\lambda)$ means $a(\lambda)/b(\lambda) \to 1$ as $\lambda \to \infty$, then it follows from (5.18) and the definitions that

(5.19) $\qquad \phi_0(\lambda) \sim \dfrac{\beta}{\lambda} \quad$ and $\quad \gamma(\lambda) - 1 \sim \dfrac{-\ln\beta}{\ln\lambda}.$

Introduce the following r.v.'s (extend the underlying probability space, where necessary, to allow these additions).

(5.20)
- τ_0 is a r.v. which is independent of $(\epsilon_i)_{i\geq 1}$ and geometric distributed with parameter $\beta_0 = \alpha\beta$, that is $P_0(\tau_0 \geq k) = \beta_0^k$, $k = 0, 1, \ldots$
- $M_0 := \max\{\epsilon_0 + \cdots + \epsilon_j : 0 \leq j \leq \tau_0\}$, and is geometric distributed with parameter $\phi_0 = \lambda^{-\gamma} = \psi^{-1}$, and $E_0\lambda^{M_0} = (\lambda^\gamma - 1)/(\lambda^\gamma - \lambda)$.
- $\tau^* = \tau^*(\lambda^*, r, \lambda, \beta)$ is the 'level-crossing' stopping time defined by

$$\tau^* := \inf\{n \geq 0 : x + \epsilon_0 + \cdots + \epsilon_n \geq \log_\lambda E_0\lambda^{M_0}\} \quad \text{if this set is nonempty,}$$
$$= \infty \qquad \qquad \qquad \qquad \qquad \qquad \qquad \qquad \text{otherwise.}$$

PROPOSITION 5.5. *In the setting of* Example 5.1, *assume also that* $0 < \beta < 1$. *Then* $v_C(s, r, \lambda, \beta) = \sup\{E_0(\beta^\tau(S_\tau - 1)^+) : \tau$ *is a finite stopping time*$\}$ *satisfies*

(5.21) $\quad v_C(s, r, \lambda, \beta) = \dfrac{E_0((s\lambda^{M_0} - E_0\lambda^{M_0})^+)}{E_0\lambda^{M_0}} = E_0(\beta_0^{\tau^*}(s\lambda^{\epsilon_0 + \cdots + \epsilon_{\tau^*}} - 1)^+);$

(5.22) $\quad v_C(s, r, \lambda, \beta) = \begin{cases} s - 1 & \text{for } s \geq E_0\lambda^{M_0} \\ s(\lambda^{1-\gamma})^{k(s)} - \lambda^{-\gamma k(s)} & \text{for } 0 < s \leq E_0\lambda^{M_0} \end{cases}$

where $k(s) = \lceil \log_\lambda(E_0\lambda^{M_0}/s) \rceil$; *and for all integers* x, *with* $s = \lambda^x$,

(5.23) $\quad v_C(s, r, \lambda, \beta) = \begin{cases} s - 1 & \text{for } s = \lambda^x \geq \lambda^{\hat{m}(\lambda)} \\ C(\lambda)s^\gamma & \text{for } s = \lambda^x \leq \lambda^{\hat{m}(\lambda)} \end{cases}$

where $\hat{m} = \hat{m}(\lambda) = \hat{m}(r, \lambda, \beta) := \lceil \log_\lambda E_0\lambda^{M_0} \rceil$ *and* $C(\lambda) = C(r, \lambda, \beta) := \lambda^{-\gamma\hat{m}}(\lambda^{\hat{m}} - 1)$. *The function* $v_C(s)$ *is continuous, piecewise-linear, increasing, with discontinuities in its derivative at points* $\{E_0\lambda^{M_0}/\lambda^k : k = 0, 1, \ldots,\}$, *and*

(5.24) $\quad v_C(s) = \lambda^{-(\gamma-1)l}s - \lambda^{-\gamma l} \quad \text{for } \dfrac{E_0\lambda^{M_0}}{\lambda^l} \leq s < \dfrac{E_0\lambda^{M_0}}{\lambda^{l-1}}, \quad l = 1, 2, \ldots.$

and has upper bound

(5.25) $\qquad\qquad v_C(s, r, \lambda, \beta) \leq K_{r,\beta}s^\gamma \qquad \text{for all } s > 0,$

where

(5.26) $$K_{r,\beta} := \frac{1}{\beta}\left(\frac{r+1+\sqrt{(r+1)^2-\beta^2}}{r+1+\sqrt{(r+1)^2-1}}\right)$$

for $r > 0$, $0 < \beta < 1$.

PROOF. The representations of (5.21) are verified through a straightforward modification of the results in Darling, Liggett and Taylor [**DLT**]. The representations (5.22) and (5.23) follow from an expectation calculation based on (5.20) and (5.21). Note that (5.23) is consistent with the representation for v_C derived in Shiryaev, et al. [**SKKM**]; however, for a correct expression one should replace 'min' by 'max' in the fifth line of page 51 of [**SKKM**]. It can be shown that the function $C(\lambda)$ has upper bound $C(\lambda) < 1$, but this only gives an upper bound for $v_C(s)$ at $s = \lambda^*$ for integers x. To obtain the upper bound, use (5.24) and verify that for $l = 1, 2, \ldots$, with $s \in [\lambda^{-l} E_0 \lambda^{M_0}, \lambda^{-l+1} E_0 \lambda^{M_0}]$,

$$\frac{s^\gamma}{v_C(s)} \geq \lambda^{\gamma l}\left(\frac{s^\gamma}{s\lambda^\gamma-1}\right) \geq \frac{(\gamma/(\gamma-1))^{\gamma-1}}{\lambda^{\gamma-l}-((\gamma-1)/\gamma)}$$

$$\geq \frac{1}{\lambda^{\gamma-l}} \geq \lambda^{1-\gamma} = \lambda\phi_0(\lambda) \geq \beta\left(\frac{r+1+\sqrt{(r+1)^2-1}}{r+1+\sqrt{(r+1)^2-\beta^2}}\right)$$

where calculus gives the second inequality, and the last inequality follows from (5.14); this proves (5.25). □

Now one can show that within these symmetric Binomial markets, the limiting lookback option prices $m_0(s, r, \lambda, \beta) = E_0(\sup_{n\geq 0} \beta_0^n (S_n - 1)^+)$ can be arbitrarily large in comparison to the prices of the American perpetual call options $v_C(s, r, \lambda, \beta) = \sup\{E_0(\beta^\tau (S_\tau - 1)^+) : \tau \text{ is a finite stopping time}\}$, by letting $\lambda \to \infty$ and thus trading in 'long shot' markets.

THEOREM 5.6. *Given the symmetric Binomial market of* Example 5.1 *with $0 < \lambda^{-1} < r+1 < \lambda$, $r > 0$ and dividend factor $0 < \beta < 1$. For $s = \lambda^x$ for integers x, $m_0(\lambda) = m_0(\lambda^x, r, \lambda, \beta)$ and $v_C(\lambda) = v_C(\lambda^*, r, \lambda, \beta)$ satisfy for $x < 0$,*

$$\liminf_{\lambda\to\infty} \frac{m_0(\lambda)}{v_C(\lambda)} = \infty \quad \text{for each } r > 0, \quad 0 < \beta < 1; \text{ and}$$

for $x \geq 0$

$$\liminf_{\lambda\to\infty} \frac{m_0(\lambda)}{v_C(\lambda)} \geq \frac{\beta^x}{K_{r,\beta}(1-\beta)} \quad \text{for each } r > 0, \quad 0 < \beta < 1; \text{ and}$$

$$\liminf_{\beta\nearrow 1} \liminf_{\lambda\to\infty} \frac{m_0(\lambda)}{v_C(\lambda)} = \infty \quad \text{for each } r > 0.$$

For each constant $s > 0$,

$$\liminf_{\lambda\to\infty} \frac{m_0(\lambda)}{v_C(\lambda)} \geq \frac{1}{K_{r,\beta}}\left(\frac{1}{1-\beta} - \frac{1}{s}\right) \quad \text{for each } r > 0, \quad 0 < \beta < 1 \text{ and}$$

$$\liminf_{\beta\nearrow 1} \liminf_{\lambda\to\infty} \frac{m_0(\lambda)}{v_C(\lambda)} = \infty \quad \text{for each } r > 0.$$

PROOF. Use inequalities (5.16), (5.17) and (5.25) with asymptotics from the definitions, such as (5.19). □

This brings up the interesting question of how the perpetual American lookback option price v_L compares to the perpetual American call option price v_C and to the limiting lookback options prices m_0. We are able to show that for 'long shot' symmetric Binomial markets v_L can be arbitrarily large compared to v_C, and v_L and m_0 can be very close. For this verification, we use results of Kramkov and Shiryaev [**Ka-S**] on characterization of the solution to the optimal stopping problem (associated with the 'Russian option') $w_L(s, r, \lambda, \beta) := \sup\{E_0(\beta_0^\tau \max_{0 \leq n \leq \tau} S_n) : \tau$ is a finite stopping time$\}$ within the data of the symmetric Binomial market of Example 5.1 with dividend factor $0 < \beta < 1$ ($\beta_0 = \alpha\beta = \beta/(r+1)$). We record results from [**Ka-S**] and then establish the comparison results. This necessitates introduction of the following parameters:

$$(5.27) \qquad \tilde{p} := \frac{\lambda - \alpha}{\lambda - \lambda^{-1}} = \alpha\lambda p_0$$

(so $\tilde{p} \to 1$ as $\lambda \to \infty$, whereas $p_0 \to 0$ as $\lambda \to \infty$); $\gamma_1 < 0$ and $\gamma_2 > 1$ are defined through

$$(5.28) \qquad y_1 = \lambda^{\gamma_1} = \frac{\hat{A}}{2} - \sqrt{\frac{\hat{A}^2}{4} - \hat{B}} \quad \text{and} \quad y_2 = \lambda^{\gamma_2} = \frac{\hat{A}}{2} + \sqrt{\frac{\hat{A}^2}{4} - \hat{B}}$$

where y_1 and y_2 are the solutions of equation $(1-\tilde{p})\beta y^2 - y + \tilde{p}\beta = 0$ (compare this to (5.11) and (5.18)), where

$$(5.29) \qquad \hat{A} = \frac{1}{(1-\tilde{p})\beta} \quad \text{and} \quad \hat{B} = \frac{\tilde{p}}{1-\tilde{p}};$$

and b_0, $\bar{x}_0$, C_1 and C_2 are defined by

$$(5.30) \qquad b_0 := \frac{(1-\tilde{p})\lambda^{\gamma_2} + \tilde{p} - \beta^{-1}}{(1-\tilde{p})(\lambda^{\gamma_2} - \lambda^{\gamma_1})}$$

$$(5.31) \qquad \bar{x}_0 := \left(\frac{b_0}{1-b_0} \frac{1-\gamma_1}{\gamma_2 - 1}\right)^{1/(\gamma_2 - \gamma_1)}$$

$$(5.32) \quad C_1 := \frac{\lambda^\mu}{b_0\lambda^{\gamma_1\mu} + (1-b_0)\lambda^{\gamma_2\mu}} \quad \text{and} \quad C_2 := \frac{\lambda^{\mu+1}}{b_0\lambda^{\gamma_1(\mu+1)} + (1-b_0)\lambda^{\gamma_2(\mu+1)}}$$

where $\mu := \lfloor \log_\lambda \bar{x}_0 \rfloor$. Kramkov and Shiryaev [**Ka-S**] prove that

$$(5.33) \qquad w_L = s \cdot \max\{C_1, C_2\}.$$

Observe this value w_L is related to $v_L = \sup_\tau E_0(\beta_0^\tau(\max_{0 \leq n \leq \tau} S_n - 1))$ by

$$(5.34) \qquad w_L \geq v_L \geq w_L - 1.$$

THEOREM 5.7. *Given the symmetric Binomial market of* Example 5.1 *with* $0 < \lambda^{-1} < r + 1 < \lambda$, $r > 0$ *and dividend factor* $0 < \beta < 1$. *Then for each* $S_0 = s > 0$, $v_L(\lambda) = v_L(s, r, \lambda, \beta)$ *and* $v_C(\lambda) = v_C(s, r, \lambda, \beta)$ *satisfy*

$$(5.35) \qquad \begin{aligned} \liminf_{\lambda \to \infty} \frac{v_L(\lambda)}{v_C(\lambda)} &\geq \frac{1}{K_{r,\beta}}\left(\frac{1}{1-\beta} - \frac{1}{s}\right) \quad \text{for each } r > 0, 0 < \beta < 1, \text{ and,} \\ \liminf_{\beta \nearrow 1} \liminf_{\lambda \to \infty} \frac{v_L(\lambda)}{v_C(\lambda)} &= \infty \text{ for each } r > 0; \end{aligned}$$

and for $s > 1 - \beta$, $v_L(\lambda) = v_L(s, r, \lambda, \beta)$ *and* $m_0(\lambda) = m_0(s, r, \lambda, \beta)$ *satisfy*

(5.36)
$$\frac{s/(1-\beta)}{(s/(1-\beta))-1} \geq \limsup_{\lambda\to\infty}\frac{m_0(\lambda)}{v_L(\lambda)} \geq \liminf_{\lambda\to\infty}\frac{m_0(\lambda)}{v_L(\lambda)}$$
$$\geq 1 - \left(\frac{1-\beta}{s}\right)$$
for each $r > 0, 0 < \beta < 1$, and
$$\lim_{\beta\to 1}\liminf_{\lambda\to\infty}\frac{m_0(\lambda)}{v_L(\lambda)} = 1 = \lim_{\beta\to 1}\limsup_{\lambda\to\infty}\frac{m_0(\lambda)}{v_L(\lambda)}$$

PROOF. First, observe that as $\lambda \to \infty$, we have

(5.37) $\tilde{p} \sim 1, \quad 1 - \tilde{p} \sim \alpha/\lambda, \quad \lambda^{\gamma_1} \sim \beta, \quad \lambda^{\gamma_2} \sim \lambda/\beta_0, \quad \gamma_1 \sim \frac{\ln\beta}{\ln\lambda},$

$\gamma_2 - 1 \sim \frac{-\ln\beta_0}{\ln\lambda},$

$b_0 \sim \beta, \quad \bar{x}_0 \sim \frac{\beta}{(1-\beta)(-\ln\beta_0)}\ln\lambda, \quad C_1 \sim \frac{1}{1-\beta}$ and

$C_2 \sim \frac{\beta_0}{1-\beta}.$

and thus from (5.33) and (5.34)
$$w_L(s, r, \lambda, \beta) \sim \frac{s}{1-\beta}$$
and as $\lambda \to \infty$

(5.38) $$\frac{s}{1-\beta} \geq v_L(s, r, \lambda, \beta) \geq \frac{s}{1-\beta} - 1.$$

The results (5.35) and (5.36) now follow from the asymptotic inequality (5.38) and from (5.16), (5.17), (5.25) and (5.19). □

Of course, for market modelling, the infinite horizon model can only be used as an approximation to finite horizon models. With this point in mind, it is important that, under the conditions of Theorems 5.6 and 5.7, the finite horizon counterparts of v_C, v_L and m_0 converge to the infinite horizon values v_C, v_L and m_0 respectively, as the horizon increases to infinity, so finite horizon approximations for fair prices are possible (see Chapter 4 of [**CRS**]). This topic of approximating finite-lived options by perpetual American options was discussed from another point of view by Carr and Faguet [**C-F**].

6. Other 'prophet-type' problems in finance

In this Section we mention one other use of prophet problems in the option pricing literature. There are probability ordering comparisons which are in the spirit of the prophet inequality comparisons, which can be used to prove expectation-based prophet inequalities, and which have been applied to problems within mathematical finance. One result of this nature is given here, for its proof, related results and mathematical applications see Blackwell and Dubins [**B-D**], Dubins and Gilat [**D-G**], Gilat [**G**], Kertz and Rösler [**K-R1**], [**K-R2**], [**K-R3**], and Rogers [**R**]).

(6.1) Let μ be any p.m. on $\mathbb{R}$ with $\int |x|d\mu(x) < \infty$ and let μ^* denote its associated Hardy-Littlewood maximal p.m. Then
$\{\nu$ is a p.m. on $\mathbb{R}: \mu \prec \nu \prec \mu^*\} =$
$\{\nu :$ there is a martingale $(Z_t)_{0 \leq t \leq 1}$ with $\sup_{0 \leq t \leq 1} Z_t \stackrel{\mathcal{D}}{=} \nu$ and $Z_1 \stackrel{\mathcal{D}}{=} \mu\}$

where '$\prec$' is the stochastic order of p.m.'s and '$\stackrel{\mathcal{D}}{=}$' means the r.v. has this p.m.

Hobson [**Hob**] uses this type of result to price and hedge lookback options in certain types of continuous-time markets in which European call options are the underlying instruments.

References

[B-D] Blackwell, D. and Dubins, L. E. (1963) *A converse to the dominated convergence theorem.* Illinois J. Math. **7**, 508–514.

[C-F] Carr, P. and Faguet, D. (1996) *Valuing finite-lived options as perpetual.* Working paper, Johnson Graduate School of Management, Cornell Univ.

[CRS] Chow, Y.S., Robbins, H. and Siegmund, D. (1971) *The Theory of Optimal Stopping.* Dover, New York.

[C-R] Cox, J.C. and Rubenstein, M. (1985) *Options Markets.* Prentice Hall, Englewood Cliffs, N.J.

[C-V] Conze, A. and Viswanathan (1991) *Path dependent options: The case of lookback options.* Journal of Finance **46**, 1893-1907.

[DLT] Darling, D. A., Liggett, T. and Taylor, H. M. (1972) *Optimal stopping for partial sums.* Annals Math. Statist. **43**, 1363–1368.

[D-G] Dubins, L. E. and Gilat, D. (1978) *On the distribution of maxima of martingales.* Proc. Amer. Math. Soc. **68**, 337-338.

[D-P] Dubins, L. E. and Pitman, J. (1980) *A maximal inequality for skew fields.* Z. Wahrscheinlichkeitstheorie. Verw. Gebiete **52**, 219–227.

[D-S] Dubins, L. E. and Schwarz, G. (1988) *A sharp inequality for sub-martingales and stopping times.* Soc. Math. de France Astérisque **157–158**, 129–145.

[D] Duffie, D. (1996) *Dynamic Asset Pricing Theory*, Second Edition, Princeton University Press, Princeton, NJ.

[D-H] Duffie, D. and Harrison, J. M. (1993) *Arbitrage pricing of Russian options and perpetual lookback options.* Ann. Appl. Probab **3**, 641–651.

[G] Gilat, D. (1988) *On the ratio of the expected maximum of a martingale and the L_p-norm of its last term.* Israel J. Math. **63**, 270–280.

[F] Feller, W. (1968) *An Introduction to Probability Theory and its Applications.* Wiley, New York.

[Ha-K] Harrison, J. M. and Kreps, D. M. (1979) *Martingales and arbitrage in multiperiod securities markets.* J. Econ. Theory **20**, 381–408.

[Ha-P] Harrison, J. M. and Pliska, S. R. (1981) *Martingales, stochastic integrals and continuous trading.* Stoch. Proc. Appl. **11**, 215–260.

[H-K83] Hill, T. P. and Kertz, R. P. (1983) *Stop rule inequalities for uniformly bounded sequences of random variables.* Trans. Amer. Math. Soc. **278**, 197–207.

[H-K92] Hill, T. P. and Kertz, R. P. (1992) *A survey of prophet inequalities in optimal stopping theory*, in *Strategies for Sequential Search and Selection in Real Time* (F. T. Bruss, T. S. Ferguson, and S. M. Samuels, eds.) Contemporary Mathematics Volume 125, pages 191–208. American Mathematical Society, Providence, Rhode Island.

[Hob] Hobson, D. G. (1997) *Robust hedging of the lookback option.* Preprint, School of Mathematical Sciences, University of Bath.

[Ka-S] Karatzas, I. and Shreve, S. E. (1998) *Methods of Mathematical Finance, A Research Monograph.* Springer-Verlag, New York.

[K-K] Kennedy, D. P. and Kertz, R. P. (1997) *A prophet inequality for independent random variables with finite variances.* J. Appl. Probab. **34**, 945–958.

[K] Kertz, R. P. (1987) *Prophet problems in optimal stopping: results, techniques, and variations.* Preprint.

[K-R1] Kertz, R. P. and Rösler, U. (1990) *Martingales with given maxima and terminal distributions.* Israel J. Math. **69**, 173–192.

[K-R2] Kertz, R. P. and Rösler, U. (1992) *Stochastic and convex orders and lattices of probability measures, with a martingale interpretation.* Israel J. Math. **77**, 129–164.

[K-R3] Kertz, R. P. and Rösler, U. (1992) *Hyperbolic-concave functions and the Hardy-Littlewood maximal function*, in *Stochastic Inequalities* (M. Shaked and Y. L. Tong, eds.) Institute of Mathematical Statistics Lecture Notes – Monograph Series Volume 22, pages 196–210. Hayward, California.

[Kr-Sh] Kramkov, D. O. and Shiryaev, A. N. (1994) *On the rational pricing of the "Russian option" for the symmetrical Binomial model of a (B, S)-market*. Theory Probab. Appl. **39**, 153–162.

[L-L] Lamberton, D. and Lapeyre, B. (1996) *Introduction to Stochastic Calculus Applied to Finance*. Chapman & Hall, New York.

[M] McKean, H. P. Jr. (1965) *A free-boundary problem for the heat equation arising from a problem in mathematical economics*. Industr. Manag. Rev. **6**, 32–39.

[Pl] Pliska, S. R. (1997) *Introduction to Mathematical Finance Discrete Time Models*. Blackwell Publishers Ltd., Oxford.

[Pr] Price, J.F. (1996) *Optional mathematics is not optional*. Notices Amer. Math. Soc. **43**, 964-971.

[Res] Resnick, S. (1992) *Adventures in Stochastic Processes*. Birkhäuser, Boston.

[R] Rogers, L. C. G. (1993) *The joint law of the maximum and terminal value of a martingale*. Probab. Theory Relat. Fields **95**, 451–466.

[S] Schachermayer, W. (1994) *Martingale measures for discrete-time processes with infinite horizon*. Math. Finance **4**, 25–55.

[S-S] Shepp, L. A. and Shiryaev, A. N. (1993) *The Russian option: reduced report*. Ann. Appl. Probab. **3**, 631–640.

[SKKM] Shiryaev, A. N., Kabanov, Y. M., Kramkov, O. D. and Mel'nikov, A. V. (1994) *Toward the theory of pricing of options of both European and American types. I. Discrete time*. Theory Probab. Appl. **39**, 14–60.

[W] Williams, D. (1991) *Probability with Martingales*. Cambridge University Press, Cambridge.

[Z] Zhang, P. (1997) *Exotic Options - A Guide to Second Generation Options*. World Scientific Publ., Singapore.

SCHOOL OF MATHEMATICS, GEORGIA INSTITUTE OF TECHNOLOGY, ATLANTA, GA 30332-0160
E-mail address: kertz@math.gatech.edu

A correlation inequality for stable random vectors

Alexander Koldobsky

ABSTRACT. Let $X_1, ..., X_n$ and $Y_1, ..., Y_n$ be jointly q-stable symmetric random variables, $0 < q \leq 2$, so that, for some $k \in \mathbb{N}$, $1 \leq k < n$, the vectors $(X_1, ..., X_k)$ and $(X_{k+1}, ..., X_n)$ have the same distributions as $(Y_1, ..., Y_k)$ and $(Y_{k+1}, ..., Y_n)$, respectively, but Y_i and Y_j are independent for every choice of $1 \leq i \leq k$, $k+1 \leq j \leq n$. Let $(\mathbb{R}^n, \|\cdot\|)$ be an n-dimensional normed space such that $\|(u,v)\| = \|(u,-v)\|$ for every $u \in \mathbb{R}^k$, $v \in \mathbb{R}^{n-k}$. We prove that, for every $p \in [n-3, n)$, $\mathbb{E}(\|X\|^{-p}) \geq \mathbb{E}(\|Y\|^{-p})$.

1. Introduction

Let $X_1, ..., X_n$ and $Y_1, ..., Y_n$ be jointly q-stable symmetric random variables, $0 < q \leq 2$, so that, for some $k \in \mathbb{N}$, $1 \leq k < n$, the vectors $(X_1, ..., X_k)$ and $(X_{k+1}, ..., X_n)$ have the same distributions as $(Y_1, ..., Y_k)$ and $(Y_{k+1}, ..., Y_n)$, respectively, but Y_i and Y_j are independent for every choice of $1 \leq i \leq k$, $k+1 \leq j \leq n$. Let $B = (\mathbb{R}^n, \|\cdot\|)$ be an n-dimensional normed space.

The following result was established in [K, Th. 4] and later proved by Houdré [H, Remark 2.4] by different methods:

THEOREM A. *If $0 < p < n$, $\|x\|^{-p}$ is a positive definite distribution and the norm satisfies a symmetry condition $\|(u,v)\| = \|(u,-v)\|$ for every $u \in \mathbb{R}^k$, $v \in \mathbb{R}^{n-k}$, then $\mathbb{E}(\|X\|^{-p}) \geq \mathbb{E}(\|Y\|^{-p})$.*

Here we consider $\|x\|^{-p}$ as a tempered distribution. Recall that by L.Schwartz's generalization of Bochner's theorem (see [GV, p. 152]), a tempered distribution $f \in \mathcal{S}'(\mathbb{R}^n)$ is positive definite if and only if its Fourier transform $\hat{f}$ is a positive distribution. The latter means that $\langle \hat{f}, \phi \rangle \geq 0$ for every non-negative test function $\phi \in \mathcal{S}(\mathbb{R}^n)$.

It was shown in [K, Corollary 2(ii)] that if B is a subspace of L_r with $0 < r \leq 2$ then $\|x\|^{-p}$ is positive definite for every $0 < p < n$. However, if $B = \ell_r^n$, $2 < r \leq \infty$, $n \geq 3$ then $\|x\|^{-p}$ is positive definite if and only if $p \in [n-3, n)$. In particular, for every $p \in [n-3, n)$, $n \geq 3$

$$\mathbb{E}(\max_{i=1,...,n} |X_i|^{-p}) \geq \mathbb{E}(\max_{i=1,...,n} |Y_i|^{-p}).$$

1991 *Mathematics Subject Classification.* 60E07.
Part of this work was done when the author was visiting the Weizmann Institute of Science. Research supported in part by the NSF Grant DMS-9531594.

In this article we show that the latter inequality is a part of a more general result:

THEOREM 1. *For every $p \in [n-3, n)$ and every n-dimensional normed space $B = (\mathbb{R}^n, \|\cdot\|)$, $n \geq 3$, whose norm satisfies the symmetry condition $\|(u,v)\| = \|(u,-v)\|$ for each $u \in \mathbb{R}^k$, $v \in \mathbb{R}^{n-k}$, we have $\mathbb{E}(\|X\|^{-p}) \geq \mathbb{E}(\|Y\|^{-p})$.*

The proof is based on the fact that, for $p \in [n-3, n)$, the distribution $\|x\|^{-p}$ is positive definite for every n-dimensional normed space $B = (\mathbb{R}^n, \|\cdot\|)$. Theorem 1 follows immediately from this result and Theorem A.

2. Proof of Theorem 1

We use methods of convex geometry to prove positive definiteness of powers of the norm. Let $K = \{x \in \mathbb{R}^n : \|x\| \leq 1\}$ be the unit ball of the space B. For every unit vector $\xi \in S^{n-1}$ the parallel section function A_ξ in the direction of ξ is defined as a function on $\mathbb{R}$ so that for each $t \in \mathbb{R}$, $A_\xi(t)$ is the $(n-1)$-dimensional volume of the section of K by the hyperplane perpendicular to ξ and located at the distance t from the origin. We say that the space B is infinitely smooth if the restriction of the norm of B to the unit sphere S^{n-1} belongs to the space $C^\infty(S^{n-1})$ of infinitely differentiable functions on the sphere. If B is infinitely smooth then, for every $\xi \in S^{n-1}$, A_ξ is an infinitely differentiable function in a neighborhood of zero. For $\beta \in (-1, 0)$, the fractional derivative of order β of the function A_ξ at zero is defined by

$$(1) \qquad A_\xi^{(\beta)}(0) = \frac{1}{\Gamma(-\beta)} \int_0^\infty t^{-1-\beta} A_\xi(t) \, dt.$$

If $\beta \in (0, 2)$, $\beta \neq 1$ then

$$(2) \qquad A_\xi^{(\beta)}(0) = \frac{1}{\Gamma(-\beta)} \int_0^\infty t^{-1-\beta}(A_\xi(t) - A_\xi(0)) \, dt$$

(note that A_ξ is an even function so its first derivative at zero is equal to zero; for more on fractional derivatives see, for example, [GKS, Section 3]).

Our main tool is the following theorem, which was proved in [GKS, Th.2] in a more general form (for every $\beta \in \mathbb{C}$, $\Re(\beta) > -1$, $\beta \neq n-1$).

THEOREM B. *Let B be an infinitely smooth n-dimensional normed space, K is the unit ball of B, $\beta \in (-1, 2)$, β is not an integer. Then for every $\xi \in S^{n-1}$*

$$(3) \qquad A_\xi^{(\beta)}(0) = \frac{\cos\frac{\beta\pi}{2}}{\pi(n-\beta-1)} (\|x\|^{-n+\beta+1})^\wedge(\xi).$$

Note that this result in its general form was used in [GKS] as one of the major ingredients of the solution to the Busemann-Petty problem on sections of convex bodies.

THEOREM 2. *Let $B = (\mathbb{R}^n, \|\cdot\|)$ be an n-dimensional normed space. Then for every $p \in [n-3, n)$ the distribution $\|x\|^{-p}$ is positive definite.*

PROOF. First assume that B is infinitely smooth and p is not an integer. Put $\beta = n - p - 1 \in (-1, 2)$. We are going to show that $(\|x\|^{-n+\beta+1})^\wedge$ is a non-negative continuous function on S^{n-1}. Since this function is also homogeneous of degree $-\beta - 1$ on $\mathbb{R}^n$, $n \geq 3$, we deduce that it is non-negative and locally integrable on

$\mathbb{R}^n$. This would mean, in particular, that $(\|x\|^{-n+\beta+1})^\wedge = (\|x\|^{-p})^\wedge$ is a positive distribution, and $\|x\|^{-p}$ is positive definite.

Since the restriction of the norm to S^{n-1} is infinitely smooth and the volume of every section can be expressed in terms of the norm, it is easily seen that $A_\xi^{(\beta)}(0)$ is a continuous function of $\xi \in S^{n-1}$. By Theorem B, the restriction of the function $(\|x\|^{-n+\beta+1})^\wedge$ to the sphere is continuous on S^{n-1}.

Let us show that $(\|x\|^{-n+\beta+1})^\wedge$ is a non-negative function. First let $p \in (n-1, n)$. Then $\beta \in (-1, 0)$, so $\Gamma(-\beta) > 0$ and, by (1), $A_\xi^{(\beta)}(0) > 0$ for every $\xi \in S^{n-1}$. Also $\cos \frac{\beta\pi}{2} > 0$, so (3) implies non-negativity.

If $p \in (n-2, n-1)$ then $\beta \in (0, 1)$, so $\Gamma(-\beta) < 0$. But, since the unit ball K of the space B is a convex body, the function A_ξ has maximum at zero (the central section has maximal volume among all sections perpendicular to ξ; this follows for example from the Brunn-Minkowski theorem, see [S, Th. 6.1.1].) Therefore, the integral in (2) is less or equal to zero, and again $A_\xi^{(\beta)}(0) \geq 0$ for every $\xi \in S^{n-1}$. Also $\cos \frac{\beta\pi}{2} > 0$, so the result follows from (3).

If $p \in (n-3, n-2)$ then $\beta \in (1, 2)$, so $\Gamma(-\beta) > 0$. The integral in (2) is less or equal to zero for the same reason as in the case $\beta \in (0, 1)$, so $A_\xi^{(\beta)}(0) \leq 0$ for every $\xi \in S^{n-1}$. But now $\cos \frac{\beta\pi}{2} < 0$.

Now we have to free ourselves from the restrictions imposed in the beginning of the proof.

Suppose that B is not infinitely smooth. We can approximate the unit ball K of B in the Hausdorff metric by infinitely smooth convex bodies K_m, $m \in \mathbb{N}$ so that $K_m \subset K$ for every m. Let $\|\cdot\|_m$ be the norm on $\mathbb{R}^n$ with the unit ball K_m. Since $p < n$, the functions $\|x\|_m^{-p}$ are locally integrable on $\mathbb{R}^n$. Hence, for every test function ϕ, the functions $\|x\|_m^{-p}|\hat{\phi}(x)|$ are integrable on $\mathbb{R}^n$. Also these functions are majorated by an integrable function $\|x\|^{-p}|\hat{\phi}(x)|$. By definition of the Fourier transform of distributions and the dominated convergence theorem, for every non-negative test function ϕ and every $p \in [n-3, n)$,

$$\langle (\|x\|^{-p})^\wedge, \phi \rangle = \int_{\mathbb{R}^n} \|x\|^{-p}\hat{\phi}(x)\,dx =$$

$$\lim_{m \to \infty} \int_{\mathbb{R}^n} \|x\|_m^{-p}\hat{\phi}(x)\,dx = \lim_{m \to \infty} \langle (\|x\|_m^{-p})^\wedge, \phi \rangle \geq 0$$

because we have already proved that the distributions $\|x\|_m^{-p}$ are positive definite.

Finally, let us show that the statement of Theorem 2 is true for $p = n-3, n-2, n-1$. Suppose that $0 < p < n$ and p_i is a sequence of numbers that are not integers, belong to $[n-3, n)$ and $\lim_{i \to \infty} p_i = p$. We can assume that there exists $\epsilon > 0$ so that $0 < p_i < p + \epsilon < n$ for every i. Fix a non-negative test function ϕ. Then for every $i \in \mathbb{N}$ we have $\langle (\|x\|^{-p_i})^\wedge, \phi \rangle \geq 0$. Define a function g on $\mathbb{R}^n$ by $g(x) = \|x\|^{-p-\epsilon}|\hat{\phi}(x)|$ if $\|x\| \leq 1$, and $g(x) = |\hat{\phi}(x)|$ if $\|x\| > 1$. Since $\|x\|^{-p-\epsilon}$ is a locally integrable function, the function g is integrable on $\mathbb{R}^n$ and, for every $i \in \mathbb{N}$, $x \in \mathbb{R}^n$, we have $g(x) \geq \|x\|^{-p_i}|\hat{\phi}(x)|$. By the dominated convergence theorem,

$$\langle (\|x\|^{-p})^\wedge, \phi \rangle = \int_{\mathbb{R}^n} \|x\|^{-p}\hat{\phi}(x)\,dx =$$

$$\lim_{i \to \infty} \int_{\mathbb{R}^n} \|x\|^{-p_i}\hat{\phi}(x)\,dx = \lim_{i \to \infty} \langle (\|x\|^{-p_i})^\wedge, \phi \rangle \geq 0,$$

so $(\|x\|^{-p})^\wedge$ is a positive distribution, since we have already proved that $(\|x\|^{-p_i})^\wedge$ is positive for every i. □

Theorem 1 immediately follows from Theorems A and 2. If $n = 2$ the statement of Theorem 1 remains valid for $p \in (0, 2)$, and the inequality for the expectations reverses if $p \in (-\min(1, q), 0)$. To see that, note that every two-dimensional normed space embeds isometrically in L_1, and use [K, Corollary 2(ii)] and [K, Proposition 1]. Also, note that if $p \geq n$ in Theorem 1, then the function $\|x\|^{-p}$ is not locally integrable, and the expectations do not exist.

References

[H]   C. Houdré, *Comparison and deviation from a representation formula*, Stochastic Processes and Related Topics: In Memory of Stamatis Cambanis, to appear.

[GKS]   R. J. Gardner, A. Koldobsky, Th. Schlumprecht, *An analytic solution to the Busemann-Petty problem on sections of convex bodies*, preprint.

[GV]   I. M. Gelfand, N. Ya. Vilenkin, *Generalized functions 4. Applications of harmonic analysis*, Academic Press, New York, 1964.

[K]   A. Koldobsky, *Positive definite distributions and subspaces of L_{-p} with applications to stable processes*, Canad. Math. Bull., to appear.

[S]   R. Schneider, *Convex bodies: the Brunn-Minkowski theory*, Cambridge University Press, Cambridge, 1993.

DIVISION OF MATHEMATICS AND STATISTICS, UNIVERSITY OF TEXAS AT SAN ANTONIO, SAN ANTONIO, TX 78249, U.S.A.

E-mail address: koldobskmath.utsa.edu

A note on the maximal inequalities for VC classes

Rafał Latała

ABSTRACT. We investigate generalizations of Lévy and Lévy-Octaviani maximal inequalities. A general conjecture is stated and proved in several particular cases.

Introduction

The famous inequality due to Paul Lévy states that for any a.s. convergent series $\sum_{i=1}^{\infty} X_i$ of independent symmetric r.v.'s with values in some separable Banach space and $t > 0$ we have

$$(0.1) \qquad P(\max_n \|\sum_{i=1}^{n} X_i\| \geq t) \leq 2P(\|\sum_{i=1}^{\infty} X_i\| \geq t).$$

The generalization of the Lévy inequality to a nonsymmetric case is frequently called Lévy-Octaviani inequality. It states that for any a.s. convergent series $\sum_{i=1}^{\infty} X_i$ of independent Banach-space valued r.v.'s and $t > 0$

$$(0.2) \qquad P(\max_n \|\sum_{i=1}^{n} X_i\| \geq 3t) \leq 3 \max_n P(\|\sum_{i=1}^{n} X_i\| \geq t).$$

Both Lévy and Lévy-Octaviani inequalities have numerous applications (e.g. see [**KW**]). Roughly speaking they often enable one to reduce an almost sure statement to a statement in probability (like for example in the Itô-Nisio theorem).

However sometimes one has to consider more complicated sets of indices and methods of convergence of sums of random variables. Therefore it would be very useful to have suitable versions of the maximal inequalities (0.1) and (0.2) in a more general setting. The purpose of this article is to propose some version of the maximal inequality and to collect known facts and conjectures about it.

Some part of this paper consists of well known facts, which are already part of the folklore. Theorem 1.7 and parts of Proposition 1.1 were communicated to the author by S. Kwapień. However we were unable to find suitable references in the existing literature (Theorem 1.7 is stated in [**Kr**], but only with an idea of the

---

1991 *Mathematics Subject Classification.* 60G50, 60E15.
Key words and phrases. sums of independent random variables, maximal inequality, VC class.
Partially supported by KBN Grant (Poland) 2 P301 022 07 .

proof). Therefore, for completeness, we decided to include these statements in our paper together with the proofs.

Notation

We will denote by (ε_i) the Bernoulli sequence, i.e. a sequence of i.i.d. symmetric r.v.'s taking on values ± 1. A sequence of independent standard $\mathcal{N}(0,1)$ Gaussian random variables will be denoted by (g_i).

If (T,d) is a compact metric space and $\varepsilon > 0$ then $N(T,d,\varepsilon)$ will denote the minimal number of closed balls of radius ε that covers T.

1. General results

PROPOSITION 1.1. *Let $\mathcal{C}$ be a countable class of subsets of I and $(F, \|.\|)$ be a fixed separable Banach space. Then the following conditions are equivalent*

a): *There exists $K_1 < \infty$ such that for any sequence (X_i) of independent symmetric r.v.'s with values in F satisfying $\#\{i : X_i \neq 0\} < \infty$ a.s.*

$$\forall_{t>0} \ P(\max_{C \in \mathcal{C}} \|\sum_{i \in C} X_i\| \geq K_1 t) \leq K_1 P(\|\sum_{i \in I} X_i\| \geq t).$$

b): *There exists $K_2 < \infty$ such that for any sequence (X_i) of independent symmetric r.v.'s with values in F satisfying $\#\{i : X_i \neq 0\} < \infty$ a.s.*

$$E \max_{C \in \mathcal{C}} \|\sum_{i \in C} X_i\| \leq K_2 E \|\sum_{i \in I} X_i\|.$$

c): *There exists $K_3 < \infty$ such that for any sequence (v_i) of vectors in F satisfying $\#\{i : v_i \neq 0\} < \infty$*

$$\forall_{t>0} \ P(\max_{C \in \mathcal{C}} \|\sum_{i \in C} v_i \varepsilon_i\| \geq K_3 t) \leq K_3 P(\|\sum_{i \in I} v_i \varepsilon_i\| \geq t).$$

d): *There exists $K_4 < \infty$ such that for any sequence (v_i) of vectors in F satisfying $\#\{i : v_i \neq 0\} < \infty$*

$$E \max_{C \in \mathcal{C}} \|\sum_{i \in C} v_i \varepsilon_i\| \leq K_4 E \|\sum_{i \in I} v_i \varepsilon_i\|.$$

e): *There exists $K_5 < \infty$ such that for any sequence (X_i) of independent r.v.'s with values in F satisfying $\#\{i : X_i \neq 0\} < \infty$ a.s.*

$$\forall_{t>0} \ P(\max_{C \in \mathcal{C}} \|\sum_{i \in C} X_i\| \geq K_5 t) \leq K_5 \max_{C \in \mathcal{C} \cup \{I\}} P(\|\sum_{i \in C} X_i\| \geq t).$$

f): *There exists $K_6 < \infty$ such that for any sequence (X_i) of independent r.v.'s with values in F satisfying $\#\{i : X_i \neq 0\} < \infty$ a.s.*

$$E \max_{C \in \mathcal{C}} \|\sum_{i \in C} X_i\| \leq K_6 \max_{C \in \mathcal{C} \cup \{I\}} E \|\sum_{i \in C} X_i\|.$$

PROOF. Implications a)$\Rightarrow$c), b)$\Rightarrow$d) and c)$\Rightarrow$d) are obvious. By Fubini's Theorem easily follows that c)$\Rightarrow$a) and d)$\Rightarrow$b). Moreover for symmetric r.v.'s and $C \subset I$ we have $P(\|\sum_{i \in C} X_i\| \geq t) \leq 2P(\|\sum_{i \in I} X_i\| \geq t)$ and $E\|\sum_{i \in C} X_i\| \leq E\|\sum_{i \in I} X_i\|$, so e)$\Rightarrow$a) and f)$\Rightarrow$b). Thus to prove Proposition 1 it is enough to show that d)$\Rightarrow$c), a)$\Rightarrow$e) and b)$\Rightarrow$f).

d)$\Rightarrow$c). The result of Dilworth and Montgomery-Smith [**DM**] states that for any sequence of vectors w_i in some Banach space E

(1.1) $$P(\|\sum \varepsilon_i w_i\| \geq 2E\|\sum \varepsilon_i w_i\| + 6K_{1,2}^w((w_i), u)) \leq 4e^{-u^2/8}.$$

Here for a sequence of real numbers (a_i)

$$K_{1,2}((a_i), u) = \inf\{\sum |b_i| + t(\sum |c_i|^2)^{1/2} : a_i = b_i + c_i\}$$

and for a sequence of vectors (w_i)

$$K_{1,2}^w((w_i), u) = \sup\{K_{1,2}((w^*(w_i)), u) : w^* \in \text{Ext}(B_{E^*})\},$$

where $\text{Ext}(B_{E^*})$ denotes the set of extremal points in the unit ball of the dual space E^*. On the other side by [**MS**] there exists some universal constant $c > 0$ such that for any $w^* \in E^*$ and $u > 0$ we have

(1.2) $$P(|\sum \varepsilon_i w^*(w_i)| \geq cK_{1,2}((w^*(w_i)), u)) \geq ce^{-u^2/c}$$

Since $K_{1,2}((a_i), \lambda u) \leq \lambda K_{1,2}((a_i), u)$ for any $\lambda > 1, u > 0$, we get by (1.1) and (1.2)

(1.3) $$P(\|\sum \varepsilon_i w_i\| \geq K(E\|\sum \varepsilon_i w_i\| + t))$$
$$\leq K \max\{P(w^*(\sum \varepsilon_i w_i) \geq t) : w^* \in \text{Ext}(B_{E^*})\}$$

for some absolute constant K.

Let us notice that $\max_{C \in \mathcal{C}} \|\sum_{i \in C} \varepsilon_i v_i\| = \|\sum_{i \in I} \varepsilon_i w_i\|_E$ for a suitable choice of $w_i \in E := l^\infty(\mathcal{C}; F)$. Hence (1.3) implies that

(1.4) $$P(\max_{C \in \mathcal{C}} \|\sum_{i \in C} \varepsilon_i v_i\| \geq K(E\max_{C \in \mathcal{C}} \|\sum_{i \in C} \varepsilon_i v_i\| + t)) \leq KP(\|\sum_{i \in I} \varepsilon_i v_i\| \geq t).$$

We will show that c) holds for $K_3 = \max(8, (2K_4 + 1)K)$. If $t \geq \frac{1}{2}E\|\sum_{i \in I} \varepsilon_i v_i\|$ then by d) $E\max_{C \in \mathcal{C}} \|\sum_{i \in C} \varepsilon_i v_i\| + t \leq (2K_4 + 1)t$ and c) follows by (1.4). For $t \leq \frac{1}{2}E\|\sum_{i \in I} \varepsilon_i v_i\|$, by the Paley-Zygmund inequality (see [**Ka**], p.8) we get

$$P(\|\sum_{i \in I} \varepsilon_i v_i\| \geq t) \geq \frac{1}{4}\frac{(E\|\sum_{i \in I} \varepsilon_i v_i\|)^2}{E\|\sum_{i \in I} \varepsilon_i v_i\|^2} \geq \frac{1}{8}$$

and the inequality in c) is obvious.

a)$\Rightarrow$e) and b)$\Rightarrow$f). Let X_i' be an independent copy of X_i, then the variables $X_i - X_i'$ are symmetric, $E\|\sum_{i \in I}(X_i - X_i')\| \leq 2E\|\sum_{i \in I} X_i\|$ and $P(\|\sum_{i \in I}(X_i - X_i')\| \geq 2t) \leq 2P(\|\sum_{i \in I} X_i\| \geq t)$. Thus both implications are simple consequences of the next Lemma. □

LEMMA 1.2. *If $\max_{C \in \mathcal{C}} P(\|\sum_{i \in C} X_i\| \geq t/2) \leq 1/2$ then*

$$P(\max_{C \in \mathcal{C}} \|\sum_{i \in C} X_i\| \geq t) \leq 2P(\max_{C \in \mathcal{C}} \|\sum_{i \in C}(X_i - X_i')\| \geq \frac{t}{2}).$$

PROOF. Suppose that $\mathcal{C} = \{C_1, C_2, \ldots\}$ and for simplifying the notation let $Y_k = \sum_{i \in C_k} X_i, Y_k' = \sum_{i \in C_k} X_i'$ for $k = 1, 2, \ldots$. We have

$$P(\|Y_k\| \geq t, \|Y_i\| < t \text{ for } i < k, \max_k \|Y_k - Y_k'\| \leq \frac{t}{2})$$

$$\leq P(\|Y_k\| \geq t, \|Y_i\| < t \text{ for } i < k, \|Y_k'\| \geq \frac{t}{2})$$

$$\leq P(\|Y_k\| \geq t, \|Y_i\| < t \text{ for } i < k) \max_k P(\|Y_k\| \geq \frac{t}{2}).$$

Hence summing the above inequalities over k we get

$$P(\max_k \|Y_k\| \geq t) \leq P(\max_k \|Y_k - Y_k'\| \geq \frac{t}{2}) + P(\max_k \|Y_k\| \geq t) \max_k P(\|Y_k\| \geq \frac{t}{2})$$

and Lemma 1.2 follows. □

REMARK 1.3. The proof above shows that condition a)-d) are equivalent without the assumption of countability of the class $\mathcal{C}$. If the set I is countable then

$$P(\max_{C \in \mathcal{C}} \|\sum_{i \in C} X_i\| \geq t) = \lim_{n \to \infty} P(\max_{C \in \mathcal{C}} \|\sum_{i \in C \cap I_n} X_i\| \geq t),$$

where $I_1 \subset I_2 \subset \ldots$ is a sequence of finite subsets of I such that $\bigcup_{n=1}^{\infty} I_n = I$. Thus in the case of the countable set I, Proposition 1.1 holds for any class $\mathcal{C}$.

DEFINITION 1.4. In the sequel we will say that the class $\mathcal{C}$ of subsets of I *satisfies the maximal inequality in F* if any of conditions a)-f) of Proposition 1 holds true. If this is true for any separable Banach space F we will say that $\mathcal{C}$ *satisfies the maximal inequality* or that it is the *MI-class*.

It is therefore of interest to solve the following

Main Problem. Determine all classes $\mathcal{C}$ that satisfy the maximal inequality.

It has turned out that the following definition plays the crucial role for this problem

DEFINITION 1.5. We say that a class $\mathcal{C}$ of subsets of I *shatters* the set $A \subset I$ if

$$\{A \cap C : C \in \mathcal{C}\} = 2^A.$$

A class $\mathcal{C}$ is called a *Vapnik-Chervonenkis class* (or in short a *VC class*) *of order n* if it does not shatter any set of cardinality $n+1$ and it shatters some set of cardinality n. A class will be called a *VC class* if it is a VC class of some finite order.

For some properties and examples of VC classes see e.g. [**D1, D2, SY**].

PROPOSITION 1.6. *If $\mathcal{C}$ satifies the maximal inequality in some Banach space F then $\mathcal{C}$ is a VC class.*

PROOF. Obviously it is enough to prove Proposition for $F = \mathbb{R}$. Suppose that $\mathcal{C}$ shatters the set $A \subset I$ of cardinality n. Let

$$v_i = \begin{cases} 1 & \text{for} \quad i \in A \\ 0 & \text{for} \quad i \in I \setminus A \end{cases}.$$

Then

$$E|\sum \varepsilon_i v_i| \leq (E|\sum \varepsilon_i v_i|^2)^{1/2} = \sqrt{n}$$

and

$$E \max_{C \in \mathcal{C}} |\sum_{i \in C} \varepsilon_i v_i| = E \max_{B \subset A} |\sum_{i \in B} \varepsilon_i|$$

$$= E \max(\#\{i \in A : \varepsilon_i = 1\}, \#\{i \in A : \varepsilon_i = -1\}) \geq \frac{n}{2}.$$

Therefore if condition d) of Proposition 1.1 is satisfied then $\mathcal{C}$ does not shatter any set of cardinality $> 4K_4^2$. □

THEOREM 1.7. *The class $\mathcal{C}$ of subsets of I satisfies the maximal inequality in $\mathbb{R}$ if and only if $\mathcal{C}$ is a VC class.*

In the proof of this theorem we will use the following two results of Dudley (see [**LT**], Theorems 11.1 and 14.12)

THEOREM A. *Let $\psi_2(x) = e^{x^2} - 1$ and (X_t) be a random process on (T, d) such that*

$$E\psi_2(|X_t - X_s|/d(t,s)) \leq 1 \text{ for any } t, s \in T.$$

Then

$$E \sup_{s,t \in T} |X_t - X_s| \leq 12 \int_0^\infty \ln^{1/2} N(T, d, \varepsilon) d\varepsilon.$$

THEOREM B. *Let Q be a probability measure on I and $d_Q(A, B) = \sqrt{Q(A \triangle B)}$ for $A, B \subset I$. Then for any VC class $\mathcal{C}$ of order $\leq n$ and $\varepsilon \in (0, 1)$*

$$\ln N(\mathcal{C}, d_Q, \varepsilon) \leq K_B n(1 - \ln \varepsilon),$$

where K_B is an absolute constant.

PROOF OF THEOREM 1.7. One implication follows by Proposition 1.6. To prove the second, assume that $\mathcal{C}$ is a VC class of order $\leq n$ and we will prove the condition d) of Proposition 1.1. We may also assume that $\emptyset \in \mathcal{C}$. Let v_i be fixed real numbers with $\sum v_i^2 = 1$ and $X_A = \sum_{i \in A} \varepsilon_i v_i$ for $A \subset I$. Let us also define the probability measure Q on I by the formula $Q(A) = \sum_{i \in A} v_i^2$ and a distance d on $\mathcal{C}$ by $d(A, B) = (Q(A \div B))^{1/2}$. Then $N(\mathcal{C}, d, \varepsilon) = 1$ for $\varepsilon > 1$. By the properties of Rademacher sums (see [**LT**], sect.4.1) there exists universal constant K such that $\|X_A\|_{\psi_2} \leq K(\sum_{i \in A} v_i^2)^{1/2}$, so $E\psi_2((X_A - X_B)/Kd(A, B)) \leq 1$. Therefore by Theorem A and B

$$E \sup_{C \in \mathcal{C}} |\sum_{i \in C} \varepsilon_i v_i| \leq E \sup_{C, C' \in \mathcal{C}} |X_C - X_{C'}| \leq 12 \int_0^\infty \ln^{1/2} N(\mathcal{C}, d, \varepsilon/K) d\varepsilon$$

$$\leq 12 \sqrt{K_B} \sqrt{n} \int_0^K (1 - \ln \varepsilon + \ln K)^{1/2} d\varepsilon = \tilde{K} \sqrt{n}.$$

The Theorem 1.7 follows if we notice that $E|\sum_{i \in I} \varepsilon_i v_i| \geq (\sum v_i^2)^{1/2}/\sqrt{2}$ by the Khinchine inequality. □

Theorem 1.7 and Proposition 1.6 suggest that the following conjecture is reasonable

CONJECTURE 1.8. *A class $\mathcal{C}$ satisfies the maximal inequality if and only if $\mathcal{C}$ is a VC class.*

2. Some partial results for VC and MI-classes

Using Theorem 1.7 and Talagrands majorizing measure theorem, L. Krawczyk proved in [**Kr**] that if $\mathcal{C}$ is a VC class then conditions (a) and (b) holds if we additionaly assume that X_i are Gaussian vectors. This was slightly generalized in [**L**] to the following Theorem.

THEOREM 2.1. *Let $(X_i)_{i \in I}$ be a sequence of symmetric real random variables with logarithmically concave tails i.e. such that the functions $N_i(t) = -\ln P(|X_i| > t)$ are convex on $[0, \infty)$ and such that*

$$\forall_{t>0} \ N_i(2t) \leq A N_i(t)$$

for some constant $A < \infty$. Then for any VC class $\mathcal{C}$ of subsets of I of order $\leq n$ there exists a constant K, which depends only on A and n such that for any sequence of vectors v_i in some Banach space for which the sum $\sum v_i X_i$ is a.e. convergent, the following inequality holds

$$E \sup_{C \in \mathcal{C}} \| \sum_{i \in C} v_i X_i \| \leq K E \| \sum_{i \in I} v_i X_i \|.$$

REMARK 2.2. Using concentration properties of logconcave measures one may prove in the similar way as in the proof of implication d)⇒c) of Proposition 1 that under the assumptions of Theorem 2.1

$$P(\sup_{C \in \mathcal{C}} \| \sum_{i \in C} v_i X_i \| \geq \tilde{K} t) \leq \tilde{K} P(\| \sum_{i \in I} v_i X_i \| \geq t)$$

for any $t > 0$, where $\tilde{K}$ is a constant depending only on A and n.

COROLLARY 2.3. *Let F be a separable Banach space with finite cotype. Then every VC class $\mathcal{C}$ satisfies the maximal inequality in F.*

PROOF. Let $v_i \in F$ be as in condition (d) of Proposition 1.1. Then since F has finite cotype

$$E\| \sum_{i \in I} v_i g_i \| \leq A E \| \sum_{i \in I} v_i \varepsilon_i \|,$$

where A is a constant depending only on F. By the contraction principle

$$E \max_{C \in \mathcal{C}} \| \sum_{i \in C} v_i \varepsilon_i \| \leq \sqrt{\frac{\pi}{2}} E \max_{C \in \mathcal{C}} \| \sum_{i \in C} v_i g_i \|$$

and condition (d) immediately follows by the result of Krawczyk. □

REMARK 2.4. Proofs of the result of Krawczyk and Theorem 2.1 are based on general theorems about geometric conditions for subsets T of l^2 equivalent to the boundedness of processes $(\sum t_i X_i)_{t \in T}$. For Rademacher processes an important conjecture (for some partial results see [**T2**]) states that if for some $T \subset l^2$, $E \sup_{t \in T} \sum \varepsilon_i t_i < \infty$ then $T \subset U + K B_1$ for some $K < \infty$, where B_1 denotes a ball in l^1 and U is such that $E \sup_{t \in U} \sum t_i g_i < \infty$. It is not hard to check that the above conjecture immediately implies our conjecture about VC classes.

DEFINITION 2.5. Let $\mathcal{C}_1$ and $\mathcal{C}_2$ be two families of subset of I. Then we may define the following families

$$\mathcal{C}_1^c = \{ I \setminus C_1 : C_1 \in \mathcal{C}_1 \}$$

$$\mathcal{C}_1 \wedge \mathcal{C}_2 = \{ C_1 \cap C_2 : C_1 \in \mathcal{C}_1, C_2 \in \mathcal{C}_2 \}$$

and

$$\mathcal{C}_1 \vee \mathcal{C}_2 = \{ C_1 \cup C_2 : C_1 \in \mathcal{C}_1, C_2 \in \mathcal{C}_2 \}.$$

PROPOSITION 2.6. *Suppose that $\mathcal{C}_1$ and $\mathcal{C}_2$ are MI-classes. Then also the families $\mathcal{C}_1^c$, $\mathcal{C}_1 \wedge \mathcal{C}_2$ and $\mathcal{C}_1 \vee \mathcal{C}_2$ satisfy the maximal inequality.*

PROOF. Since $\|\sum_{i \in I \setminus C} X_i\| \leq \|\sum_{i \in I} X_i\| + \|\sum_{i \in C} X_i\|$ by the triangle inequality, we immediately get that $\mathcal{C}_1^c$ is a MI-class. Moreover $\mathcal{C}_1 \vee \mathcal{C}_2 = (\mathcal{C}_1^c \wedge \mathcal{C}_2^c)^c$, so it is enough to prove that $\mathcal{C}_1 \wedge \mathcal{C}_2$ satisfies the maximal inequality. We will check the condition (b). Let $(F, \|.\|)$ be a given Banach space and $\tilde{F} = l^\infty(\mathcal{C}, F)$. Suppose that $\mathcal{C}_1$ and $\mathcal{C}_2$ satisfy (b) in F and $\tilde{F}$ respectively with constants K and $\tilde{K}$. Let $\tilde{X}_i$ be independent r.v.'s with values in $\tilde{F}$ defined by the formula

$$\tilde{X}_i(C) = \begin{cases} X_i & \text{for } i \in C \\ 0 & \text{for } i \notin C. \end{cases}$$

Then for $A \subset I$, we have

$$\|\sum_{i \in A} \tilde{X}_i\|_{\tilde{F}} = \sup_{C_1 \in \mathcal{C}_1} \|\sum_{i \in A \cap C_1} X_i\|.$$

Thus

$$E \max_{C \in \mathcal{C}_1 \wedge \mathcal{C}_2} \|\sum_{i \in C} X_i\| = E \max_{C_2 \in \mathcal{C}_2} \|\sum_{i \in C_2} \tilde{X}_i\|_{\tilde{F}} \leq \tilde{K} E \|\sum_{i \in I} \tilde{X}_i\|_{\tilde{F}}$$

$$= \tilde{K} E \max_{C_1 \in \mathcal{C}_1} \|\sum_{i \in C_1} X_i\| \leq K\tilde{K} E \|\sum_{i \in I} X_i\|.$$

$\square$

PROPOSITION 2.7. *Every VC class of order 1 satisfies the maximal inequality.*

PROOF. Following the notation of [S] we will call the family $\mathcal{F}$ of subsets of I a chain if it is linearly ordered by the inclusion, i.e. for each $A, B \in \mathcal{F}$ either $A \subset B$ or $B \subset A$. Families of the form $\mathcal{F}_1 \wedge \mathcal{F}_2$, where $\mathcal{F}_1$ and $\mathcal{F}_2$ are chains, will be called 2-chains. Smoktunowicz in [S] proved that if $\mathcal{C}$ is a VC class of order 1 then $\mathcal{C} \subset \mathcal{G}_1 \vee \mathcal{G}_2^c$ for some 2-chains $\mathcal{G}_1$ and $\mathcal{G}_2$. Since every chain is a MI-class by the Lévy inequality, Proposition 2.7 follows by Proposition 2.6. $\square$

By Propositions 2.6 and 2.7 we immediately get the following

COROLLARY 2.8. *Suppose that $\mathcal{C}'$ is a family of subsets of I that is obtained from some VC classes of order 1 by finitely many operations ^{c}, $\vee$ and $\wedge$. Then any subfamily $\mathcal{C} \subset \mathcal{C}'$ satisfies the maximal inequality.*

Unfortunately even very simple VC classes are not of the form described in the above Corollary. A. Smoktunowicz in [S] showed that the family of all lines in Z^2 does not have such form. As follows from Corollary 2.12 below, the lattest family is an MI-class, so Corollary 2.8 does not describe all families that satisfy the maximal inequality.

DEFINITION 2.9. In the last part of the paper we will consider the classes $\mathcal{C}$ of subsets of I, which satisfy the additional condition

(2.1) $$\forall_{A,B \in \mathcal{C}} \; A \neq B \Rightarrow \#(A \cap B) \leq 1.$$

Such classes will be called *1-disjoint*.

We will also denote for a fixed sequence of vectors v_i and $A \subset I$ by X_A the variable $\sum_{i \in A} v_i \varepsilon_i$.

LEMMA 2.10. *If $M \geq 2\max_i \|v_i\|$, class $\mathcal{C}$ is 1-disjoint and $A_1,\ldots,A_n \in \mathcal{C}$ are such that for some $t > 0$*

$$P(\|X_{A_k}\| \geq M) \geq t \text{ for } k = 1,\ldots,n,$$

then there exist disjoint subsets $B_1,\ldots,B_m \subset I$ such that $m \geq \sqrt[3]{n}$ and

$$P(\|X_{B_k}\| \geq M/2) \geq t/2 \text{ for } k = 1,\ldots,m.$$

PROOF. In this proof we will say that the set A_k is *good* if

(2.2) $\qquad\qquad \forall_{C \subset A_k} \#C \leq \sqrt[3]{n} \Rightarrow P(\|X_C\| \geq M/2) \leq t/2.$

Let us notice that the condition (2.2) also implies by the triangle inequality that $P(\|X_{A_k \setminus C}\| \geq M/2) \geq t/2$. We will consider 3 cases

CASE 1. Among $A_1,\ldots,A_n$ there are $m \geq \sqrt[3]{n}$ good sets, say $A_1,\ldots,A_m$. Without loss of generality we may assume that $m < \sqrt[3]{n}+1$. If we put $B_1 = A_1$ and $B_i = A_i \setminus (\bigcup_{j<i} A_j)$ for $1 < i \leq m$ we get by (2.1) that $\#(A_i \setminus B_i) \leq i-1 \leq \sqrt[3]{n}$. Hence we get the result in this case by the definition of good sets.

CASE 2. There exists $i \in I$ such that $\#\{k : i \in A_k\} \geq \sqrt[3]{n}$. Without loss of generality we may assume that $i \in A_1 \cap \ldots \cap A_m$ with $m \geq \sqrt[3]{n}$. We put in this case $B_k = A_k \setminus \{i\}$ and notice that $\|X_{B_k}\| \geq \|X_{A_k}\| - \|v_i\| \geq \|X_{A_k}\| - M/2$. Sets B_k are disjoint by the property (2.1).

CASE 3. There are less then $\sqrt[3]{n}$ good sets A_k and $\#\{k : i \in A_k\} < \sqrt[3]{n}$ for all $i \in I$. We have more than $n - \sqrt[3]{n}$ not good sets, let A_{i_1} be one of them. We may then find $B_1 \subset A_{i_1}$ with $\#B_1 \leq \sqrt[3]{n}$ and $P(\|X_{B_1}\| \geq M/2) \geq t/2$. At most $\sqrt[3]{n}\#B_1$ sets A_i have nonempty intersection with B_1. So we have more than $n - \sqrt[3]{n} - \sqrt[3]{n^2}$ not good sets disjoint from B_1, let A_{i_2} be one of them. Then we may find $B_2 \subset A_{i_2}$ with $\#B_2 \leq \sqrt[3]{n}$ and $P(\|X_{B_2}\| \geq M/2) \geq t/2$. Continuing in this way completes the proof. $\square$

COROLLARY 2.11. *Suppose that $M \geq 8E\|X_I\|$ and class $\mathcal{C}$ is 1-disjoint, then*

$$\sum_{A \in \mathcal{C}} (P(\|X_A\| \geq M))^4 \leq 2^{14} P(\|X_I\| \geq M/2).$$

PROOF. Suppose that there exist $A_1,\ldots,A_n \in \mathcal{C}$ such that $P(\|X_{A_k}\| \geq M) \geq t$ for all k. Then by Lemma 2.10 we may find disjoint subsets $B_1,\ldots,B_m \subset I$ with $m \geq \sqrt[3]{n}$ and $P(\|X_{B_k}\| \geq M/2) \geq t/2$. But by the Lévy inequality

$$P(\max \|X_{B_k}\| \geq M/2) \leq 2P(\|X_I\| \geq M/2) \leq 1/2,$$

so

$$t\sqrt[3]{n}/2 \leq \sum P(\|X_{B_k}\| \geq M/2) \leq 4P(\|X_I\| \geq M/2)$$

and $\sqrt[3]{n} \leq 8P(\|X_I\| \geq M/2)/t$. Therefore we obtain

$$\sum_{A \in \mathcal{C}} (P(\|X_A\| \geq M))^4 \leq 16 \sum_{n=1}^{\infty} 2^{-4n} \#\{A \in \mathcal{C} : P(\|X_A\| \geq M) \geq 2^{-n}\}$$

$$\leq 2^{13} P(\|X_I\| \geq M/2) \sum_{n=1}^{\infty} 2^{-4n} 2^{3n} \leq 2^{14} P(\|X_I\| \geq M/2).$$

$\square$

COROLLARY 2.12. *There exists a universal constant C such that for any 1-disjoint class $\mathcal{C}$*

$$\sum_{A \in \mathcal{C}} E\|X_A\| I_{\{\|X_A\| \geq CE\|X_I\|\}} \leq CE\|X_I\|.$$

In particular

$$E \max_{A \in \mathcal{C}} \|X_A\| \leq 2CE\|X_I\|,$$

so the maximal inequality holds for any VC class satisfying (2.1).

PROOF. By the properties of Rademacher sums (cf [**Ka**]) we have

$$P(\|X_A\| \geq 4M) \leq C_1(P(\|X_A\| \geq M))^4$$

for some constant $C_1 < \infty$. Therefore by the previous Corollary we obtain for $M \geq 8E\|X_I\|$

$$\sum_{A \in \mathcal{C}} P(\|X_A\| \geq 4M) \leq 2^{14} C_1 P(\|X_I\| \geq M/2).$$

Corollary follows by integration the above inequality with respect to M. □

REMARK 2.13. All the above results remain true (with a change of constants) if we substitute (2.1) by the more general condition

$$\forall_{A,B \in \mathcal{C}} \ A \neq B \Rightarrow \#(A \cap B) \leq m,$$

where m is a fixed positive integer.

ACKNOWLEDGMENTS. The main part of this research was carried out when the author was working in the Institute of Mathematics Polish Academy of Science. The author would like to thank Prof. Stanisław Kwapień for introducing him into the subject and many useful discussion and suggestions.

References

[DM] S. J. Dilworth and S. J. Montgomery-Smith, *The distribution of vector-valued Rademacher series*, Ann. Prob. 21 (1993), 2046-2052.

[D1] R. M. Dudley, *A course on empirical processes*, École d'été de probabilités de Saint-Flour, XII–1982, 1-142, Lecture Notes in Math. 1097, Springer, Berlin-New York, 1984

[D2] ———, *The structure of some Vapnik-Červonenkis classes*, Proceedings of the Berkeley conference in honor of Jerzy Neyman and Jack Kiefer (Berkeley, Calif. 1983), 495-508, Wadsworth, Belmont, Calif., 1985

[Ka] J.-P. Kahane, *Some random series of functions*, Cambridge University Press, 1985, 2nd ed.

[Kr] L. Krawczyk *Maximal Inequality for Gaussian Vectors*, Bull. Acad. Polon. Sci. Math. 44 (1996), 157-160.

[KW] S. Kwapień, W. A. Woyczyński, *Random Series and Stochastic Integrals: Single and Multiple*, Birkhauser, Boston 1992.

[L] R. Latała, *Sudakov minoration principle and supremum of some processes*, Geom. and Funct. Anal. 7 (1997), 936-953.

[LT] M. Ledoux, M. Talagrand, *Probability in Banach Spaces*, Springer, Berlin Heidelberg 1991.

[MS] S. J. Montgomery-Smith, *The distribution of Rademacher sums*, Proc. Amer. Math. Soc. 109 (1990), 517-522

[S] A. Smoktunowicz, *A remark on Vapnik-Chervonenkis classes*, Colloq. Math. 74 (1997), 93-98.

[SY] G. Stengle, J. E. Yukich *Some new Vapnik-Chervonenkis classes*, Ann. Statist. 17 (1989), 1441-1446.

[T1] M. Talagrand, *Regularity of Gaussian processes*, Acta Math. 159 (1987), 99-149.

[T2] ——, *Construction of majorizing measures, Bernoulli processes and cotype*, Geom. and Funct. Anal. 4 (1994), 660-717.

INSTITUTE OF MATHEMATICS, WARSAW UNIVERSITY, BANACHA 2, 02-097 WARSZAWA, POLAND
E-mail address: rlatala@mimuw.edu.pl

Comparison of moments via Poincaré-type inequality

Krzysztof Oleszkiewicz

ABSTRACT. In this paper we establish some inequalities between moments of sums of independent random vectors. To prove these results we attribute to each random vector some linear operator and from its spectral properties we deduce a Poincaré-type inequality. In some cases we get best possible constants.

Notation and preliminary results

We begin with some notation and simple observations, then we formulate our main theorem and we prove it in its most abstract setting. Finally we use this theorem in some concrete cases to obtain some new inequalities and generalize ones known before.

General assumptions. Let X be a symmetric random vector with values in a separable Banach space F and with finite second moment i.e. such that $E\|X\|^2 < \infty$. We denote the distribution of X by μ. In many cases it is convenient to assume that X is simple i.e. it has only finite number of values or more generally that it takes values in some finite-dimensional linear space and is bounded. As we can approximate any random vector by simple ones such reduction will usually suit our needs. Usually we can treat all Banach spaces as if they were finite-dimensional and the constants obtained will not depend on dimension. All considered Banach spaces have real scalar fields as well as L^p-spaces; application to the complex Banach spaces making use of their complex structure was not investigated. Let us recall that functions from L^p are identified if they differ only on the set of measure zero - the word **countable** seems necessary whenever it appears, also to assure measurability. It should be also pointed out that we use symmetry of X only once in our argumentation; except this only moment assumption that $EX = 0$ suffices. Standard symmetrization techniques allow us to deduce moment inequalities for mean-zero random vectors from the inequalities for symmetric ones (although the constants obtained for mean-zero vectors by this method are slightly worse than in the symmetric case).

1991 *Mathematics Subject Classification.* primary 47A57, 60G50, secondary 46B09, 47B25.

Key words and phrases. comparison of moments, sums of independent random variables, Poincaré inequality, random vectors, positivity preserving property.

The work was partially supported by KBN Grant 2 P301 022 07 and it was completed during the author's post-doctoral fellowship at the Weizmann Institute of Science.

© 1999 American Mathematical Society

Class $\mathcal{L}(X)$. Let us define a class of linear operators $\mathcal{L}(X)$. For a linear operator L defined on a dense linear subspace $Dom(L)$ of $L^2(F,\mu)$, positive definite (i.e. $\langle f, Lf \rangle \geq 0$ for any $f \in Dom(L)$) and **self-adjoint** with respect to the standard inner product we will say that $L \in \mathcal{L}(X)$ if and only if it satisfies the following four conditions:
1. constant functions belong to $Dom(L)$ and they are eigenfunctions of L with eigenvalue 0,
2. all linear bounded functionals on F belong to $Dom(L)$ and they are eigenfunctions of L with eigenvalue 1,
3. $f \in Dom(L^{1/2})$ implies $|f| \in Dom(L^{1/2})$
and $\langle L^{1/2}f, L^{1/2}f \rangle \geq \langle L^{1/2}|f|, L^{1/2}|f| \rangle$,
4. $L(\tilde{f}) = (\widetilde{Lf})$ and $f \in Dom(L)$ implies $\tilde{f} \in Dom(L)$, where $\tilde{f}(x) = f(-x)$ for any function $f \in Dom(L)$.

In the case of a simple random vector instead of the third condition we can require that $(Lf)(a) \geq 0$ in each point a in which f has the global maximum.

Alternatively one can consider a semigroup $T_t = e^{-tL}$ ($t \geq 0$) of self-adjoint contractions defined on whole $L^2(F,\mu)$ and require that
1. $T_t \mathbf{1} = \mathbf{1}$,
2. $T_t x^* = e^{-t} x^*$ for any $x^* \in F^*$,
3. $f \geq 0$ implies $T_t f \geq 0$
4. $T_t(\tilde{f}) = (\widetilde{T_t f})$ for $t \geq 0$.

For the non-trivial part of the proof of equivalence of these two sets of conditions see for example Theorem 1. 3. 2 of [**D**].

Preliminary results. Let us note some consequences of our assumptions: for $f_1, f_2, \ldots \in L^2(F,\mu)$ we have $T_t \sup_i(f_i) \geq \sup_i(T_t f_i)$ and

$$E(T_t f)(X) = \langle T_t f, \mathbf{1} \rangle = \langle f, T_t \mathbf{1} \rangle = E f(X)$$

for any $f \in L^2(F,\mu)$. A simple observation that the norm of the space F is a supremum of some countable family of bounded functionals on F yields the following

COROLLARY 0.1. *If X is a random vector with values in a separable Banach space F and its second moment is finite then the function $g : F \longrightarrow \mathbf{R}$ given by $g(x) = \|x\|$ satisfies the inequality*

$$(T_t g)(X) \geq e^{-t} g(X)$$

(with probability 1) for any $t \geq 0$, where $T_t = e^{-tL}$ for some $L \in \mathcal{L}(X)$.

In a similar way we can prove that $E\phi((T_t f)(X)) \leq E\phi(f(X))$ for any convex function $\phi : \mathbf{R} \longrightarrow \mathbf{R}_+$ - note that ϕ is a supremum of a countable family of affine functions and that for any affine function φ we have $T_t(\varphi(f)) = \varphi(T_t(f))$; therefore T_t is a family of contractions also in L^p-norm for $1 \leq p \leq \infty$.

Let us introduce for $\delta > 0$ a new non-negative function $g_\delta = e^{-\delta L} g$. It has some nice properties. As indicated above,

$$e^{-\delta} \|g\|_p \leq \|g_\delta\|_p \leq \|g\|_p$$

for $p \geq 1$. Moreover $g_\delta \in Dom(L)$ because $Le^{-\delta L}$ is a bounded operator ($\lambda e^{-\delta \lambda} \leq \frac{1}{e\delta}$ for $\lambda \geq 0$). From Corollary 0.1 and the fact that $e^{-\delta L}$ preserves non-negativity of functions we infer that $e^{-tL} g_\delta \geq e^{-t} g_\delta$. Differentiating at $t=0$ we get

COROLLARY 0.2. *Under assumptions of Corollary 0.1 the inequality*

$$Lg_\delta \leq g_\delta$$

holds (pointwise, $\mu-$ almost everywhere) for any $\delta > 0$.

We also need a simple lemma:

LEMMA 0.3. *If $p \geq \frac{1}{2}$ and $x, y \geq 0$ then*

$$x^p y^p \geq \frac{p^2}{4p-2}(xy^{2p-1} + x^{2p-1}y) - \frac{(p-1)^2}{4p-2}(x^{2p} + y^{2p}).$$

PROOF. We have to show that $u(t) = (p-1)^2 t^{2p} - p^2 t^{2p-1} + (4p-2)t^p - p^2 t + (p-1)^2 \geq 0$ for $t \geq 1$. To prove this it is enough to show that $u(1) = 0, u'(1) = 0$, and $u''(t) \geq 0$ for $t \geq 1$ which follows easily from Young's inequality (the case $p < 1$ needs a separate but very similar treatment). □

Poincaré inequality and spectral properties of operator L. Let $\lambda_2(L)$ be the infimum of the spectrum of operator L **restricted** to the subspace of $L^2(F, \mu)$ consisting of functions orthogonal to all functionals from F^* and to the constant function **1**. In other words let $\lambda_2(L)$ be the greatest constant c such that a Poincaré-type inequality

$$D^2 f(X) = Ef(X)^2 - (Ef(X))^2 \leq \frac{1}{c}Ef(X)(Lf)(X)$$

is satisfied for any $f \in Dom(L)$ such that $Ex^*(X)f(X) = 0$ for all $x^* \in F^*$. Then we define $\lambda_2(X) = \sup_{L \in \mathcal{L}(X)} \lambda_2(L)$ and $w(X) = \min(\lambda_2(X), 2)$.

Let us note that thanks to the assumption $L(\tilde{f}) = (\widetilde{Lf})$, if L has a complete orthogonal system of eigenfunctions then we can also find an orthogonal basis of $L^2(F, \mu)$ consisting of symmetric (even) and anti-symmetric (odd) eigenfunctions of operator L only; therefore we will assume that all bases considered throughout the rest of this paper consist of symmetric and anti-symmetric functions only. In the general case for $L \in \mathcal{L}(X)$ we can split $L^2(F, \mu)$ into three orthogonal invariant subspaces C_L, A_L and S_L consisting of constant functions, anti-symmetric functions and symmetric mean-zero functions respectively.

Notice that for a symmetric function $g \in L^2(F, \mu)$ also $e^{-\delta L}g$ will be symmetric.

1. Main result

Now we are in position to state our main theorem:

THEOREM 1.1. *Let $X_1, X_2, \ldots, X_n$ be independent symmetric random vectors with values in Banach spaces $F_1, F_2, \ldots, F_n$ respectively, having finite second moments and let $A_i : F_i \longrightarrow B$ be bounded linear transformations into some separable Banach space B for $i = 1, 2, \ldots, n$. Let $S = \|\sum_{i=1}^n A_i(X_i)\|$ and let $w = \min_{1 \leq i \leq n} w(X_i)$. If $w > 1$ then for $p \in [1, w + \sqrt{w^2 - w})$ the inequality*

$$\|S\|_{2p} \leq C_{2p,p}(w)\|S\|_p,$$

holds true, where $C_{2p,p}(w) = \left(\frac{(2p-1)w}{(2p-1)w - p^2}\right)^{\frac{1}{2p}}$.

PROOF. Let us consider a new random vector $X = (X_1, X_2, \ldots, X_n)$ with values in $F = F_1 \times F_2 \times \ldots \times F_n$. We introduce a natural product measure μ on F and a seminorm function $g : F \longrightarrow \mathbf{R}_+$ given by the formula $g(x_1, \ldots, x_n) = \|A_1(x_1) + \ldots + A_n(x_n)\|$. Since $L_i \in \mathcal{L}(X_i)$ for $i = 1, 2, \ldots, n$, we can define $L \in \mathcal{L}(X)$ by $L = L_1 \otimes Id_2 \otimes \ldots \otimes Id_n + Id_1 \otimes L_2 \otimes Id_3 \otimes \ldots \otimes Id_n + \ldots + Id_1 \otimes Id_2 \otimes \ldots Id_{n-1} \otimes L_n$. It is a well-known fact that $\sigma(L) = \sigma(L_1) + \ldots + \sigma(L_n)$. If L has a complete orthogonal system of eigenfunctions then eigenvalues $\lambda_1 + \ldots + \lambda_n$ correspond to eigenfunctions of form $f_1 \otimes \ldots \otimes f_n$, where f_i are eigenfunctions of L_i with eigenvalues respectively λ_i. Therefore we see that a **symmetric** (even) function $h \in L^2(F, \mu)$ given by the formula $h(x_1, \ldots, x_n) = g_\delta(x_1, \ldots, x_n)^p$ can be written down in the form

$$h = Eh(X) + \sum_\alpha h_\alpha,$$

where h_α are symmetric and $Lh_\alpha = \lambda_\alpha h_\alpha$ with all $\lambda_\alpha \geq \hat{w} = \min_{1 \leq i \leq n}(2, \lambda_2(L_i))$ because we have either to tensorize at least two non-constant eigenfunctions or to take a symmetric non-constant one to get a non-constant symmetric function. In the general case similar reasoning works. We will sketch it for $n = 2$. It is easy to see that $C_L = C_{L_1} \otimes C_{L_2}$ whereas

$$A_L = (A_{L_1} \otimes C_{L_2}) \oplus (A_{L_1} \otimes S_{L_2}) \oplus (C_{L_1} \otimes A_{L_2}) \oplus (S_{L_1} \otimes A_{L_2})$$

and

$$S_L = (C_{L_1} \otimes S_{L_2}) \oplus (S_{L_1} \otimes C_{L_2}) \oplus (S_{L_1} \otimes S_{L_2}) \oplus (A_{L_1} \otimes A_{L_2}).$$

Now it suffices to tensorize Poincaré inequality on each of the nine pieces separately and deduce lower bounds

$$\inf \sigma(L|_{S_L}) \geq \min(2, \inf \sigma(L_1|_{S_{L_1}}), \inf \sigma(L_2|_{S_{L_2}}))$$

and

$$\inf \sigma(L|_{A_L}) \geq 1.$$

Hence we can proceed by induction on n to arrive at

$$\inf \sigma(L|_{S_L}) \geq \min_{1 \leq i \leq n}(2, \inf \sigma(L_i|_{S_{L_i}})) = \hat{w}.$$

We leave details to the reader.

Anyway,

$$\|L^{1/2}h\|_2^2 \geq \hat{w}Eh(X)^2 - \hat{w}(Eh(X))^2 =$$
$$= \hat{w}Eg_\delta(X)^{2p} - \hat{w}(Eg_\delta(X)^p)^2.$$

Note that the expression $\|L^{1/2}h\|_2^2$ is always definite (because L is positive definite) although *a priori* it can take on value $+\infty$ if h does not belong to $Dom(L^{1/2})$. On the other hand from Lemma 0.3 we deduce that

$$\|L^{1/2}(g_\delta^p)\|_2^2 = Eg_\delta(X)^p(L(g_\delta^p))(X) \leq \frac{p^2}{2p-1}Eg_\delta(X)^{2p-1}(Lg_\delta)(X) \quad (*)$$

(the middle term makes no sense if g_δ^p does not belong to $Dom(L)$ but it is not necessary to consider it; it was included just to underline the fact that $(*)$ is linear with respect to L). To see this note that in the cone of symmetric operators L such that e^{-tL} preserves (for all $t \geq 0$) non-negativity of functions the operators of form

$$(L_{v,u}f)(x) = E((\mathbf{1}_v(x)\mathbf{1}_u(X) + \mathbf{1}_u(x)\mathbf{1}_v(X))(f(x) - f(X))) =$$

$$\delta_{v,x}(f(v) - f(u))P(X = u) + \delta_{u,x}(f(u) - f(v))P(X = v),$$

where v and u are values of X, are extremal. Therefore, since each $L \in \mathcal{L}(X)$ is a linear combination of operators of such form with non-negative coefficients, it is enough to prove the inequality $(*)$ for them - and this follows from Lemma 0.3 if we put $x = g_\delta(v)$ and $y = g_\delta(u)$. This argument works for a simple random vector X; for the proof of $(*)$ in a general situation see the classical Stroock-Varopoulos inequality [S], [V].

Now we use Corollary 0.2 (which was formulated for the function g given by the norm but works also for a continuous seminorm) to obtain

$$\|L^{1/2}h\|_2^2 = \|L^{1/2}(g_\delta^p)\|_2^2 \leq$$

$$\leq \frac{p^2}{2p-1} Eg_\delta(X)^{2p}.$$

Putting together two inequalities involving $\|L^{1/2}h\|_2^2$ we finish the proof of Theorem 1.1 (the operators L_i can be chosen to make $\hat{w}$ arbitrarily close to w; then upon letting $\delta \longrightarrow 0$, we conclude that $\|g_\delta\|_p \longrightarrow \|S\|_p, \|g_\delta\|_{2p} \longrightarrow \|S\|_{2p}$). □

REMARK 1.2. Note that, in the general case, the constant $\frac{p^2}{2p-1}$ appearing in the proof cannot be replaced by any smaller constant. However for our special tensor form of operator L we can tensorize the inequality $(*)$ and therefore if we prove it with some better constant for all L_i's, we will have also a better constant in $(*)$ for L.

REMARK 1.3. Let us note that the proof is correct if $Eg(X)^{2p}$ is finite. To show that $\|S\|_{2p} < \infty$ it suffices to prove that $\|X_i\|_{2p} < \infty$ for $1 \leq i \leq n$. We postpone the formal proof till the end of subsection 2.1.

2. Applications

Since now on we concentrate on the case $p = 1$. However it should be pointed out that if $w = 2$ then for $p = 2$ we get constant $\sqrt[4]{3}$ in the inequality between second and fourth moments; this is best possible because $\sqrt[4]{3}$ is equal to the ratio of fourth and second norm of real symmetric gaussian random variable (so that if we consider $S = \frac{X_1 + \ldots + X_n}{\sqrt{n}}$ with n tending to infinity for a seminorm given by the absolute value of some linear functional then we obtain asymptotic equality).

Formulation of Theorem 1.1 does not suggest how to compute $\lambda_2(X)$. We need to know how to choose an appropriate operator L and how to find $\lambda_2(L)$. Now we present several special cases in which we are able to give natural and simple answers to these questions.

2.1. Connections with hypercontractivity. First we should make sure that the class of operators $\mathcal{L}(X)$ is not empty. To see this let us consider the operator $L = Id - E$ defined by $(Lf)(x) = f(x) - Ef(X)$. It is easy to prove that this operator satisfies all required conditions. However, it should be pointed out that usually for this operator $\lambda_2(L) = 1$ and we cannot deduce anything from Theorem 1.1 (the only exceptions are random vectors equal to the product of a fixed vector and Bernoulli random variable because in this case $\lambda_2(X) = \infty$). Moreover, there exists a symmetric random vector X having finite second moment and such that $\lambda_2(X) = 1$ i.e. $\lambda_2(L) \leq 1$ for all $L \in \mathcal{L}(X)$. To see this, let us consider a symmetric real random variable X such that $EX^2 < \infty$ and X is not $(2+\epsilon, 2)$-hypercontractive for any $\epsilon > 0$ (an easy example is X such that $P(|X| > t) = 1/(t \ln t)^2$ for $t > 2$; then $E|X|^{2+\epsilon} = \infty$ for all $\epsilon > 0$). Let us assume that $\lambda_2(X) > 1$. Then from Theorem 1.1 we deduce that there exist some positive constants C and ϵ such that for any vectors $v_1, v_2, \ldots, v_n$ of any normed linear space we have

$$\|\sum_{i=1}^n X_i v_i\|_{2+\epsilon} \le C \|\sum_{i=1}^n X_i v_i\|_{1+\epsilon/2},$$

where $X_1, X_2, \ldots, X_n$ are independent copies of X. Therefore Theorem 2. 2 of [**KS**] implies that X is $(2+\epsilon, 2)$-hypercontractive and we get a contradiction which proves that $\lambda_2(X) = 1$. Note that it follows from the results of [**HKLSSZ**] that X is $(2+\epsilon, 2)$-hypercontractive for some $\epsilon > 0$ if and only if X is $(2, 1\frac{1}{2})$-hypercontractive. Perhaps the following equivalence holds true:

CONJECTURE 2.1. *A symmetric real random variable X with finite second moment is $(2, 1\frac{1}{2})$-hypercontractive if and only if $\lambda_2(X) > 1$.*

There exists an equivalent tail condition (see [**KS**]). A symmetric (and more generally zero-mean, see [**La**]) real random variable is $(2, 1\frac{1}{2})$-hypercontractive if and only if there exist positive numbers s and C such that for any $t > s$ the inequality

$$E|X|^2 \mathbf{1}_{|X| \ge t} \le C t^2 P(|X| \ge t)$$

holds true or equivalently ([**HKLSSZ**])

$$\limsup_{D \to \infty} \limsup_{t \to \infty} \frac{D^2 P(|X| > Dt)}{P(|X| > t)} = 0.$$

Now we are in position to complete the proof of Theorem 1.1. We need to show that if X_i has finite second moment then $\|A_i(X_i)\|_{2p} < \infty$ for $p \in [1, w + \sqrt{w^2 - w})$, where $w = w(X) = \min(\lambda_2(X), 2)$.

PROOF. Let $Y = A_i(X_i)$. Let us start with the simplest case: Y being a symmetric real random variable. Let $p_{\max} = \sup\{p \in [1, w + \sqrt{w^2 - w}) : \|Y\|_{2p} < \infty\}$. If $p_{\max} = w + \sqrt{w^2 - w}$ then we are done. Assume the opposite. We have already proved that Theorem 1.1 is valid for $p < p_{\max}$. Therefore we know that

$$\|Y\|_{2p} \le C_{2p,p}(w) \|Y\|_p \le (\sup_{p \in [1, p_{\max}]} C_{2p,p}(w)) \|Y\|_{p_{\max}} < \infty$$

for any $p \in [1, p_{\max})$. Hence $E|Y|^{2p} \mathbf{1}_{|Y| \ge 1}$ is uniformly bounded and from the Lebesgue's Monotone Convergence Lemma we deduce that also $E|Y|^{2p_{\max}} < \infty$. Hence we can use Theorem 1.1 for $p_{\max}$ and we prove that there exists $C > 0$ such that

$$\|\sum_{i=1}^n Y_i v_i\|_{2p_{\max}} \le C \|\sum_{i=1}^n Y_i v_i\|_{p_{\max}}$$

for any vectors $v_1, v_2, \ldots, v_n$ of any normed space, where Y_i's are independent copies of Y. Theorem 2. 2 of [**KS**] yields that Y is $(2p_{\max}, p_{\max})$-hypercontractive and therefore ([**HKLSSZ**]) it is also $(2p_{\max} + \epsilon, p_{\max})$-hypercontractive for some $\epsilon > 0$. In particular $\|Y\|_{2p_{\max} + \epsilon} < \infty$ in contradiction with the definition of $p_{\max}$. Thus the proof in the case of real random variable is finished. The case of Y with values in some finite-dimensional normed space is very similar - it is enough to show that $x^*(Y)$ has finite $2p$-th moment for $p < w+\sqrt{w^2-w}$ and for any linear functional x^*, with essentially the same proof. In the general case $A_i(X_i)$ take values in some separable Banach space B. The norm of this space is a supremum of some countable set of bounded linear functionals $\{x_1^*, x_2^*, \ldots\}$. Put $\sum^{(k)} = (x_1^*(\sum), x_2^*(\sum), \ldots, x_k^*(\sum))$, where $\sum = \sum_{i=1}^n A_i(X_i)$. Then $\sum^{(k)}$ takes on values in $(\mathbf{R}^k, \|\cdot\|_{\sup})$ which is a finite-dimensional space. Therefore $\|\sum^{(k)}\|_{2p} \leq C_{2p,p}(w)\|\sum^{(k)}\|_p$. Letting k tend to infinity and applying Lebesgue's Monotone Convergence Lemma we deduce Theorem 1.1. □

2.2. Continuous distributions on real line. To a symmetric real random variable X with finite second moment and positive, continuous density function m on the interval $I = (-a, a)$, where $0 < a \leq \infty$, we attribute functions $p(x) = \int_x^a sm(s)ds$ and $u(x) = \frac{p(x)}{m(x)}$. For $f \in C_0^\infty(I)$ let us define the differential operator $(Lf)(x) = xf'(x) - u(x)f''(x) = -\frac{1}{m(x)}(pf')'(x)$. The general theory (Sturm-Liouville's) says that the operator L can be extended to some self-adjoint operator, which we denote by the same symbol, and all its eigenvalues except 0 and 1 are greater than 1. Therefore $L \in \mathcal{L}(X)$. There exist many concrete distributions of X for which $\lambda_2(L)$ can be computed, including symmetric gaussian, exponential and beta distributions, and some more general classes of distributions. These and some connected results can be found in [**KLO**]; the methods of that paper inspired the present work.

REMARK 2.2. Let us underline that the assumption of self-adjointness of operator L is crucial not only for our proof of Theorem 1.1 (which makes use of the spectral theorem) but also for the validity of Theorem 1.1 itself. Symmetry of L does not suffice as the example of $(L_c f)(x) = xf'(x) - u(x)f''(x) - cf''(x)/m(x)$ (where c is any positive constant) indicates. To see this, take $p = 2$ and $m(x) = \frac{1}{2}e^{-|x|}$ for $x \in \mathbf{R}$. Then L is well defined and obviously symmetric on the subspace of polynomials, dense in $L^2(\mathbf{R}, m(x)dx)$ but $\lim_{c \to \infty} \lambda_2(L_c) = \infty$ (use Lemma 4. 1 of [**L**] to get

$$2\int_\mathbf{R} f'(x)^2 dx \geq 4\int_\mathbf{R} f'(x)^2 m(x)dx \geq \int_\mathbf{R} f(x)^2 m(x)dx - (\int_\mathbf{R} f(x)m(x)dx)^2$$

which means that $\langle -cf''/m, f\rangle = c\int_\mathbf{R} f'(x)^2 dx \geq \frac{1}{2}c\langle f, f\rangle$ for all f orthogonal to 1) and Theorem 1.1 would give the inequality

$$\|\varepsilon\|_4 \leq \sqrt[4]{3}\|\varepsilon\|_2$$

for the random variable ε having a symmetric exponential distribution. This is not true since $\|\varepsilon\|_p = (\Gamma(p+1))^{1/p}$. The obtained contradiction proves that the generalized "**symmetric**" version of Theorem 1.1 is not valid.

2.3. Rotation invariant random vectors. M. Solomyak suggested to the author that another approach could help to generalize the method. Instead of self-adjoint L one can consider a positive definite closed symmetric bilinear form B (such that $B(f, f) < \infty$ on some dense subspace $Dom(B)$ of $L^2(F, \mu)$). The equivalent set of conditions could look as follows:
1. $B(1, 1) = 0$,
2. $B(x^*, x^*) = Ex^*(x)^2$,

3. $B(|f|,|f|) \leq B(f,f)$ (and $f \in Dom(B)$ implies $|f| \in Dom(B)$),
4. $B(\tilde{f},g) = B(f,\tilde{g})$.

Most of methods used in this paper can be translated into the language of bilinear forms by use of Friedrichs' scheme and other standard techniques. These ideas were not presented here for the sake of clarity and also because the author does not feel competent enough, however at least one potential application should be mentioned. Let X be a random vector taking on values in the Euclidean space $\mathbf{R}^n$ having a finite second moment and a continuous, strictly positive density m concentrated on some open ball with centre at zero and radius a (or on whole $\mathbf{R}^n$). Moreover, assume that m is rotation invariant, i.e. $|x| = |y|$ implies $m(x) = m(y)$. Then for f smooth enough we can define the symmetric operator $(Lf)(x) = \langle x, \nabla f(x) \rangle - u(x) \Delta f(x)$. Its essential self-adjointness is very problematic (and some counterexamples are known) whereas the closedness of a related symmetric bilinear form $B(f,g) = \int \langle \nabla f(x), \nabla g(x) \rangle p(x) dx$ can be usually easily checked. In the above formulas $p(x) = \int_{|x|}^{a} sm(s)ds$ and $u(x) = \frac{p(x)}{m(x)}$.

2.4. Discrete distributions on real line. Now we study a related case which was investigated in the author's Ph.D. Thesis.

Let X be a symmetric real random variable taking on a finite number of values $x_1 < x_2 < \ldots < x_n$ ($n \geq 2$) with probabilities respectively $p_1, p_2, \ldots, p_n > 0$, in other words let distribution μ of X be given by a formula $\sum_{k=1}^{n} p_k \delta_{x_k}$. Let L be given by the formula

$$(Lf)(x_k) = \frac{1}{p_k} \left(\frac{f(x_k) - f(x_{k-1})}{x_k - x_{k-1}} s_k - \frac{f(x_{k+1}) - f(x_k)}{x_{k+1} - x_k} s_{k+1} \right),$$

if $2 \leq k \leq n-1$, and

$$(Lf)(x_1) = x_1 \frac{f(x_2) - f(x_1)}{x_2 - x_1}$$

and

$$(Lf)(x_n) = x_n \frac{f(x_n) - f(x_{n-1})}{x_n - x_{n-1}},$$

where $s_k = \sum_{i=k}^{n} x_i p_i$ (it is easy to see that $s_k > 0$ for $k = 2, \ldots, n$, since $EX = 0$). Simple calculation shows that L is symmetric and therefore self-adjoint, because $L^2(\mathbf{R}, \mu)$ is finite-dimensional. Other conditions can also be easily verified so that $L \in \mathcal{L}(X)$. Let us investigate eigenvalues of L. Certainly they are onefold, i.e. respective eigenspaces are one-dimensional and orthogonal. To see this assume that $Lf = \lambda f$. Then

$$\lambda f(x_1) = \frac{x_1}{x_2 - x_1}(f(x_2) - f(x_1)),$$

and

$$f(x_2) = \frac{(x_2 - x_1)\lambda + x_1}{x_1} f(x_1).$$

Since all other values of f can be obtained by a linear recurrence from $f(x_1)$ and $f(x_2)$, it is clear that an eigenfunction f respective to the eigenvalue λ is determined (up to a multiplicative constant). We know that all eigenvalues of L are non-negative. Now we will prove that $\lambda_2(L) > 1$. Since all eigenfunctions orthogonal to the constant function and to the function $f(x) = x$ correspond to eigenvalues different from 0 and 1, it will be enough to prove that none of eigenvalues of L lies in interval (0,1). Assume that $Lf = \lambda f$, where $\lambda \in (0,1)$. We will not lose generality if we assume that $f(x_n) \geq 0$. For these k's for which $x_k > 0$ we put $a_k = \frac{f(x_k)}{x_k}$. Of course $a_n \geq 0$. Moreover,

$$\lambda f(x_n) = (Lf)(x_n) = x_n \frac{f(x_n) - f(x_{n-1})}{x_n - x_{n-1}},$$

so that

$$a_{n-1} = \frac{x_n - \lambda(x_n - x_{n-1})}{x_{n-1}} a_n \geq \frac{x_n - (x_n - x_{n-1})}{x_{n-1}} a_n = a_n$$

(if only $x_{n-1} > 0$). We will show inductively that for $x_n > x_{n-1} > \ldots > x_m > 0$ inequalities $0 \leq a_n \leq a_{n-1} \leq \ldots \leq a_m$ are valid. In fact, for $n > k > m$ we have

$$p_k a_k x_k \geq \lambda p_k a_k x_k = \lambda p_k f(x_k) = p_k (Lf)(x_k) =$$

$$= \frac{f(x_k) - f(x_{k-1})}{x_k - x_{k-1}} s_k - \frac{f(x_{k+1}) - f(x_k)}{x_{k+1} - x_k} s_{k+1} =$$

$$= \frac{x_k a_k - x_{k-1} a_{k-1}}{x_k - x_{k-1}} s_k - \frac{x_{k+1} a_{k+1} - x_k a_k}{x_{k+1} - x_k} s_{k+1} \geq$$

(inductional assumption says that $a_{k+1} \leq a_k$)

$$\geq \frac{x_k a_k - x_{k-1} a_{k-1}}{x_k - x_{k-1}} s_k - a_k s_{k+1} =$$

$$= \frac{x_k a_k - x_{k-1} a_{k-1}}{x_k - x_{k-1}} s_k - a_k s_k + p_k a_k x_k.$$

Hence $a_k(x_k - x_{k-1}) \geq x_k a_k - x_{k-1} a_{k-1}$, i.e. $a_{k-1} \geq a_k$ and the induction is finished. Hence for $x_n > x_{n-1} > \ldots > x_m > 0$ values $f(x_n), f(x_{n-1}), \ldots, f(x_m)$ are of the same sign. In the similar way we show that for $0 \geq x_{m-1} > \ldots > x_2 > x_1$ values $f(x_{m-1}), \ldots, f(x_2), f(x_1)$ are of the same sign. On the other hand we know that $\sum_{k=1}^{n} f(x_k) = 0$ and $\sum_{k=1}^{n} x_k f(x_k) = 0$ (because an eigenspace corresponding to the eigenvalue λ is orthogonal to eigenspaces corresponding to 0 and 1) and this implies $f = \mathbf{0}$. We have proved that operator L does not have eigenvalues between 0 and 1 and therefore $\lambda_2(L) > 1$.

As a concrete example let us consider a four-valued symmetric real random variable X with the distribution equal to $\alpha(\delta_{-x} + \delta_x) + (\frac{1}{2} - \alpha)(\delta_{-y} + \delta_y)$ for some $\alpha \in (0, \frac{1}{2})$ and $x > y > 0$. One can easily check that $\lambda_2(L) = \frac{1}{1-2\alpha} \frac{x}{x-y}$ corresponds to the eigenfunction $f(x) = f(-x) = 1 - 2\alpha, f(y) = f(-y) = -2\alpha$. For this particular kind of random variable $\lambda_2(L) = \lambda_2(X)$ (we omit long but elementary computation), i.e. the operator L introduced above seems to be optimal for our purposes. However, we should realize that it is usually hard to compute λ_2 even for this concrete operator. It is worth effort to find operators with λ_2 maybe a bit less but easier to compute. We make one step in this direction.

2.5. Isotropic position method. If X is a symmetric bounded random vector taking values in some finite-dimensional vector space, we can consider the minimal linear space F carrying X. Then there exists the unique inner product in F such that for any vector $x \in F$ we have $E\langle x, X\rangle \cdot X = x$. This simple fact is connected with the notion of so-called isotropic position. Let us define for x being a value of X

$$(Lf)(x) = E\Big(\frac{\|X\|_\infty^2 + \langle x, X\rangle}{\|X\|_\infty^2 - 1}(f(x) - f(X))\Big),$$

where $\|X\|_\infty = ess\sup(\langle X, X\rangle)^{\frac{1}{2}}$. Note that $\|X\|_\infty^2 \geq E\langle X, X\rangle = dim\, F$. Such an operator has exactly three eigenvalues: 0 for the constant function, 1 corresponding to all linear functionals and $\lambda_2 = \frac{\|X\|_\infty^2}{\|X\|_\infty^2 - 1}$ for all functions orthogonal to the linear span of $\mathbf{1}$ and F^*. It is easy to check that $L \in \mathcal{L}(X)$. Hence, in the inequality between first and second moments of the sum of (linear images of) such vectors, we obtain the constant equal to $\max_i(\sqrt{2}, \|X_i\|_{\infty,i})$. Let us notice three immediate consequences of this fact:

COROLLARY 2.3. *If* $X_1, X_2, \ldots, X_n$ *are independent simple symmetric random vectors in a Banach space* F *and* $S = \|\sum_{i=1}^n X_i\|$ *then*

$$\|S\|_2 \leq \max(\sqrt{2}, \frac{1}{a^{\frac{1}{2}}})\|S\|_1,$$

*where a is the least of all probabilities of symmetric atoms of $X'_i s$, i.e. the least of **non-zero** numbers* $P(X_i \in \{v, -v\}); v \in F$.

PROOF. Without loss of generality we can assume that values of each X_i are linearly symmetric-independent i.e. their linear combination is equal to zero if and only if for any value v of X_i coefficients assigned to v and $-v$ are equal. Certainly by an easy approximation argument this case would imply the general one. Now it is easy to find proper inner product for each X_i; we leave simple computation to the reader. □

Let us note that this result is quite sharp. If $v_1, \ldots, v_k$ are linearly independent vectors and X is a symmetric random vector with values $\pm v_1, \ldots, \pm v_k$ then we can consider a seminorm given by the formula $|x_i^*(\cdot)|$, where x_i^* is a linear functional such that $x_i^*(v_j) = \delta_{i,j}$. It is easy to check that $\frac{\|x_i^*(X)\|_1}{\|x_i^*(X)\|_2} = \sqrt{P(X \in \{v_i, -v_i\})}$. Note also that from this corollary and from the next one we can deduce quite sharp inequality between p–th and q–th moments of sum, where $2 \geq p > q \leq 1$, by simple use of Hölder inequality (see [**LO**]).

COROLLARY 2.4. *If* $X_1, X_2, \ldots, X_n$ *are independent bounded symmetric real random variables and* $v_1, v_2, \ldots, v_n$ *are vectors of linear normed space* F *then for* $S = \|\sum_{i=1}^n X_i v_i\|$ *the following inequality holds true:*

$$\|S\|_2 \leq \max_{1 \leq i \leq n}(\sqrt{2}, \frac{\|X_i\|_\infty}{\|X_i\|_2})\|S\|_1.$$

PROOF. On the real line there is no problem with choice of an appropriate inner product. Constants in this inequality are optimal in the case of linear combination of independent copies of a three-valued symmetric real random variable (but only if probability atom in zero is greater than $1/2$) with vector coefficients. □

COROLLARY 2.5. *Let K be a convex, centrally symmetric body in $\mathbf{R}^n$ and $Id = \sum c_i x_i \otimes x_i$ be the symmetric John's decomposition of identity related to this body (x_i's are contact points of the maximal volume ellipsoid contained in K and coefficients assigned to $x_i \otimes x_i$ and $(-x_i) \otimes (-x_i)$ are equal). Then for any norm $\|\cdot\|$ in $\mathbf{R}^n$ the following inequality holds true:*

$$\sum c_i \|x_i\|^2 \leq (\sum c_i \|x_i\|)^2.$$

PROOF. Introduce in $\mathbf{R}^n$ such a Euclidean structure in which the maximal volume ellipsoid is a ball and consider a random vector taking on values $\sqrt{n} x_i$ with probabilities c_i/n respectively ($\sum c_i = tr(Id) = n$), which appears to be in the isotropic position with respect to this Euclidean structure. $\square$

2.6. Uniform distribution on the sphere. If X is uniformly distributed on the unit sphere S^{m-1} then we can define $T_t f = e^{-tL} f$ for $f \in C(S^{m-1})$ in the following way. Let $\hat{f}$ be the unique harmonic extension of f on the unit ball. Then put $(T_t f)(x) = \hat{f}(e^{-t}(x))$. It is easy to check that $L \in \mathcal{L}(X)$, except the self-adjointness. Although it is not very easy to prove, it is well-known that L defined in such a way is essentially self-adjoint and its eigenvalues are $0, 1, 2, \ldots$ with respective spherical harmonics as corresponding eigenspaces of eigenfunctions, so that we get $\lambda_2(L) = 2$. We see that this constant and hence constants in comparison of (low) moments inequalities do not depend on dimension m. Note that the method used in the subsection 2.5 would give worse constants, depending on dimension. The hypercontractive properties of the same linear operator were investigated in [**B**].

3. Log-Sobolev inequalities and comparison of higher moments

3.1. Discrete cube. The simplest problem connecting our methods with hypercontractive ones and yielding comparison of higher moments can be formulated as follows:

Consider family $\mathcal{F}$ of **symmetric** functions $f : \{-1,1\}^n \longrightarrow \mathbf{R}_+$ such that $Lf \leq f$, where $(Lf)(x) = \frac{1}{2} \sum_y (f(x) - f(y))$ with summing over all n neighbours of x, i.e. such y's that they differ from x on exactly one coordinate. Take $p > q > 0$. We are looking for the best constants $C_{p,q}$ such that for any $f \in \mathcal{F}$ the moment inequality $\|f\|_p \leq C_{p,q} \|f\|_q$ holds true (on $\{-1,1\}^n$ we introduce normalized counting measure).

The basic example of function from $\mathcal{F}$ is any function of the following type: $f(x) = \|\sum_{i=1}^n x_i v_i\|$, where v_i's are vectors of some linear normed space, but the family $\mathcal{F}$ is much larger. To see this, consider the case $n = 5$ and $f(x) = \mathbf{1}_{|x_1+x_2+x_3+x_4+x_5|=1}$. It is easy to check that $f \in \mathcal{F}$. Assume that $f(x) = \|\sum_{i=1}^5 x_i v_i\|$. Since a linear span of $(1,1,1,1,-1)$, $(1,1,1,-1,1)$, $(1,1,-1,1,1)$, $(1,-1,1,1,1)$ and $(-1,1,1,1,1)$ is equal to $\mathbf{R}^5$ and f takes on zero value in all these points the seminorm must be identically equal to zero. We get a contradiction because $f(1,1,-1,-1,-1) \neq 0$.

The methods presented in this paper give upper estimates on $C_{2p,p}$ for $p < 2 + \sqrt{2}$ (and exact values of $C_{4,2}$ and $C_{p,q}$ for $2 \geq p > q \leq 1$). To get comparison of higher moments let us use log-Sobolev inequality (this idea was suggested by S. Kwapień)

$$Ef^2 \ln(f^2) - Ef^2 \ln(Ef^2) \leq 2EfLf$$

valid for all real functions f on $\{-1,1\}^n$ (see for example Chapter 2. 2 of [**L**] in which the inequality is given in the equivalent form). Putting f^p in place of f we get

$$Ef^{2p}\ln(f^{2p}) - Ef^{2p}\ln(Ef^{2p}) \leq 2Ef^p L(f^p) \leq \frac{2p^2}{2p-1} Ef^{2p-1} Lf \leq \frac{2p^2}{2p-1} Ef^{2p}.$$

The middle inequality is a consequence of the Stroock-Varopoulos inequality used in the proof of Theorem 1.1 and it works for all $p > \frac{1}{2}$. Hence

$$\frac{d}{dp}\left(\frac{1}{2p}\ln(Ef^{2p})\right) \leq \frac{1}{2p-1}$$

and therefore

$$\ln(\|f\|_{2p}) - \ln(\|f\|_{2q}) \leq \frac{1}{2}(\ln(2p-1) - \ln(2q-1)).$$

This gives estimate $C_{p,q} \leq \sqrt{\frac{p-1}{q-1}}$ for any $p > q > 1$. The method seems essentially different from usual probabilistic hypercontractive methods (see [**KW**] for a nice review) which use the values of the norm also in other points, not only for $x \in \{-1,1\}^n$.

Note that we did **not** use symmetry of f in the last proof while it was strongly used for the comparison of low moments (however having this kind of result in the symmetric case it is easy to prove it also in the non-symmetric case with a slightly worse constant).

3.2. Sums of independent random vectors. Note that the log-Sobolev ideas can be used also for comparison of higher moments of sums of independent random vectors - all we need is log-Sobolev inequality for each $L_i \in \mathcal{L}(X_i)$ separately, with common upper bound on the constants in log-Sobolev inequality. Standard tensorization techniques show then that L also satisfies log-Sobolev inequality with the same upper bound on the constant. Since we can use Stroock-Varopoulos inequality for L, the proof given above can be completely transferred into this generalized setting. In this way we obtain generalization of some results of the second part of [**KLO**].

Let us define $\hat{\mathcal{L}}(X)$ in the similar way as $\mathcal{L}(X)$ but omitting the fourth condition. Let $\kappa(L)$ be the least constant κ such that log-Sobolev inequality

$$Ef(X)^2 \ln(f(X)^2) - Ef(X)^2 \ln(Ef(X)^2) \leq \kappa Ef(X)(Lf)(X)$$

is satisfied for any $f \in Dom(L)$ and let $\kappa(X) = \inf_{L \in \hat{\mathcal{L}}(X)} \kappa(L)$. Now we can formulate the following theorem (its proof was sketched above, we leave details to the reader).

THEOREM 3.1. *Let $X_1, X_2, \ldots, X_n$ be independent random vectors with values in Banach spaces $F_1, F_2, \ldots, F_n$ respectively, having finite second moments and let $A_i : F_i \longrightarrow B$ be bounded linear transformations into some separable Banach space B for $i = 1, 2, \ldots, n$. Let $S = \|\sum_{i=1}^n A_i(X_i)\|$ and let $\kappa = max_{1 \leq i \leq n} \kappa(X_i)$. Then for $p > q > 1$ the inequality*

$$\|S\|_p \leq \left(\frac{p-1}{q-1}\right)^{\kappa/4} \|S\|_q$$

holds true.

The author does not know any application of Theorem 3.1 for operators other than those described in subsection 2.2. It is of interest if this particular kind of operators minimizes $\kappa(L)$ among $L \in \hat{\mathcal{L}}(X)$.

4. Pełczyński's conjecture

We will finish with a partial proof of the following conjecture of A. Pełczyński.

CONJECTURE 4.1. Let $r_1, r_2, \ldots, r_n$ be independent symmetric Bernoulli random variables. Then for arbitrary vectors $(v_{i,j})_{1 \leq i < j \leq n}$ of any normed linear space
$$\|S\|_2 \leq 2\|S\|_1,$$
where $S = \|\sum_{1 \leq i < j \leq n} r_i r_j v_{i,j}\|$.

We will sketch the proof of the conjecture for $n = 6$ (and therefore for $n \leq 6$) omitting long but elementary calculations. The author is quite sceptical about possibility of generalization of the presented method. Let us mention that the conjecture is well-known to hold with some positive constant. It follows from the hypercontractive methods for multilinear forms but also from the fact that for the function g defined as below the inequality $Lg \leq 2g$ holds true. Therefore, by similar use of log-Sobolev inequality as in subsection 3.1, one can prove that $\|S\|_p \leq \frac{p-1}{q-1}\|S\|_q$ for $p > q > 1$ and deduce the inequality between first and second moments with some absolute constant.

PROOF. Let us consider the natural product probabilistic measure and Hamming's metric d on $\{-1, 1\}^6$. For any symmetric real function f on $\{-1, 1\}^6$ let us define
$$(Kf)(x) = \sum_{y: d(x,y)=1} f(y)$$
and
$$(Mf)(x) = \sum_{y: d(x,y)=3} f(y).$$

It is easy to check that operators K and M are self-adjoint and their eigenfunctions are Walsh functions. The eigenvalues of operators K, M and $2K + M$ for Walsh functions of order k are given below:

k	K	M	2K + M
0	6	20	32
2	2	-4	0
4	-2	4	0
6	-6	-20	-32

Therefore $\langle f, (2K + M)f \rangle \leq 32\|f\|_1^2$ for any symmetric real function f on $\{-1, 1\}^6$. In particular $\langle g, (2K + M)g \rangle \leq 32\|g\|_1^2$ for
$$g(x) = \|\sum_{1 \leq i < j \leq 6} x_i x_j v_{i,j}\|.$$

On the other hand norm inequalities yield that $Kg \geq 2g, Mg \geq 4g$ and therefore $(2K+M)g \geq 8g$. Hence $\langle g, (2K+M)g \rangle \geq 8\|g\|_2^2$. Putting together two inequalities involving $\langle g, (2K+M)g \rangle$ we finish the proof. □

ACKNOWLEDGEMENTS. The main idea of this paper, attribution of a self-adjoint operator to the random variable in the context of Poincaré inequality, is due to Prof. Stanisław Kwapień. Lemma 0.3 and some other ideas evolved from the joint unpublished work with Rafał Latała. Prof. Michael Solomyak turned my attention to the Stroock-Varopoulos inequality and suggested some extensions of the method, as mentioned in the paper. His remarks enabled me to generalize results proved previously only in the case of L having complete system of eigenfunctions and helped to organize the proof of Theorem 1.1. I owe most of my knowledge and intution about self-adjoint operators to Jan Herczyński.

Main part of the paper was written during my work in the Institute of Mathematics of Polish Academy of Sciences. The work was completed and some new results included during my post-doctoral stage in the Weizmann Institute of Science. I am very grateful to my WIS host, Prof. Gideon Schechtman for his time and encouragement, and creating excellent research conditions.

References

[B]   W. Beckner *Sobolev Inequalities, the Poisson Semigroup and Analysis on the Sphere S^n*, Proc. Nat. Acad. Sci. USA 89 (1992), 4816-4819.
[D]   E. B. Davies *Heat Kernels and Spectral Theory*, Cambridge Tracts in Mathematics, 92, Cambridge University Press, 1989.
[HKLSSZ] P. Hitczenko, S. Kwapień, W. V. Li, G. Schechtman, T. Schlumprecht and J. Zinn *Hypercontractivity and comparison of moments of iterated maxima and minima of independent random variables*, Electr. J. Prob. 3 (1998), Paper no. 2, 1-26 (http://www.math.washington.edu/~ejpecp/EjpVol3/paper2.abs.html).
[KLO] S. Kwapień, R. Latała and K. Oleszkiewicz *Comparison of moments of sums of independent random variables and differential inequalities*, J. Funct. Anal. 136 (1996), 256-268.
[KS]   S. Kwapień, J. Szulga *Hypercontraction methods in moment inequalities for series of independent random variables in normed spaces*, Ann. Prob. 19 (1991), 369-379.
[KW]   S. Kwapień and W. A. Woyczyński *Random Series and Stochastic Integrals: Single and Multiple*, Birkhäuser, Boston-Basel-Berlin, 1992.
[L]   M. Ledoux *Concentration of Measure and Logarithmic Sobolev Inequalities*, to appear.
[La]   R. Latała *Hypercontractive random variables*. Master degree thesis, Department of Mathematics, Warsaw University, 1994 (in Polish).
[LO]   R. Latała and K. Oleszkiewicz *On the best constant in the Khintchin-Kahane inequality*, Studia Math. 109 (1994), 101-104.
[S]   D. W. Stroock *An Introduction to the Theory of Large Deviations*, Springer-Verlag, New York, 1984.
[V]   N. Th. Varopoulos *Hardy-Littlewood theory for semigroups*, J. Funct. Anal. 63 (1985), 240-260.

INSTITUTE OF MATHEMATICS, WARSAW UNIVERSITY, BANACHA 2, 02-097 WARSZAWA, POLAND
Current address: Department of Theoretical Mathematics, the Weizmann Institute of Science, 76100 Rehovot, Israel
E-mail address: koles@mimuw.edu.pl

Contemporary Mathematics
Volume **234**, 1999

Fractional sums and integrals of r-concave tails and applications to comparison probability inequalities

Iosif Pinelis

ABSTRACT. A function $q > 0$ is said to be r-concave if $q^{-1/r}$ is convex. It is shown that α-fractional integrals of r-concave tail functions are $(r-\alpha)$-concave if $0 < \alpha < r$. A "discrete" counterpart of this result is also obtained. Applications to comparison probability inequalities are given. Among the results is the following: If $a_1^2 + \cdots + a_n^2 = n$ and $\varepsilon_1, \ldots, \varepsilon_n$ are independent Rademacher random variables, then for all $x \in 2\mathbb{Z} - n$,
$$\Pr(a_1\varepsilon_1 + \cdots + a_n\varepsilon_n \geq x) \leq c\Pr(\varepsilon_1 + \cdots + \varepsilon_n \geq x),$$
where $c := 2e^3/9 = 4.46\ldots$.

1. Introduction

In Pinelis (1994) [**13**], a multidimensional version of the following inequality was obtained:

(1.1) $\quad \Pr(b_1\varepsilon_1 + \cdots + b_n\varepsilon_n \geq x) \leq c\Pr(\xi \geq x) = c(1 - \Phi(x)) \quad \forall x \in \mathbb{R},$

where the ε_i's are independent Rademacher random variables (r.v.'s), ξ is a standard normal r.v., Φ is its distribution function, the b_i's are any real numbers such that $b_1^2 + \cdots + b_n^2 = 1$, and $c = 2e^3/9$.

The version of (1.1) – only for $x > \sqrt{2}$ and with $\dfrac{1}{x\sqrt{2\pi}}\exp\left(-\dfrac{x^2}{2}\right)$ in place of $1 - \Phi(x)$ – had for 20 years been an unproved conjecture by Eaton (1974) [**9**].

Analysis of the proof of (1.1) in [**13**] can show the crucial role of the fact that however many times integrated normal tails are log-concave. This idea was further developed in Pinelis (1998) [**14**] in the "continuous" case.

In this paper, we treat mainly the more fundamental and hard "discrete" case. "Continuous" results follow easily from the "discrete" counterparts. E.g., it is easy to deduce (1.1) from the inequality mentioned in the Abstract
$$\Pr(a_1\varepsilon_1 + \cdots + a_n\varepsilon_n \geq x) \leq c\Pr(\varepsilon_1 + \cdots + \varepsilon_n \geq x),$$
where $a_1^2 + \cdots + a_n^2 = n$ and $x \in 2\mathbb{Z} - n$.

1991 *Mathematics Subject Classification.* Primary: 60E15; Secondary: 05A20, 26A12, 26A33, 26A51, 26D07, 26D10, 26D15, 33E20, 39A12, 51M16, 52A01, 52A40.

Key words and phrases. Fractional summation and integration, r-concave tails, α-convex functions, probability inequalities.

© 1999 American Mathematical Society

2. r-concave functions and fractional integrals and sums: definitions and examples

DEFINITION 1. Let I be a convex subset of a linear space E and $r \in (0, \infty)$. A function $q \colon E \to [0, \infty]$ is called r-*concave* on I if $q^{-1/r}$ is convex on I (we assume $0^{-1/r} = \infty$ and $\infty^{-1/r} = 0$); q is called ∞-*concave* or, interchangeably (see Remark 1 below), *log-concave* if it is s-concave for all $s \in (0, \infty)$. Similarly, we can define "discrete" r-concave and ∞-concave on $I \cap \mathbb{Z}$ or at $n \in \mathbb{Z}$ functions $q \colon \mathbb{Z} \to [0, \infty]$, where I is a convex subset of $\mathbb{R}$, if we say that a function $f \colon \mathbb{Z} \to (-\infty, \infty]$ is convex on $I \cap \mathbb{Z}$ iff it is the restriction to $\mathbb{Z}$ of a function $g \colon \mathbb{R} \to (-\infty, \infty]$ which is convex on I and if we say that a function $f \colon \mathbb{Z} \to (-\infty, \infty]$ is convex at $n \in \mathbb{Z}$ iff if is convex on $\{n-1, n, n+1\}$; the latter is equivalent to $f(n+1) + f(n-1) \geq 2f(n)$. Note that if a function $f \colon \mathbb{Z} \to (-\infty, \infty]$ is either finite on $I \cap \mathbb{Z}$ or monotone on $I \cap \mathbb{Z}$, then it is convex on $I \cap \mathbb{Z}$ iff f is convex at each $n \in \mathbb{Z}$ such that $\{n-1, n+1\} \subset I$; the condition that $f \colon \mathbb{Z} \to (-\infty, \infty]$ is either finite or monotone is essential here, as can be seen from the following example, due to an anonymous referee: $I = \mathbb{R}$, $f(0) = f(1) = \infty$, $f(z) = 0 \; \forall z \in \mathbb{Z} \setminus \{0, 1\}$. Let us say that $f \colon \mathbb{Z} \to (-\infty, \infty]$ is *strictly convex* at an $n \in \mathbb{Z}$ if $f(n+1) + f(n-1) > 2f(n)$. Then let us say that a function $q \colon \mathbb{Z} \to [0, \infty]$ is *strictly r-concave* at an $n \in \mathbb{Z}$ if $q^{-1/r}$ is strictly convex at n.

REMARK 1. If $q \geq 0$ is r-concave, then it is r'-concave for all $r' \in (0, r)$ (because the function $q^{-1/r} \mapsto (q^{-1/r})^{r/r'} = q^{-1/r'}$ is increasing and convex); hence, q is ∞-concave (which is the same as being log-concave) iff it is r-concave for all large enough $r > 0$. A *finite* $q \geq 0$ is ∞-concave iff $(-\ln q)$ is convex (assuming $\ln 0 = -\infty$); this is so because $r(q^{-1/r} - 1) \to (-\ln q)$ as $r \to \infty$, and, on the other hand, the function $(-\ln q) \mapsto e^{\frac{1}{r}(-\ln q)} = q^{-1/r}$ is increasing and convex.

Here are some possible variations.

DEFINITION 2. For a function $q \colon \mathbb{R} \to [0, \infty]$ and an $r \in (0, \infty]$, let us say that q is *eventually r-concave* if q is r-concave on $[N, \infty)$ for some $N \in \mathbb{R}$; let us say that q is $(r-0)$-*concave* if $\forall r' \in (0, r) \; \exists N \in \mathbb{R} \quad q$ is r'-concave on $[N, \infty)$; let us say that q is *asymptotically r-concave* if there exists an r-concave function $\hat{q}$ such that $\lim_{x \to \infty} q(x)/\hat{q}(x) = 1$. Let us assume similar definitions in the "discrete" case as well, when $q \colon \mathbb{Z} \to [0, \infty]$.

REMARK 2. An important characterization of r-concavity in terms of a generalized Brunn-Minkowski inequality follows from Theorem 3.2 of Borell (1975) [3].

REMARK 3. For any $r \in (0, \infty)$, a nonnegative function is r-concave iff it can be represented as the pointwise infimum of a family of power functions of the form $x \mapsto (ax+b)_+^{-r}$; for $r = \infty$, the enveloping functions are exponential, rather than power ones. As usual, $x_+ := \max(x, 0)$. It follows that any decreasing r-concave function cannot decrease slower than a power function. On the other hand, not too irregular tails dominated by a power tail should be normally expected to be at least eventually r-concave for some $r > 0$.

Almost all standard statistical distributions possess such a property.

Say, in the "discrete" case, a great number of types of distributions fall into the very broad family of generalized hypergeometric distributions, characterized by

$$\frac{\Pr(X = m+1)}{\Pr(X = m)} = \frac{(a_1 + m) \ldots (a_A + m)\lambda}{(b_1 + m) \ldots (b_B + m)}, \quad m \in \{N, N+1, \ldots\} \subset \mathbb{Z};$$

cf. formula (2.70) and Table 2.4 in Johnson, Kotz, and Kemp (1992) [**11**]. For this formula to define a non-zero distribution tail, one must have $\lambda > 0$ and $A \le B$; also, if $A = B$, then $0 < \lambda \le 1$; if $A = B$ and $\lambda = 1$, then $a := a_1 + \cdots + a_A - b_1 - \cdots - b_B \le 0$; etc. Then such a probability mass function (p.m.f.) is eventually ∞-concave if $A < B$; $(\infty - 0)$-concave if $A = B$ and $0 < \lambda < 1$; $(|a| - 0)$-concave if $A = B$, $\lambda = 1$, and $a < 0$.

In particular, all binomial, hypergeometric, and Poisson p.m.f.'s are ∞-concave. The negative binomial p.m.f. with parameters k and P (see (5.1) in [**11**]) is $(\infty - 0)$-concave when $k < 1$ and ∞-concave when $k \ge 1$; in particular, the geometric distribution is obviously ∞-concave.

In the "continuous" case, the probability distribution function (p.d.f.) of the Student t distribution with k degrees of freedom is $(k+1)$-concave. In particular, the p.d.f. of the normal distribution is ∞-concave; moreover, the same holds for the distribution of $\chi_d := \sqrt{\chi_d^2}$, where χ_d^2 is a r.v. having the χ^2 distribution with $d \ge 2$ degrees of freedom.

If $k \ge 2$, then the p.d.f. of the F distribution with k and m degrees of freedom is $(m/2 + 1)$-concave. In particular, the p.d.f. of the χ^2 distribution with $k \ge 2$ degrees of freedom is ∞-concave. So is the logistic p.d.f. $e^x(1 + e^x)^{-2}$.

More examples can be found in Borell (1975) [**3**].

DEFINITION 3. The fractional – or, more exactly, the α-fractional – (tail) integral of a measurable function $q \colon \mathbb{R} \to [0, \infty]$ is defined for any $\alpha > 0$ by the formula
$$\mathcal{I}^\alpha q(x) = \frac{1}{\Gamma(\alpha)} \int_0^\infty u^{\alpha-1} q(x+u) du, \quad x \in \mathbb{R}.$$

Similarly, the α-fractional (tail) sum of a function $q \colon \mathbb{Z} \to [0, \infty]$ is defined for any $\alpha > 0$ by the formula
$$\mathcal{J}^\alpha q(n) = \sum_{k=0}^\infty \binom{\alpha + k - 1}{k} q(n+k), \quad n \in \mathbb{Z},$$
where, as usual, $\binom{\beta}{k} := \dfrac{\beta(\beta-1)\ldots(\beta-k+1)}{k!}$, so that

(2.1) $\quad \binom{\alpha + k - 1}{k} = (-1)^k \binom{-\alpha}{k} = \dfrac{\alpha(\alpha+1)\ldots(\alpha+k-1)}{k!} [> 0 \text{ if } \alpha > 0].$

REMARK 4. At least if restricted to regular enough q, $\mathcal{I}_\alpha$ can be obtained from $\mathcal{J}_\alpha$ by a Riemann type of limit transition: setting $q_\varepsilon(m) := \varepsilon^\alpha q(\varepsilon m) \; \forall m \in \mathbb{Z}$, one has $(\mathcal{J}^\alpha q_\varepsilon)(n) \to (\mathcal{I}^\alpha q)(x)$ as $\varepsilon \downarrow 0$, $\varepsilon n \to x$, and $n \in \mathbb{Z}$. This follows since
$$\binom{\alpha + k - 1}{k} \frac{\Gamma(\alpha)}{k^{\alpha-1}} \to 1 \quad \text{as} \quad k \to \infty.$$

We shall not pursue here most general conditions under which this limit transition from $\mathcal{J}^\alpha$ to $\mathcal{I}^\alpha$ can be done. For the specific purposes of this work, suffice it to say that it can done at least for the r-concave q and $\alpha \in (0, r)$. Indeed, then Remark 3 provides for the uniform (in ε) summability of the series representing $(\mathcal{J}^\alpha q_\varepsilon)(n)$ if $q(x) \to 0$ as $x \to \infty$; otherwise, both $\mathcal{J}^\alpha q_\varepsilon$ and $\mathcal{I}^\alpha q$ are identically ∞.

REMARK 5. Note that each of the families $(\mathcal{I}^\alpha)_{\alpha > 0}$ and $(\mathcal{J}^\alpha)_{\alpha > 0}$ constitutes a semigroup with $\mathcal{I}^1 q(x) = \int_{[0,\infty)} q(x+u) du$ and $\mathcal{J}^1 q(n) = \sum_{k=0}^\infty q(n+k)$. The semigroup property of $(\mathcal{J}^\alpha)_{\alpha>0}$ follows from (2.1) and the well-known identity

$$\sum_{k=0}^{n} \binom{-\alpha}{k}\binom{-\beta}{n-k} = \binom{-\alpha-\beta}{n},$$

which in turn is easy to obtain using generating functions. The semigroup property of $(\mathcal{I}^\alpha)_{\alpha>0}$ can be deduced from that of $(\mathcal{J}^\alpha)_{\alpha>0}$ and Remark 4; alternatively, it follows easily from the well-known identity

$$\mathrm{B}(\alpha, \beta) := \int_0^1 x^{\alpha-1}(1-x)^{\beta-1}\mathrm{d}x = \frac{\Gamma(\alpha)\Gamma(\beta)}{\Gamma(\alpha+\beta)}.$$

It is natural to complete these semigroups by setting $(\mathcal{I}^0 q)(x) := \lim_{\alpha \downarrow 0}(\mathcal{I}^\alpha q)(x) = q(x+0)$ for q with finite variation and $\mathcal{J}^0 q := q$. Fractional integration $\mathcal{I}^\alpha$ is known as a Weyl transform.

REMARK 6. Note that fractional integration arises quite naturally in [14] (cf. eq. (16) therein), since for any $\alpha > 0$ and any r.v. η with the tail function $q(t) := \Pr(\eta \geq t) \ \forall t \in \mathbb{R}$, one has

$$\mathcal{I}^\alpha q(t) = \frac{\mathsf{E}\,(\eta - t)_+^\alpha}{\Gamma(\alpha + 1)}, \quad t \in \mathbb{R}.$$

3. $(r - \alpha)$-concavity of α-fractional sums and integrals of r-concave tails: results

THEOREM 1. *Let $0 < \alpha < r \leq \infty$ and let $q \colon \mathbb{Z} \to [0, \infty]$ be r-concave and non-increasing. Then $\mathcal{J}^\alpha q$ is $(r - \alpha)$-concave; we assume $r - \alpha = \infty$ if $r = \infty > \alpha$.*

A proof of this result – and others, whenever the proofs we know are comparatively long – is given in Section 5 below.

If $\alpha \geq 1$, then q need not be non-increasing in Theorem 1, so that the following theorem holds.

THEOREM 2. *Let $1 \leq \alpha < r \leq \infty$ and let $q \colon \mathbb{Z} \to [0, \infty]$ be r-concave. Then $\mathcal{J}^\alpha q$ is $(r - \alpha)$-concave.*

By Remark 4, the "discrete" Theorems 1 and 2 imply the following "continuous" statement.

THEOREM 3. *Let $0 < \alpha < r \leq \infty$ and let $q \colon \mathbb{R} \to [0, \infty]$ be r-concave. Assume also that either q is non-increasing or $\alpha \geq 1$. Then $\mathcal{I}^\alpha q$ is $(r - \alpha)$-concave.*

REMARK 7. The condition in Theorems 1, 2, and 3 that either q is non-increasing or $\alpha \geq 1$ is essential. Indeed, if, for instance, $q(x) := \mathrm{I}\{0 < x < 1\}$, then q is obviously ∞-concave; on the other hand, $\mathcal{I}^\alpha q(x)$ behaves in this case like $|x|^{\alpha-1}$ as $x \to -\infty$, and so, $\mathcal{I}^\alpha q$ is not ∞-concave when $\alpha < 1$.

REMARK 8. It should be clear that q, together with $\mathcal{J}^\alpha q$ and $\mathcal{I}^\alpha q$, can be considered in Theorems 1, 2, and 3 as defined and possessing the respective properties only on an interval of the form $[a, \infty) \cap \mathbb{Z}$ or $[a, \infty)$, rather than on $\mathbb{Z}$ or $\mathbb{R}$.

REMARK 9. In the special case when $\alpha = 1$ and q is a basic r-concave function of the form $q(x) = (ax + b)_+^{-r}$, the statements of Theorems 1 and 2 are equivalent to the following inequality: for any $r > 1$ and $b > 0$,

$$\left(b^{-r} + (b+1)^{-r} + (b+2)^{-r} + (b+3)^{-r} + \ldots\right)^{-1/(r-1)}$$
$$+ \left((b+2)^{-r} + (b+3)^{-r} + \ldots\right)^{-1/(r-1)}$$
(3.1)
$$\geq 2 \left((b+1)^{-r} + (b+2)^{-r} + (b+3)^{-r} + \ldots\right)^{-1/(r-1)}.$$

Even in this very special case, Theorems 1 and 2 seem to be non-trivial. We shall see that the semigroup property of the fractional summation is very helpful, as it provides a continuous bridge over the gap between r and $r - 1$ or, in general, between r and $r - \alpha$. One may note that a stronger version of (3.1) with the three entries of "$+\ldots$" replaced by "$+ \cdots + (b+n)^{-r}$" with $n = 2, 3, \ldots$ also follows from Theorem 1 or 2.

REMARK 10. In the "continuous" setting of Theorem 3, the special case $\alpha = 1$ is well known at least for $r = \infty$; see Barlow, Marshall, and Proshan (1963) [**1**] or Hall, Kanter, and Perlman (1980) [**10**]. This case may be dealt with by means of the celebrated Prékopa-Leindler theorem when $r = \infty$ and by the Borell-Brascamp-Lieb extension of the Prékopa-Leindler theorem when $r < \infty$ [**3, 6**], since the function $\mathbb{R}^2 \ni (t, x) \mapsto I\{t \geq x\} q(t)$ is r-concave whenever q is r-concave. This "continuous" case with $\alpha = 1$ can also be treated by more elementary means. Relations between the Prékopa-Leindler-Borell-Brascamp-Lieb theorems and generalized Brunn-Minkowski theorems were demonstrated by Prékopa (1971, 1973) [**15, 16**], Leindler (1972) [**12**], Borell (1975) [**3**], Brascamp and Lieb (1975, 1976) [**5, 6**], Rinott (1976) [**18**], Das Gupta (1980) [**7**], et al.

REMARK 11. If, in accordance with Remark 3, the r-concavity is considered as a formalization of the idea of "decrease no slower than that of $0 < x \mapsto x^{-r}$", then the corresponding meaning of the statements of Theorems 1, 2, and 3 is transparent: If $q(x)$ "decreases no slower than x^{-r}", then its αth tail sum $\mathcal{J}^\alpha q(x)$ or integral $\mathcal{I}^\alpha q(x)$ "decreases no slower than $x^{-(r-\alpha)}$".

REMARK 12. Theorems 1, 2, and 3 hold if instead of the *bona fide* notion of r-concavity, any of its variations described in Definition 2 is used. Thus, if q is asymptotically r-concave [alternatively, $(r-0)$-concave or eventually r-concave] and $r > \alpha > 0$, then $\mathcal{J}^\alpha q$ or (depending on the context) $\mathcal{I}^\alpha q$ is asymptotically $(r-\alpha)$-concave [respectively, $(r-\alpha-0)$-concave or eventually $(r-\alpha)$-concave]. In these "asymptotic" settings, the condition in Theorems 1, 2, and 3 that either $\alpha \geq 1$ or q is non-increasing is not needed; indeed, if a function q is r-concave either *bona fide* or in any of the above broader senses, then either q is non-increasing in a neighborhood of ∞ or, for each $\alpha > 0$, the α-fractional sum or integral of q is identically ∞, as well as that of any other function asymptotically equivalent to q at ∞.

REMARK 13. One can apply Theorems 1, 2, or 3 to any of the r-concave distributions mentioned in Section 2, say. E.g., it follows that for a binomial p.d.f. p_{bin}, the "α times integrated" tail $\mathcal{J}^\alpha p_{\text{bin}}$ is ∞-concave on $\mathbb{Z}$ if $\alpha \geq 1$ and is so on $\mathbb{Z} \cap [m_{\text{bin}}, \infty)$ if $\alpha \in (0, 1)$, where m_{bin} is the mode of the binomial distribution. Or, as another example, $\mathcal{I}^1 p_F$ is $(m/2)$-concave on $\mathbb{R}$, where p_F is the p.d.f. of the F distribution with $k \geq 2$ and m degrees of freedom.

4. Applications

Theorem 3 was stated without proof as Proposition 3.8 in [14]; the condition that either q is non-increasing or $\alpha \geq 1$ (satisfied in all applications in [14]) was missing there.

To state Theorem 4 below, we need to introduce the following notions of α-convexity, which are closely related to fractional integration.

From now on, unless otherwise specified, assume that I is either $\mathbb{R}$ or $[-\infty, \infty)$; $\alpha \in (0, \infty)$; and ξ and η are r.v.'s with values in I.

DEFINITION 4. Let $\mathcal{E} := \mathcal{E}_I$ stand for the set of all continuous non-decreasing functions $f: I \to \mathbb{R}$ with $\lim_{x \downarrow (-\infty)} f(x) = 0$. Let

$$\mathcal{F}_\alpha := \mathcal{F}_{\alpha,I} := \{f \in \mathcal{E}_I : f^{1/\alpha} \text{ is convex on } \mathbb{R}\};$$

$$\mathcal{G}_\alpha := \mathcal{G}_{\alpha,I} := \left\{f \in \mathcal{E}_I : \exists \mu \in M \; \forall u \in I \; f(u) = \int (u-t)_+^\alpha \mu(dt)\right\},$$

where M stands for the set of all nonnegative Borel measures on $\mathbb{R}$;

$$\mathcal{H}_\alpha := \mathcal{H}_{\alpha,I} := \{I \ni u \mapsto (u-t)_+^\alpha : t \in \mathbb{R}\};$$

assume $(-\infty - t)_+ = 0 \; \forall t \in \mathbb{R}$. Functions belonging either to $\mathcal{F}_{\alpha,I}$ or $\mathcal{G}_{\alpha,I}$ may be referred to as α-convex on I. Let us endow $\mathcal{E}_I$, $\mathcal{F}_{\alpha,I}$, and $\mathcal{G}_{\alpha,I}$ with the topologies induced by the topology of the uniform convergence on every set of the form $I \cap [-\infty, b]$, $b \in \mathbb{R}$, in the space $\mathbb{R}^I$ of all real functions on I. Note that then $\mathcal{E}_I$ is closed in $\mathbb{R}^I$.

We also give the following definition of majorization.

DEFINITION 5. For any $\mathcal{C} \subseteq \mathcal{E}_I$, let us write $\xi \preceq_\mathcal{C} \eta$ iff $\mathsf{E} f(\xi) \leq \mathsf{E} f(\eta)$ for all $f \in \mathcal{C}$; the expectations here always exist but may be infinite.

Relations between the notions introduced in the last two definitions are summarized in the following proposition. We write $A \subset B$ iff $A \subseteq B$ but $A \neq B$.

PROPOSITION 1.

(i): $\mathcal{H}_\alpha \subseteq \mathcal{F}_\alpha \cap \mathcal{G}_\alpha$.
(ii): $\alpha < \beta$ implies $\mathcal{F}_\beta \subseteq \mathcal{F}_\alpha$ and $\mathcal{G}_\beta \subseteq \mathcal{G}_\alpha$.
(iii): $\mathcal{G}_\alpha$ is a convex cone, which is closed in $\mathbb{R}^I$.
(iv): $\mathcal{F}_\alpha$ is a convex cone iff $\alpha \geq 1$.
(v): If $\alpha < 1$, then $\mathcal{F}_\alpha \subset \mathcal{G}_\alpha$.
(vi): $\mathcal{F}_1 = \mathcal{G}_1$.
(vii): If $\alpha > 1$, then $\mathcal{G}_\alpha \subset \mathcal{F}_\alpha$.
(viii): $\xi \preceq_{\mathcal{G}_\alpha} \eta$ iff $\xi \preceq_{\mathcal{H}_\alpha} \eta$ iff $\mathcal{I}^\alpha p \leq \mathcal{I}^\alpha q$ on $\mathbb{R}$,
where $p(x) := \Pr(\xi \geq x)$ and $q(x) := \Pr(\eta \geq x)$.
(ix): If $\alpha > 1$, then $\xi \preceq_{\mathcal{F}_\alpha} \eta$ implies $\xi \preceq_{\mathcal{G}_\alpha} \eta$.
(x): If $\alpha \leq 1$, then $\xi \preceq_{\mathcal{F}_\alpha} \eta$ iff $\xi \preceq_{\mathcal{G}_\alpha} \eta$ iff $\xi \preceq_{\mathcal{H}_\alpha} \eta$.
(xi): If card $(\mathbb{R} \cap \text{supp}\,\xi) \leq 1$, then $\xi \preceq_{\mathcal{F}_\alpha} \eta$ iff $\xi \preceq_{\mathcal{G}_\alpha} \eta$ iff $\xi \preceq_{\mathcal{H}_\alpha} \eta$.
(xii): Yet, in contrast to (ix)–(xi), for any $\alpha > 1$ and $\beta > 1$, there exist ξ and η such that card supp$\,\xi \leq 2$ and $\xi \preceq_{\mathcal{G}_\beta} \eta$, while $\xi \npreceq_{\mathcal{F}_\alpha} \eta$.

Proposition 1, especially its Part (xii), stems from a comment by the referee.

Based on Theorem 3, one can obtain Theorem 4 below, which somewhat generalizes Theorem 3.11 [14].

THEOREM 4. *Let $0 \leq \beta < \alpha < r \leq \infty$. Suppose that*

$$\xi \preceq_{\mathcal{G}_\alpha} \eta$$

and the tail function of η, $q(x) := \Pr(\eta \geq x)$, is r-concave. Then

(4.1) $$\mathsf{E}\, g(\xi) \leq c(r;\alpha,\beta)\, \mathsf{E}\, g(\eta) \quad \forall g \in \mathcal{G}_\beta,$$

which is equivalent – cf. Part (viii) of Proposition 1 – to

$$\mathsf{E}\,(\xi - x)_+^\beta \leq c(r;\alpha,\beta)\, \mathsf{E}\,(\eta - x)_+^\beta \quad \forall x \in \mathbb{R}$$

and also to

$$\mathcal{I}^\beta p \leq c(r;\alpha,\beta)\mathcal{I}^\beta q,$$

where

$$c(r;\alpha,\beta) := \frac{\Gamma(\alpha+1)\Gamma(r-\alpha)\alpha^{-\alpha}(r-\alpha)^{\alpha-r}}{\Gamma(\beta+1)\Gamma(r-\beta)\beta^{-\beta}(r-\beta)^{\beta-r}}$$

if $r < \infty$, and

$$c(\infty;\alpha,\beta) := \lim_{r \to \infty} c(r;\alpha,\beta) = \frac{\alpha^{-\alpha}\Gamma(\alpha+1)e^\alpha}{\beta^{-\beta}\Gamma(\beta+1)e^\beta}.$$

In particular,

$$\Pr(\xi \geq x) \leq c(r;\alpha,0)\Pr(\eta \geq x) \quad \forall x \in \mathbb{R},$$

and

$$c(r;\alpha,0) := \frac{\Gamma(\alpha+1)\Gamma(r-\alpha)\alpha^{-\alpha}(r-\alpha)^{\alpha-r}}{\Gamma(r)r^{-r}}$$

if $r < \infty$ and

$$c(\infty;\alpha,0) = \alpha^{-\alpha}\Gamma(\alpha+1)e^\alpha.$$

The constant $c(r;\alpha,\beta)$ is the best possible in (4.1).

Inequality (4.1) remains true if the condition $\xi \preceq_{\mathcal{G}_\alpha} \eta$ is replaced by $\xi \preceq_{\mathcal{F}_\alpha} \eta$; the constant $c(r;\alpha,\beta)$ remains the best possible in (4.1) in this case if $I = [-\infty,\infty)$.

REMARK 14. The differences between Theorem 4 above and Theorem 3.11 [14] are that here, first, the condition $\alpha \geq 1$ for optimality of $c(r;\alpha,\beta)$ is now removed and, second, the case $\xi \preceq_{\mathcal{G}_\alpha} \eta$ is treated, along with $\xi \preceq_{\mathcal{F}_\alpha} \eta$. In addition, g, ξ, and η for which (4.1) turns into an equality are explicitly constructed in the proof of Theorem 4.

In [14], a number of applications of Theorem 4 were given. In this article, we concentrate on the "discrete" case. The following is the main new application in this paper.

THEOREM 5. *Let $\eta_1,\ldots,\eta_n$ be independent zero-mean r.v.'s such that $\Pr(|\eta_i| \leq 1) = 1$ for all i. Let $\varepsilon_1,\ldots,\varepsilon_n$ be independent Rademacher r.v.'s, such that $\Pr(\varepsilon_i = \pm 1) = 1/2$ for all i. Let $a_1,\ldots,a_n$ be any real numbers such that $a_1^2 + \cdots + a_n^2 = n$. Then for any $x \in 2\mathbb{Z} - n := \{2z - n\colon z \in \mathbb{Z}\}$*

(4.2) $$\Pr(a_1\eta_1 + \cdots + a_n\eta_n \geq x) \leq c\Pr(\varepsilon_1 + \cdots + \varepsilon_n \geq x),$$

where $c := \dfrac{2e^3}{9} = c(\infty;3,0)$.

By conditioning on $\xi_1^2,\ldots,\xi_n^2$, one deduces the following corollary from Theorem 5.

COROLLARY 1. *Let $\xi_1,\ldots,\xi_n$ be any independent symmetric r.v.'s such that $\Pr(\xi_1 = 0)\ldots\Pr(\xi_n = 0) = 0$. Then for any $x \in 2\mathbb{Z} - n$*

$$\Pr\left(\frac{\xi_1 + \cdots + \xi_n}{\sqrt{\xi_1^2 + \cdots + \xi_n^2}} \geq \frac{x}{\sqrt{n}}\right) \leq c\,\Pr(\varepsilon_1 + \cdots + \varepsilon_n \geq x), \tag{4.3}$$

where $c := \dfrac{2e^3}{9}$.

REMARK 15. The condition $x \in 2\mathbb{Z} - n$ in Theorem 5 and Corollary 1 is essential. Indeed, let $n \geq 2$, $a_1 = \cdots = a_{n-1} = 1 - \delta$, $0 < \delta < 1$, and $a_n := \sqrt{n - (n-1)(1-\delta)^2}$, so that $a_1^2 + \cdots + a_n^2 = n$ and $a_n = 1 + (n-1)\delta + o(\delta)$ as $\delta \downarrow 0$. Hence, $\exists \delta \in (0,1)$ $(n-3)(1-\delta) + a_n > n - 2$. Then

$$\Pr(a_1\varepsilon_1 + \cdots + a_n\varepsilon_n > n - 2) \geq$$

$$\sum_{i=1}^{n-1} \Pr(\varepsilon_i = -1;\ \varepsilon_1 = \cdots = \varepsilon_{i-1} = \varepsilon_{i+1} = \cdots = \varepsilon_n = 1) = (n-1)2^{-n},$$

while

$$\Pr(\varepsilon_1 + \cdots + \varepsilon_n > n - 2) = 2^{-n}.$$

Therefore, there is no $c \in \mathbb{R}$ such that (4.2) holds for all $x \in \mathbb{R}$ $\Big($since $\Pr(\xi > n - 2) = \lim_{x\downarrow(n-2)} \Pr(\xi \geq x)$ for any r.v. $\xi\Big)$.

Using Corollary 1 with $\xi_{m+1} = \xi_{m+2} = \cdots = \xi_n = 0$ and letting $n \to \infty$, one obtains the following generalization of inequality (1.1) in the Introduction.

COROLLARY 2. *Let $\xi_1,\ldots,\xi_m$ be any independent symmetric r.v.'s such that $\Pr(\xi_1 = 0)\ldots\Pr(\xi_m = 0) = 0$. Then for any $u \in \mathbb{R}$*

$$\Pr\left(\frac{\xi_1 + \cdots + \xi_m}{\sqrt{\xi_1^2 + \cdots + \xi_m^2}} \geq u\right) \leq c(1 - \Phi(u)), \tag{4.4}$$

where $c := \dfrac{2e^3}{9}$.

REMARK 16. Inequalities (4.2) and (4.3) turn into the trivial equality $0 = 0$ when $x > n$. However, it can be seen from the proof, given below, that both inequalities (4.2) and (4.3) are strict when $x = n, n-2, n-4, \ldots$. Using some other considerations (cf. [13]), it can also be shown that inequality (4.4) is strict for all $u \in \mathbb{R}$.

5. Proofs

PROOF OF THEOREM 1. By Definition 1, it suffices to consider only $r < \infty$. It suffices to prove that $\mathcal{J}^\alpha q$ is $(r - \alpha)$-concave at each $n \in \{1, 2, \ldots\}$ – recall also Definition 3 and note that the r-concavity is invariant with respect to the shift $q(\cdot) \mapsto q(\cdot + m)$, $m \in \mathbb{Z}$. Next, without loss of generality, we shall assume that $\exists N \in \{1, 2, \ldots\}$ $\forall n \in (N, \infty) \cap \mathbb{Z}$ $q(n) = 0$ $\Big($otherwise, consider the r-concave function $q_N(n) := q(n)\mathrm{I}\{n \leq N\}$ and let $N \to \infty$; then $\mathcal{J}^\alpha q_N \to \mathcal{J}^\alpha q$ pointwise; note also that the pointwise convergence preserves the β-concavity for every $\beta > 0\Big)$. Then, on the set $\{N+1, N+2, \ldots\}$, $\mathcal{J}^\alpha q$ is zero and hence it is β-concave at each

$n \in \{N+1, N+2, \ldots\}$ for all $\beta > 0$. Thus, it suffices to prove that $\mathcal{J}^\alpha q$ is $(r-\alpha)$-concave for all $\alpha \in (0, r)$ at each $n \in \{1, \ldots, N\}$.

Introduce now the set

$$A := \{\alpha_1 \in [0, r) \colon \forall \alpha \in [0, \alpha_1) \ \mathcal{J}^\alpha q \text{ is } (r - \alpha) - \text{concave on } \{1, \ldots, N\}\}.$$

Obviously, $0 \in A$. Let $\rho := \sup A$. Note that $\rho \in A$ and $\rho \leq r$. It suffices to show that $\rho = r$. Hence, we may and shall assume that $\rho < r$. We may and shall also assume that $q < \infty$ on $\mathbb{Z}$; otherwise, consider the r-concave function $n \mapsto \min(C, q(n))$, letting $C \to \infty$.

Since $\rho \in A$ and $\rho < r$, it follows by the continuity of $\mathcal{J}^\alpha$ in α that $\mathcal{J}^\rho q$ is $(r - \rho)$-concave on $\{1, \ldots, N\}$. Also, $\mathcal{J}^\rho q$ is obviously non-increasing, since so is q. Therefore and in view of the mentioned semigroup property of $(\mathcal{J}^\alpha)$, we may and shall assume $\rho = 0$; otherwise, use $\mathcal{J}^\rho q$ and $r - \rho$ instead of q and r, respectively.

The continuity of $(\mathcal{J}^\alpha)$ in α also implies that if q is *strictly* r-concave at some $n \in \{1, \ldots, N\}$, then $\exists \delta_n > 0 \ \mathcal{J}^\alpha q$ is $(r - \alpha)$-concave at this n $\forall \alpha \in [0, \delta_n)$.

It remains to show that such a δ_n exists as well for each $n \in \{1, \ldots, N\}$ at which q is not strictly r-concave. Then it will follow that $\rho \geq \min\{\delta_1, \ldots, \delta_N\} > 0$, which contradicts the assumption $\rho = 0$. Thus, Theorem 1 is now reduced to Lemma 1 below. □

LEMMA 1. *Let $r \in (0, \infty)$. Suppose that $q \colon \mathbb{Z} \to [0, \infty)$ is non-increasing and such that $\exists N \in \mathbb{Z} \ \forall m \in (N, \infty) \cap \mathbb{Z} \ q(m) = 0$. Suppose also that q is r-concave on $\{1, 2, \ldots\}$ but is not strictly r-concave at some $n \in \{1, 2, \ldots\}$. Then $\exists \delta > 0 \ \forall \alpha \in (0, \delta) \ \mathcal{J}^\alpha q$ is $(r - \alpha)$-concave at this n.*

PROOF. In what follows, let us write q_m for $q(m)$, $\mathcal{J}^\alpha q_m$ for $(\mathcal{J}^\alpha q)(m)$, etc., for any $m \in \mathbb{Z}$. If $q_{n+1} = 0$ for the n in the statement of the lemma, then $\mathcal{J}^\alpha q$ being $(r - \alpha)$-concave at this n is trivial. Thus, without loss of generality, $q_{n+1} > 0$, whence $q_{n-1} \geq q_n \geq q_{n+1} > 0$. Note further that

$$\frac{d}{d\alpha}\left[(\mathcal{J}^\alpha q_m)^{-\frac{1}{r-\alpha}}\right]\bigg|_{\alpha=0} = -r^{-2} Q_m,$$

where

(5.1) $$Q_m := r\ell_m q_m^{-\frac{1}{r}-1} + q_m^{-1/r} \ln q_m,$$

$$\ell_m := \frac{d}{d\alpha}(\mathcal{J}^\alpha q_m)\bigg|_{\alpha=0} = \sum_{k=1}^{\infty} \frac{1}{k} q_{m+k}.$$

Hence, it suffices to show that

(5.2) $$\Delta^2 Q_{n+1} < 0,$$

where $\Delta^2 f := \Delta \Delta f$, and $\Delta f_m := (\Delta f)_m := f_m - f_{m-1}$, for any sequence $f := (f_m)$. Observe that for any two sequences $f := (f_m)$ and $g := (g_m)$,

(5.3) $$\Delta^2 (fg)_{n+1} = g_n \Delta^2 f_{n+1} + f_{n+1} \Delta^2 g_{n+1} + (f_{n+1} - f_{n-1}) \Delta g_n.$$

Let here $f := \ell$ and $g := q^{-\frac{1}{r}-1}$. Then

(5.4) $$g \geq 0, \quad \Delta g \geq 0, \quad \Delta^2 g \geq 0.$$

Indeed, $g \geq 0$ since $q \geq 0$, $\Delta g \geq 0$ since q is non-increasing, and $\Delta^2 g \geq 0$ since $g = (q^{-\frac{1}{r}})^{r+1}$ is convex (because $q^{-\frac{1}{r}}$ is convex and $[0,\infty) \ni x \mapsto x^{r+1}$ is increasing and convex). Also,

$$f_{n+1} = \sum_{k=2}^{\infty} \frac{1}{k-1} q_{n+k},$$

$$f_{n+1} - f_{n-1} = -q_n - \frac{1}{2} q_{n+1} + \sum_{k=2}^{\infty} \frac{2}{(k-1)(k+1)} q_{n+k},$$

$$\Delta^2 f_{n+1} = q_n - \frac{3}{2} q_{n+1} + \sum_{k=2}^{\infty} \frac{2}{(k-1)k(k+1)} q_{n+k}.$$

Observe that the coefficients of $q_{n+2}, q_{n+3}, \ldots$ in the R.H.S.'s of the last three equations are all positive. It follows from (5.1), (5.3), and (5.4) that $\Delta^2 Q_{n+1}$ is non-decreasing in q_{n+2}, in q_{n+3}, ..., since one can write

(5.5) $$\Delta^2 Q_{n+1} = c_1 + \sum_{k=2}^{\infty} c_k q_{n+k},$$

where $c_2, c_3, \ldots$ are nonnegative and, moreover, $c_1, c_2, \ldots$ depend only on r, q_{n-1}, q_n, and q_{n+1} but not on $q_{n+2}, q_{n+3}, \ldots$; in particular, c_1 contains the term $\left(\Delta^2 \left(q^{-1/r} \ln q\right)\right)_{n+1}$, which does not depend on $q_{n+2}, q_{n+3}, \ldots$.

Because $q^{-1/r}$ is convex on $\mathbb{Z}$ but *not strictly* convex at n, one has

$$q_k \leq (ak+b)^{-r} \quad \forall k \geq n-1$$

and

$$q_k = (ak+b)^{-r} \quad \forall k \in \{n-1, n, n+1\},$$

where $a = q_n^{-1/r} - q_{n-1}^{-1/r} = q_{n+1}^{-1/r} - q_n^{-1/r}$ and $b = q_n^{-1/r} - an$; recall that $q > 0$ on $\{n-1, n, n+1\}$. Note that $a \geq 0$, since q is non-increasing. Moreover, without loss of generality $a > 0$; consider otherwise, in place of q, the lowest r-concave majorant q_ε of q such that $q_\varepsilon(n-1) \geq q(n-1) + \varepsilon$ and let $\varepsilon \downarrow 0$. It also follows that $ak + b > 0$ for all $k \geq n-1$. Thus, in view of (5.5), inequality (5.2) follows from Lemma 2 below. □

LEMMA 2. *Suppose that $\exists n \in \mathbb{Z}$ $\exists a > 0$ $\exists b \in \mathbb{R}$ $a(n-1) + b > 0$ & $\forall k \in \{n-1, n, \ldots\}$ $q_k = (ak+b)^{-r}$. Then (5.2) holds.*

PROOF. Without loss of generality, $a = 1$. In this case, recalling (5.1), one has

$$\frac{1}{r} Q_m = f(x) := \sum_{k=1}^{\infty} \frac{1}{k} \frac{x^{r+1}}{(x+k)^r} - x \ln x,$$

where $x := m + b$. Hence, it suffices to show that

(5.6) $$x f''(x) = r(r+1) x^r \sum_{k=1}^{\infty} \frac{k}{(k+x)^{r+2}} - 1 < 0 \quad \forall x > 0.$$

Using the identity

$$\frac{1}{\lambda^p} = \frac{1}{\Gamma(p)} \int_0^{\infty} u^{p-1} e^{-\lambda u} du, \quad p > 0, \quad \lambda > 0,$$

with $p = r+2$, one sees that

$$xf''(x) + 1 = \frac{r(r+1)x^r}{\Gamma(r+2)} \int_0^\infty u^{r+1} e^{-xu} \sum_{k=1}^\infty k e^{-ku} du$$

$$= \frac{1}{\Gamma(r)} \int_0^\infty v^{r-1} e^{-v} h\left(\frac{v}{x}\right) dv,$$

where

$$h(u) := \frac{u^2 e^{-u}}{(1-e^{-u})^2}.$$

Now (5.6) follows from the elementary inequality $h(u) < 1$ for all $u > 0$ (in fact, for all $u \neq 0$). Thus, Lemma 2 is proved, and so are Lemma 1 and Theorem 1. □

PROOF OF THEOREM 2. We shall see that Theorem 2 can be easily reduced to Theorem 1. Note that $q \geq 0$ implies that $\mathcal{J}^1 q$ is non-increasing. Hence, using Theorem 1 and the semigroup property of $(\mathcal{J}^\alpha)$, one needs only to prove the theorem for $\alpha = 1$.

In other words, it suffices to prove that if $q \colon \mathbb{Z} \to [0, \infty)$ is r-concave for some $r \in (1, \infty)$, then

(5.7) $$(q_0 + q_1 + R)^s + R^s \geq 2(q_1 + R)^s,$$

where $R := \mathcal{J}^1 q_2 = q_2 + q_3 + \ldots$ and $s := -1/(r-1) < 0$.

If $q_1 = 0$, then the r-concavity of q implies either $q_0 = 0$ or $R = 0$, so that (5.7) is trivial in this case. Assume now that $q_1 > 0$. Then there is a function $\tilde{q}$ (say of the form $\tilde{q}(x) = (ax+b)_+^{-r}$), which is non-increasing and r-concave on the set $\{0, 1, \ldots\}$ and such that $\tilde{q}_1 = q_1$ and $\mathcal{J}^1 \tilde{q}_2 = \mathcal{J}^1 q_2 = R$. Hence, by Theorem 1, one has

(5.8) $$(\tilde{q}_0 + q_1 + R)^s + R^s \geq 2(q_1 + R)^s.$$

In view of Theorem 1, one may assume that q is *not* non-increasing on the set $\{0, 1, \ldots\}$; in other words, $q_n < q_{n+1}$ for some $n \in \{0, 1, \ldots\}$; this assumption and the r-concavity of q imply $q_0 < q_1$, so that one has $q_0 < q_1 = \tilde{q}_1 \leq \tilde{q}_0$, since $\tilde{q}$ is non-increasing. Hence, $q_0 < \tilde{q}_0$. This and (5.8), together with $s < 0$, now yield (5.7). □

PROOF OF PROPOSITION 1. (i) is trivial.

(ii). Let $\alpha < \beta$. Then $\mathcal{F}_\beta \subseteq \mathcal{F}_\alpha$ follows because the function $[0, \infty) \ni x \mapsto x^{\beta/\alpha}$ is convex and increasing; $\mathcal{G}_\beta \subseteq \mathcal{G}_\alpha$ follows because, by the Fubini Theorem, $\int (u-t)_+^\beta \mu(dt) = \int (u-z)_+^\alpha \lambda(dz)$, where

$$\lambda(dz) := \frac{\Gamma(\beta+1)}{\Gamma(\alpha+1)\Gamma(\beta-\alpha)} \left[\int (z-t)_+^{\beta-\alpha-1} \mu(dt)\right] dz.$$

(iii). That $\mathcal{G}_\alpha$ is a convex cone is trivial. It remains to show that $\mathcal{G}_\alpha$ is closed. This can be done by routine compactness argument. Suppose that (f_n) is a sequence in $\mathcal{G}_\alpha$ which converges to some $f \in \mathbb{R}^I$. Then $f \in \mathcal{E} = \mathcal{E}_I$, since $\mathcal{E}_I$ is closed in $\mathbb{R}^I$. Hence, it suffices to show that $f \in \mathcal{G}_\alpha$, because $\mathcal{E}$ has a locally countable topological base consisting, e.g., of the sets of the form $\{h \in \mathcal{E} \colon \sup_{t \leq m} |h(t) - g(t)| < 1/m\}$, where $g \in \mathcal{E}$ and $m = 1, 2, \ldots$. According to the definition of $\mathcal{G}_\alpha$, $\forall n$ $\exists \mu_n \in M$ $\forall u \in I$ $f_n(u) = \int (u-t)_+^\alpha \mu_n(dt)$. Let $\nu_n(dt) := (|t| \vee 1)^\alpha \mu_n(dt)$, so that

$$f_n(u) = \int \left(\frac{u-t}{|t| \vee 1}\right)_+^\alpha \nu_n(dt).$$

Then

$$b \geq 1 \Rightarrow \nu_n((-\infty, b]) \leq \int_{-\infty}^b \left(\frac{2b-t}{|t| \vee 1}\right)^\alpha \nu_n(dt) \leq f_n(2b),$$

$$b \leq -1 \Rightarrow 2^{-\alpha} \nu_n((-\infty, 2b]) \leq \int_{-\infty}^{2b} \left(\frac{b-t}{|t| \vee 1}\right)^\alpha \nu_n(dt) \leq f_n(b).$$

Hence, the sequence (ν_n) of measures is weakly compact on $(-\infty, b]$ for each $b \in \mathbb{R}$. Therefore, for each $b \in \{1, 2, \ldots\}$, there exists a weakly convergent on $(-\infty, b]$ subsequence $(\nu_n^{(b)})$ of (ν_n) such that $(\nu_n^{(b+1)})$ is a subsequence of $(\nu_n^{(b)})$, for each $b \in \{1, 2, \ldots\}$. Then the "diagonal" subsequence $(\nu_n^{(n)})$ weakly converges to some $\nu \in M$ on each interval $(-\infty, b]$, $b \in \{1, 2, \ldots\}$. Note also that the integrand $\left(\frac{u-t}{|t| \vee 1}\right)_+^\alpha$ is continuous in $t \in \mathbb{R}$ and bounded by $(|u|+1)^\alpha$. Thus, $\forall u \in \mathbb{R}$

$$f(u) = \lim_n f_n(u) = \int \left(\frac{u-t}{|t| \vee 1}\right)_+^\alpha \nu(dt) = \int (u-t)_+^\alpha \mu(dt),$$

where $\mu(dt) := (|t| \vee 1)^{-\alpha} \nu(dt)$. If $(-\infty) \in I$ is the case, then $f(u) = \int (u-t)_+^\alpha \mu(dt)$ is trivial at $u = -\infty$.

(iv). Suppose first that $\alpha \geq 1$; then, because the function $[0, \infty) \times [0, \infty) \ni (x, y) \mapsto (x^\alpha + y^\alpha)^{1/\alpha}$ is convex in (x, y) and also non-decreasing in x and in y, it follows that $\mathcal{F}_\alpha$ is a convex cone. If $\alpha < 1$, then taking the function $u \mapsto u_+^\alpha + (u+1)_+^\alpha$, one sees that $\mathcal{F}_\alpha$ is not a convex cone.

(v). Let $0 < \alpha < 1$. First, the same example $u \mapsto u_+^\alpha + (u+1)_+^\alpha$ shows that $\mathcal{F}_\alpha \neq \mathcal{G}_\alpha$. It remains to prove that $\mathcal{F}_\alpha \subseteq \mathcal{G}_\alpha$. Let $f \in \mathcal{F}_\alpha$. We only need to show that $f \in \mathcal{G}_\alpha$. Observe that the set of all functions in $\mathcal{F}_\alpha$ that are zero in a neighborhood of $(-\infty)$ and are of class C^2 is dense in $\mathcal{F}_\alpha$. Recall also that, according to (iii), $\mathcal{G}_\alpha$ is closed. Hence, one may assume that f itself is zero in a neighborhood of $(-\infty)$ and of class C^2. Moreover, without loss of generality, $I = \mathbb{R}$, $f = 0$ on $(-\infty, 0]$, and $f > 0$ on $(0, \infty)$. Now let $\forall v \in \mathbb{R}$

$$m_f(v) := \int_{-\infty}^v (v-t)^{-\alpha-2} [f(t) - f(v) - (t-v)f'(v)] dt.$$

Note that $m_f = 0$ on $(-\infty, 0]$. By a Taylor series formula,

(5.9) $$f(t) - f(v) - (t-v)f'(v) = \int_t^v (z-t) f''(z) dz$$

(or twice integrate the R.H.S. of (5.9) by parts, to obtain the L.H.S.). Hence, $\forall v \in \mathbb{R}$

$$m_f(v) = \int_{-\infty}^v (v-t)^{-\alpha-2} dt \int_t^v (z-t) f''(z) dz.$$

Now, using the Fubini Theorem – or, alternativiely, twice integrating by parts – one has

$$m_f(v) = \frac{1}{\alpha(\alpha+1)} \int_{-\infty}^v (v-z)^{-\alpha} f''(z) dz.$$

Integrating by parts again, one has $\forall u \in \mathbb{R}$

$$\alpha(\alpha+1)\int (u-v)_+^\alpha m_f(v)dv = \Gamma(\alpha+1)\Gamma(1-\alpha)\int_{-\infty}^u (u-z)f''(z)dz.$$

Now (5.9) with $v \to -\infty$ (or another two integrations by parts) yields $\int_{-\infty}^u (u-z)f''(z)dz = f(u)$ $\forall u \in \mathbb{R}$, and so,

$$f(u) = k\int (u-v)_+^\alpha m_f(v)dv,$$

where $k := \dfrac{\alpha+1}{\Gamma(\alpha)\Gamma(1-\alpha)}$. By definition of $\mathcal{G}_\alpha$, it remains now to show that $m_f \geq 0$ on $(0,\infty)$. But for every $v > 0$ there exists a function of the form $f_{c,d}: t \mapsto c(t-d)_+^\alpha$ for some $c := c(f,v) > 0$ and $d := d(f,v) < v$ such that

$$f_{c,d}(v) = f(v), \quad f'_{c,d}(v) = f'(v), \quad \text{and } f_{c,d} \leq f.$$

Therefore, $m_f \geq m_{f_{c,d}}$, and so, it suffices to show that $m_{f_{c,d}} = 0$ on (d,∞), for then $m_f(v) \geq m_{f_{c,d}}(v) = 0$. Without loss of generality, $c = 1$ and $d = 0$. Thus, it remains to show that $m_{f_{1,0}} = 0$ on $(0,\infty)$, which is equivalent to $\int_{-\infty}^1 (1-z)^{-\alpha-2}[z_+^\alpha - 1 - (z-1)\alpha]dz = 0$, by homogeneity. Represent the latter integral as $\int_{-\infty}^1 \cdots = \int_{-\infty}^0 \cdots + \int_0^1 \cdots$. It is easy to see that the $\int_{-\infty}^0 \cdots = \dfrac{\alpha}{\alpha+1}$. Integrating now the $\int_0^1 \cdots$ by parts, it suffices to show that $\alpha \int_0^1 (1-z)^{-\alpha-1}(z^{\alpha-1}-1)dz = 1$. But this follows from the more general formula: for all complex numbers α, β, and λ with $\operatorname{Re}\lambda < 1$, $\operatorname{Re}\alpha > 0$, and $\operatorname{Re}\beta > 0$, one has

$$(5.10) \qquad \lambda \int_0^1 \frac{z^{\alpha-1} - z^{\beta-1}}{(1-z)^{-\lambda-1}}dz = (\lambda-\alpha)\mathrm{B}(1-\lambda,\alpha) - (\lambda-\beta)\mathrm{B}(1-\lambda,\beta).$$

In turn, (5.10) is obvious when $\operatorname{Re}\lambda < 0$, since in that case the L.H.S. of (5.10) is $\lambda\mathrm{B}(-\lambda,\alpha) - \lambda\mathrm{B}(-\lambda,\beta)$. But the L.H.S. of (5.10) (as well as its R.H.S.) is analytic in λ for $\operatorname{Re}\lambda < 1$; hence, it finally remains to apply the principle of uniqueness for analytic functions. [Note that (5.10) implies the seemingly disparate Formulas 2.2.4.19 [17] $\left(\text{since } \mathrm{B}(\alpha,1-\alpha) = \dfrac{\pi}{\sin\pi\alpha} \text{ if } 0 < \operatorname{Re}\alpha < 1\right)$ and 2.2.4.20 [17] (letting $\lambda \to 0$), as well as Formulas 2.2.4.21–23 [17].] This completes the proof of Part (v).

(vi). That $\mathcal{G}_1 \subseteq \mathcal{F}_1$ is obvious. To see that every $f \in \mathcal{F}_1$ belongs to $\mathcal{G}_1$, set $\mu(dt) := df'(t)$, where f' is the (R.H.S., say) derivative of f. Since every $f \in \mathcal{F}_\alpha$ is convex, one has $(-t/2)f'(t) \leq f(t/2) - f(t)$, which implies $tf'(t) \to 0$ as $t \to -\infty$. Now integration by parts yields $f(u) = \int_{-\infty}^u f'(t)dt = \int_{-\infty}^u (u-t)df'(t) = \int (u-t)_+ \mu(dt)$.

(vii) Let $\alpha > 1$. That $\mathcal{G}_\alpha \subseteq \mathcal{F}_\alpha$ follows from (iv) and the formula

$$\int (u-t)_+^\alpha \mu(dt) = \lim_{n \to \infty} \sum_{k=-n^2}^{n^2} \left(u - \frac{k}{n}\right)_+^\alpha \mu\left(\left[\frac{k-1}{n}, \frac{k}{n}\right)\right),$$

which in turn follows from the Fatou Lemma. That $\mathcal{G}_\alpha \neq \mathcal{F}_\alpha$ follows from (xii).

(viii). The first "iff" follows from the Fubini Theorem; the second one, from Remark 6.

(ix). This is immediate from (vii).

(x). It follows from (i), (v), and (vi) that $\mathcal{H}_\alpha \subseteq \mathcal{F}_\alpha \subseteq \mathcal{G}_\alpha$ if $\alpha \leq 1$. It remains to use (viii).

(xi). Let card$(\mathbb{R} \cap \operatorname{supp} \xi) \leq 1$. Then $\operatorname{supp} \xi \subseteq \{-\infty, b\}$ for some $b \in \mathbb{R}$. In view of (viii), (ix), and (x), it suffices to show that for such ξ, $\xi \preceq_{\mathcal{H}_\alpha} \eta$ implies $\xi \preceq_{\mathcal{F}_\alpha} \eta$. Let $f \in \mathcal{F}_\alpha$. Then there exists a function $h \in \mathcal{H}_\alpha$ such that $h \leq f$ on I and $h = f$ on $\operatorname{supp} \xi \subseteq \{-\infty, b\}$. Hence, $\xi \preceq_{\mathcal{H}_\alpha} \eta$ implies $\mathsf{E} f(\xi) = \mathsf{E} h(\xi) \leq \mathsf{E} h(\eta) \leq \mathsf{E} f(\eta)$. This proves Part (xi).

(xii). A general idea of this proof was suggested by the anonymous referee. Let $\alpha > 1$ and $\beta > 1$. In view of (viii), it suffices to find a function $f \in \mathcal{F}_\alpha$ and a finite signed Borel measure γ on I such that $\int f \mathrm{d}\gamma < 0$, $\gamma(I) = 0$, $\forall h \in \mathcal{H}_\alpha \int h \mathrm{d}\gamma \geq 0$, and card $\operatorname{supp} \gamma_- \leq 2$; we denote the positive and negative parts of measure γ by γ_+ and γ_-, so that $\gamma = \gamma_+ - \gamma_-$. Indeed, then we could let ξ and η have probability distributions $2\|\gamma\|^{-1}\gamma_-$ and $2\|\gamma\|^{-1}\gamma_+$, respectively, where $\|\gamma\| := \gamma_+(I) + \gamma_-(I)$. We claim that
$$f(x) := x_+^\alpha \mathrm{I}\{x \leq n\} + [n + \lambda(x-n)]^\alpha \mathrm{I}\{x > n\}$$
and
$$\gamma := \delta_{n+2} - k\delta_{n+1} + p\delta_n + (k-p-1)\delta_{n-1},$$
where δ_x is the Dirac delta-measure at x, satisfy all above requirements if positive real numbers k, p, λ, n are chosen as described below. Indeed, it is obvious that $f \in \mathcal{F}_\alpha$ if $\lambda \geq 1$; it is also obvious that card $\operatorname{supp} \gamma_- \leq 2$. Next, fixing any $k > 2$ and $p > 0$ and then letting both λ and n/λ tend to ∞, one has
$$\int f \mathrm{d}\gamma = (n+2\lambda)^\alpha - k(n+\lambda)^\alpha + pn^\alpha + (k-p-1)(n-1)^\alpha$$
$$= \alpha n^{\alpha-1}(2\lambda - k\lambda - (k-p-1))(1 + o(1)) \to -\infty.$$
It remains to show that one can choose $k > 2$ and $p > 0$ so that $\int (u-t)_+^\beta \gamma(\mathrm{d}u) \geq 0$ $\forall t \in \mathbb{R}$, that is, $\forall x \in \mathbb{R}$

(5.11) $\qquad x_+^\beta - k(x-1)_+^\beta + p(x-2)_+^\beta + (k-p-1)(x-3)_+^\beta \geq 0.$

Inequality (5.11) is trivial for $x \leq 1$. For $x \in (1, 2]$, (5.11) is equivalent to $\left(\dfrac{x}{x-1}\right)^\beta \geq k$. Thus, (5.11) $\forall x \in (1, 2]$ is equivalent to $2^\beta \geq k$. Choose now k to be any number in the interval $(2, 2^\beta)$. It remains to choose $p > 0$ so that (5.11) holds for all $x > 2$. The last requirement is equivalent to
$$p \geq h(x) := \frac{k[(x-1)^\beta - (x-3)_+^\beta] - x^\beta + (x-3)_+^\beta}{(x-2)^\beta - (x-3)_+^\beta} \quad \forall x > 2.$$
To finish the proof of Part (xii) and thereby that of the entire proposition, it remains to notice that h is bounded from above on $(2, \infty)$, because h is continuous on $(2, \infty)$, $h(2+0) = -\infty$, and $h(\infty - 0) = 2k - 3 \in \mathbb{R}$. $\square$

PROOF OF THEOREM 4. In accordance with Remark 14, most of the statements of Theorem 4 were proved in [14].

That (4.1) remains true if the condition $\xi \preceq_{\mathcal{G}_\alpha} \eta$ is replaced by $\xi \preceq_{\mathcal{F}_\alpha} \eta$ follows from Parts (ix) and (x) of Proposition 1.

It remains to prove the optimality of the constant factor $c(r;\alpha,\beta)$ in (4.1) for all $\alpha > 0$ (rather than only for $\alpha \geq 1$ as in [**14**]), and do this not only under the condition $\xi \preceq_{\mathcal{G}_\alpha} \eta$ but also under the condition $\xi \preceq_{\mathcal{F}_\alpha} \eta$. In [**14**], the proof of optimality was based on an existence-of-minimax-duality theorem of Kemperman [**4**]. Here, we shall explicitly construct g, ξ and η for which (4.1) turns into an equality.

Consider first $r < \infty$, in which case assume that the tail function of η is

(5.12) $$q(x) := (1 \vee x)^{-r}.$$

Let

(5.13) $$\gamma_\alpha(t) := \mathsf{E}\,(\eta - t)_+^\alpha.$$

Note that

(5.14) $$\forall t > 1 \ \gamma_\alpha(t) = K_\alpha t^{\alpha-r}, \quad \text{where } K_\alpha := \frac{\Gamma(\alpha+1)\Gamma(r-\alpha)}{\Gamma(r)}.$$

Let

(5.15) $$x > 1,$$

(5.16) $$y := x\frac{r}{r - \beta},$$

(5.17) $$\pi := K_\alpha \frac{r^r}{\alpha^\alpha (r-\alpha)^{r-\alpha}} y^{-r}.$$

Assume also that x is large enough so that

(5.18) $$\pi < 1.$$

Consider first the case $I = [-\infty, \infty)$. Let

(5.19) $$\mathsf{Pr}\,(\xi = y) = \pi = 1 - \mathsf{Pr}\,(\xi = -\infty).$$

Then it follows from (5.13)–(5.17) and (5.19) that (4.1) turns into an equality for $g(u) := (u - x)_+^\beta$:

(5.20) $$\mathsf{E}\,(\xi - x)_+^\beta = c(r;\alpha,\beta)\,\mathsf{E}\,(\eta - x)_+^\beta.$$

Let us now show that for sufficiently large $x > 1$

(5.21) $$\inf_{s \in (-\infty, y)} F(s) = \pi,$$

where y and π are defined by (5.16) and (5.17), and

(5.22) $$F(s) := F_{\alpha,y}(s) := \frac{\gamma_\alpha(s)}{(y-s)^\alpha}.$$

Then it will follow that $\forall s \in (-\infty, y) \ \pi(y-s)^\alpha \leq \gamma_\alpha(s)$, which will in turn imply $\xi \preceq_{\mathcal{H}_\alpha} \eta$.

By (5.13) and Remark 6,

$$\gamma_\alpha = \Gamma(\alpha+1)\mathcal{I}^\alpha q = \Gamma(\alpha+1)\mathcal{I}^{\alpha+1}w = \mathcal{I}^1\left(\Gamma(\alpha+1)\mathcal{I}^\alpha w\right),$$

where – recall (5.12) – $w(u) := ru^{-r-1}\mathrm{I}\{u > 1\}$ is the p.d.f. of η. Next,

$$\Gamma(\alpha+1)\,(\mathcal{I}^\alpha w)\,(t) = \alpha \int_0^\infty u^{\alpha-1} w(u+t)\,du$$

is continuous in t. Hence, $\gamma := \gamma_\alpha$ has a (continuous) derivative, $\gamma' = -\Gamma(\alpha+1)\mathcal{I}^\alpha w$, and we also see that $\gamma' < 0$.

It follows that for $s \in (-\infty, y)$,

(5.23) $$\operatorname{sign} F'(s) = \operatorname{sign}(m(s) - y),$$

where

$$m(s) := s - \frac{\alpha \gamma(s)}{\gamma'(s)}$$

is continuous in s. By (5.14),

(5.24) $$m(s) = s \frac{r}{r - \alpha} \quad \forall s > 1.$$

By the Lebesgue Theorem and (5.22),

(5.25) $$F(-\infty + 0) = \lim_{s \to -\infty} \mathsf{E}\left(1 + \frac{\eta}{|s|}\right)^\alpha \frac{|s|^\alpha}{(y-s)^\alpha} = 1.$$

Note also that

(5.26) $$F(y + 0) = \infty.$$

By direct calculations and in view of (5.17) and (5.18), one has

(5.27) $$F(s_y) = \pi < 1$$

for sufficiently large y, where, according to (5.24),

(5.28) $$s_y := y \frac{r - \alpha}{r}$$

is the only root in the interval $(1, \infty)$ of the equation

(5.29) $$m(s) = y.$$

Hence and in view of (5.23), (5.25), (5.26), and (5.27), for y large enough, $\inf_{s \in (-\infty, y)} F(s)$ is achieved at one of the roots s of (5.29). Observe that if we let $y \to \infty$, then the roots of (5.29) other than s_y (if there are any such roots) must tend to $(-\infty)$, since m is continuous and (5.29) has only one root $s = s_y$ in $(1, \infty)$.

We conclude that for y large enough, $\inf_{s \in (-\infty, y)} F(s) = F(s_y) = \pi$, so that (5.21) is verified. Hence, $\xi \preceq_{\mathcal{H}_\alpha} \eta$. Now Part (xi) of Proposition 1 implies $\xi \preceq_{\mathcal{F}_\alpha} \eta$ and $\xi \preceq_{\mathcal{G}_\alpha} \eta$. Together with (5.20), this completes the proof of the optimality of $c(r; \alpha, \beta)$ in the case $I = [-\infty, \infty)$ and $r < \infty$.

The case when $I = [-\infty, \infty)$ and $r = \infty$ is quite similar; here one should consider $q(x) = 1 \wedge (e^{-x})$ (cf. (5.12)), $y := x + \beta$ (cf. (5.16)), $\pi := \Gamma(\alpha+1)\alpha^{-\alpha}e^\alpha e^{-y}$ (cf. (5.17)), and $s_y := y - \alpha$ (cf., (5.28)).

It remains to demonstrate the optimality of (4.1) under the assumptions $\xi \preceq_{\mathcal{H}_\alpha} \eta$ (which is equivalent to $\xi \preceq_{\mathcal{G}_\alpha} \eta$) and $I = \mathbb{R}$. Since here we want ξ to take on only real values, the construction given by (5.19) needs to be modified. Let us do this by replacing $(-\infty)$ in (5.19) by a $z \in (-\infty, 0)$; note that then $z < y$, in view of (5.15) and (5.16).

Then all that remains to finish the proof of the theorem is to show that $\exists z \in (-\infty, 0) \ \forall s \in (-\infty, z)$

$$[\mathsf{E}(\xi - s)^\alpha =] (1 - \pi)(z - s)^\alpha + \pi(y - s)^\alpha \leq \mathsf{E}(\eta - s)^\alpha.$$

Since $\Pr(\eta \geq 0) = 1$, the proof of the theorem is now reduced to checking of the following: $\forall y > 0 \ \forall \pi \in [0, 1) \ \exists A > 0 \ \forall v > 0$

(5.30) $$(1 - \pi)v^\alpha + \pi(A + y + v)^\alpha \leq (A + v)^\alpha;$$

here we have made the substitutions $v := z - s$ and $A := -z$. Using another substitution, $w := 1/v$, (5.30) can be further rewritten as $g \geq 0$ on $(0, \infty)$, where

$$g(w) := (1 + Aw)^\alpha - (1 - \pi) - \pi[1 + (A + y)w]^\alpha.$$

Note that $g(0) = 0$. It remains to finally notice that for all $w \geq 0$

$$\frac{g'(w)}{\alpha(A + y)(1 + Aw)^{\alpha-1}} = \frac{A}{A + y} - \pi \left[\frac{1 + (A + y)w}{1 + Aw}\right]^{\alpha-1}$$

$$\geq \frac{A}{A + y} - \pi \left[1 \vee \left(\frac{A + y}{A}\right)^{\alpha-1}\right] > 0$$

if $A > 0$ is large enough. □

PROOF OF THEOREM 5. Rewrite (4.2) as

(5.31) $$\Pr(H \geq z) \leq c \Pr(B \geq z)$$

for all $z \in \mathbb{Z}$, where

$$H := \frac{\eta_1 a_1 + \cdots + \eta_n a_n + n}{2} \quad \text{and} \quad B := \frac{\varepsilon_1 + \cdots + \varepsilon_n + n}{2}.$$

Note that $\Pr(0 \leq H \leq n) = 1$ and $\Pr(0 \leq B \leq n) = 1$; hence, it suffices to prove (5.31) for $z \in \{0, 1, \ldots, n\}$. Fix any such z.

By a result of Eaton (1970, 1974) [**8, 9**],

$$\mathsf{E}\,(H - t)_+^3 \leq \mathsf{E}\,(B - t)_+^3 \quad \forall t \in \mathbb{R};$$

cf. Pinelis (1994) [**13**]. Using now the Chebyshev inequality, one has

(5.32) $$\Pr(H \geq z) \leq \frac{\mathsf{E}\,(B - t)_+^3}{(z - t)^3}$$

for all $t \in (-\infty, z)$. Observe that for any $t \in \mathbb{Z}$,

(5.33) $$\mathsf{E}\,(B - t)_+^3 = \sum_{u=t}^{\infty}(u - t)^3 p_u = \sum_{k=1}^{\infty}(k^3 - (k - 1)^3)q_{t+k},$$

where

(5.34) $$p_u := \Pr(B = u) \text{ and } q_u := \mathcal{J}^1 p_u = \Pr(B \geq u), \quad u \in \mathbb{Z},$$

the p.m.f. and the (first-order) tail of the binomial r.v. B. By Remark 13, q is ∞-concave. Therefore, $\exists a > 0 \; \exists b \in (0, 1)$

(5.35) $$q_t \leq ab^{t-1} \; \forall t \in \mathbb{Z},$$

while

(5.36) $$q_z = ab^{z-1}.$$

It follows from (5.33) and (5.35) that

$$\mathsf{E}\,(B - t)_+^3 \leq \frac{a\,b^t\,(1 + 4b + b^2)}{(1 - b)^3}$$

$\Big($note that $k^3 - (k-1)^3 = 3k(k-1) + 1$ and $\sum_{k=1}^{\infty} k(k-1)b^{k-2} = ((1-b)^{-1})'' = 2(1-b)^{-3}\Big)$. Now (5.32) and (5.36) imply that (5.31) and hence (4.2) take place with

$$c_* := \inf\{c_m \colon m = 1, 2, \dots\}$$

instead of $c = \dfrac{2e^3}{9}$, where

$$c_m := c(m, b) := \frac{b^{1-m}\left(1 + 4b + b^2\right)}{(1-b)^3 m^3}.$$

It remains to prove that $c_* \leq c$ $\forall b \in (0, 1)$. Observe that the ratio

$$d_m := \frac{c_m}{c_{m-1}} = \frac{1}{b}\left(\frac{m-1}{m}\right)^3$$

increases from $\dfrac{1}{8b}$ to $\dfrac{1}{b} > 1$ as m increases from 2 to ∞.

Hence, in case $b < 1/8$, $c_* = c(1, b) < c(1, 1/8) = \dfrac{776}{343} < \dfrac{2e^3}{9} = c$, since $c(1, b)$ increases in $b \in (0, 1)$.

It remains to show that $c_* \leq c$ in the case $1/8 \leq b < 1$. In this case, $\exists! m \in \{2, 3, \dots\}$ $d_m \leq 1 < d_{m+1}$, i.e., $b_-(m) \leq b < b_+(m)$, where

$$b_\pm(m) := \left(\frac{m - 1/2 \pm 1/2}{m + 1/2 \pm 1/2}\right)^3.$$

Hence, it suffices to show that

$$\sup\{c(m, b) \colon m \in [2, \infty),\ b_-(m) \leq b \leq b_+(m)\} \leq c.$$

Observe that

$$\frac{\partial^2 c(m, b)}{\partial b^2} = \frac{f(m, b)}{b^{m+1}(1-b)^5 m^3},$$

where

$$f(m, b) := (14b + 44b^2 + 14b^3) + (-1 - 16b + 16b^3 + b^4)\, m + (1 + 2b - 6b^2 + 2b^3 + b^4)\, m^2$$

is a quadratic polynomial in m with the discriminant $(1-b)^2 D(b)$, where

$$D(b) := 1 - 22b - 77b^2 - 236b^3 - 77b^4 - 22b^5 + b^6.$$

Since $D(1/8) < 0$ and $D'(b) < 0$ as $b \in (0, 1)$, one has $D(b) < 0$ $\forall b \in [1/8, 1)$, which implies that $f(m, b) > 0$ for all $m \geq 2$ and $b \in [b_-(m), b_+(m)] \subset [1/8, 1)$.

Hence, $c(m, b)$ is convex in b on $[b_-(m), b_+(m)]$, and so, the theorem reduces now to the two inequalities

(5.37) $\qquad c(m, b_\pm(m)) \leq c \quad \forall m \geq 2.$

Let now

$$\ell_\pm(m) := \ln \frac{c(m, b_\pm(m))}{c}.$$

It is not hard to see that

(5.38) $\qquad \ell_\pm(m) \to 0$ as $m \to \infty$.

Next,

(5.39) $$\frac{\mathrm{d}\ell_\pm(m)}{\mathrm{d}m} = \frac{p_{1,\pm}(m)}{q_{1,\pm}(m)} - \ln b_\pm(m) \to 0 \text{ as } m \to \infty,$$

where

$$p_{1,+}(m) := -3\left(2 + 18\,m + 65\,m^2 + 132\,m^3 + 183\,m^4 + 192\,m^5 \right. \\ \left. + 147\,m^6 + 72\,m^7 + 18\,m^8\right),$$

$$q_{1,+}(m) := (1+m)\left(1 + 3\,m + 3\,m^2\right) \\ \times \left(1 + 6\,m + 15\,m^2 + 24\,m^3 + 27\,m^4 + 18\,m^5 + 6\,m^6\right),$$

$$p_{1,-}(m) := -3\left(1 - 10\,m + 44\,m^2 - 108\,m^3 + 168\,m^4 - 186\,m^5 \right. \\ \left. + 147\,m^6 - 72\,m^7 + 18\,m^8\right),$$

$$q_{1,-}(m) := m\left(1 - 3\,m + 3\,m^2\right) \\ \times \left(1 - 6\,m + 15\,m^2 - 24\,m^3 + 27\,m^4 - 18\,m^5 + 6\,m^6\right).$$

It remains to notice that

(5.40) $$\frac{\mathrm{d}^2\ell_\pm(m)}{\mathrm{d}m^2} < 0$$

for all $m \geq 2$, since

$$\frac{\mathrm{d}^2\ell_\pm(2+u)}{\mathrm{d}u^2} = -\frac{p_{2,\pm}(u)}{q_{2,\pm}(u)},$$

where the polynomials

$$p_{2,+}(u) := 3\left(481501999 + 3217818705\,u + 10039823407\,u^2 \right. \\ + 19422677441\,u^3 + 26088436326\,u^4 + 25810205607\,u^5 \\ + 19463858523\,u^6 + 11417421387\,u^7 + 5267034654\,u^8 \\ + 1917898905\,u^9 + 549605511\,u^{10} + 122685219\,u^{11} \\ + 20919447\,u^{12} + 2634786\,u^{13} + 231228\,u^{14} \\ \left. + 12636\,u^{15} + 324\,u^{16}\right),$$

$$q_{2,+}(u) := (2+u)(3+u)^2\left(19 + 15\,u + 3\,u^2\right)^2 \\ \times \left(1657 + 3810\,u + 3687\,u^2 + 1920\,u^3 + 567\,u^4 + 90\,u^5 \right. \\ \left. + 6\,u^6\right)^2,$$

$$p_{2,-}(u) := 3\left(56733 + 746746\,u + 4476412\,u^2 + 16204907\,u^3 \right. \\ + 39791424\,u^4 + 70587186\,u^5 + 93962586\,u^6 \\ + 96088041\,u^7 + 76522956\,u^8 + 47742654\,u^9 \\ + 23310720\,u^{10} + 8831079\,u^{11} + 2549277\,u^{12} + 542934\,u^{13} \\ \left. + 80568\,u^{14} + 7452\,u^{15} + 324\,u^{16}\right),$$

and
$$q_{2,-}(u) := (1+u)(2+u)^2 (7+9u+3u^2)^2$$
$$\times (97 + 342\,u + 519\,u^2 + 432\,u^3 + 207\,u^4 + 54\,u^5 + 6\,u^6)^2$$
are all positive for $u \geq 0$.

Indeed, now (5.38), (5.39), and (5.40) yield (5.37). □

Acknowledgement. The author is pleased to thank the anonymous referee, who has very carefully read the paper and made many useful comments.

References

[1] R. E. Barlow, A. W. Marshall, and F. Proshan, *Properties of probability distributions with monotone hazard rate*, Ann. Math. Statist. **34** (1963), 375–389.

[2] C. Borell, *The Brunn-Minkowski inequality in Gauss spaces*, Invent. Math. **6** (1975), 387–400.

[3] C. Borell, *Convex set functions in d-space*, Period. Math. Hungar. **6** (1975), 111-136.

[4] J. H. B. Kemperman, *On the role of duality in the theory of moments*, In: Semi-Infinite Programming and Applications. Internat. Symp. Austin, Texas, 1981; Lect. Notes Econ. Math. Systems. **215** (1983), Springer, Berlin – Heidelberg, 63–92.

[5] H. J. Brascamp and E. H. Lieb, *Some inequalities for Gaussian measures and the long-range order of the one-dimensional plasma*, Functional Integration and Its Applications (Proceedings of the International Congerence, London, 1974, A. M. Arthurs, Ed.), Clarendon Press, Oxford (1975), 1-14.

[6] H. J. Brascamp and E. H. Lieb, *On extensions of the Brunn-Minkowski and Prékopa-Leindler theorems, including inequalities for log concave functions, and with application to diffusion equation*, J. Funct. Anal. **22** (1976), 366-389.

[7] S. Das Gupta, *Brunn-Minkowski inequality and its aftermath*, J. Multivar. Anal. **10** (1980), 296-318.

[8] M. L. Eaton, *A note on symmetric Bernoulli random variables*, Ann. Math. Statist. **41** (1970), 1223–1226.

[9] M. L. Eaton, *A probability inequality for linear combinations of bounded random variables*, Ann. Statist. **2** (1974), 609–614.

[10] R. L. Hall, M. Kanter, and M. D. Perlman, *Inequalities for probability content of a rotated square and related convolutions*, Ann. Statist. **8** (1980), 802–813.

[11] N. L. Johnson, S. Kotz, and A. W. Kemp, *Univariate Dicrete Distributions, Second Ed.*, Wiley, New York, 1992.

[12] L. Leindler, *On a certain converse of Hölder's inequality II*, Acta Sci. Math. (Szeged) **33** (1972), 217-223.

[13] I. Pinelis, *Extremal probabilistic problems and Hotelling's T^2 test under a symmetry condition*, Ann. Statist. **22** (1994), 357-368.

[14] I. Pinelis, *Optimal tail comparison based on comparison of moments*, Proceedings of the Conference on High Dimensional Probability (Oberwolfach, Germany, 1996), Progress in Probability, **43**, Birkhäuser, Basel, Switzerland, (1998), 297-314.

[15] A. Prékopa, *Logarithmic concave measures with application to stochastic programming*, Acta Sci. Math. (Szeged) **32** (1971), 301-315.

[16] A. Prékopa, *On logarithmic concave measures and functions*, Acta Sci. Math. (Szeged) **34** (1973), 335-343.

[17] A. P. Prudnikov, Yu. A. Brychkov, and O. I. Marichev, *Integrals and Series, Vol. 1*, Gordon and Breach, London, 1986.

[18] Y. Rinott, *On convexity of measures*, Ann. Probab. **4** (1976), 1020-1026.

DEPARTMENT OF MATHEMATICAL SCIENCES, MICHIGAN TECHNOLOGICAL UNIVERSITY
E-mail address: `ipinelis@math.mtu.edu`

Product formula, tails and independence of multiple stable integrals

J. Rosiński and G. Samorodnitsky

ABSTRACT. We give necessary and sufficient conditions for sequences of multiple integrals of arbitrary orders with respect to a symmetric stable random measure to be independent. We also find a surprising criterion for independence of multiple Wiener–Itô integrals which follows from a work of Üstünel and Zakai (1989) (*Ann. Prob.* **17**, 1441–1453). Similarly to Wiener-Itô integrals, the main tool in establishing independence for multiple stable integrals is the product formula. The role of Itô's isometry is replaced in the stable case by the exact evaluation of the tail of stable integrals and their products.

1. Introduction and preliminaries

Stochastic independence is the basic notion in probability theory and, arguably, the one that distinguishes probability from other related areas of mathematics. Intuitively, "random objects" are independent when they have no "common causes". However, cancelations sometimes may occur. For example, if X_1 and X_2 are independent standard normal random variables then $Y_1 = X_1 + X_2$ and $Y_2 = X_1 - X_2$ are independent even though they have "common causes": X_1 and X_2. To what extent is this "cancelation" phenomenon restricted to the Gaussian world? It is natural to look at jointly infinitely divisible random variables without Gaussian components. They can be represented by stochastic integrals of deterministic functions with respect to an infinitely divisible random measure without Gaussian component. Urbanik (1967) showed that such integrals are independent if and only if the integrands have disjoint supports. Therefore, there are no common Poisson jumps in the sequence of independent stochastic integrals. Roughly speaking, Poissonian terms do not seem to cancel.

1991 *Mathematics Subject Classification*. Primary 60E07, 60H05.

Key words and phrases. multiple stochastic integral, probability tails, stable distribution, random measure, product formula, independence, non–Gaussian chaos.

The first author was partially supported by NSF grant DMS-97-04744 at University of Tennessee, Knoxville.

The second author was partially supported by NSF Grant DMS-97-04982 and NSA Grant MDA904-98-1-0041 at Cornell University.

The authors acknowledge support provided by the Tennessee Science Alliance of a visit by G. Samorodnitsky to Knoxville when this paper was completed.

© 1999 American Mathematical Society

How far does this "rigidity" of Poissonian jumps in comparison with cancellation flexibility of Gaussian random variables extend? It is interesting to look at multiple integrals with respect to Gaussian and infinitely divisible random measures. It has been proved by Üstünel and Zakai (1989) that for multiple Wiener–Itô integrals independence is equivalent to the L^2 orthogonality of one-dimensional sections of the two kernels, and so the two integrals may well be affected by the same randomness of the Gaussian random measure. On the other hand, this should not be the case for multiple integrals with respect to an infinitely divisible random measure without a Gaussian component if the Poissonian jumps did not cancel. A formal statement to this end was made by Privault (1996) who claimed that, for a particular class of infinitely divisible random measures without Gaussian component, independence is equivalent to disjoint supports of one-dimensional sections of the two kernels. Unfortunately, there appears to be an error in the proof of the necessity part of his theorem (see Example 5.5).

In this paper we consider multiple integrals with respect to symmetric α-stable random measures, $0 < \alpha < 2$, and show that in this case independence is indeed equivalent to disjoint supports of one-dimensional sections of the two kernels. It is interesting to note the difference in the approaches one takes to establish conditions for independence in the case of multiple Wiener–Itô integrals and in the case of multiple integrals with respect to a stable random measure. In the former case one uses L^2 theory, that is clearly unavailable in the heavy tailed stable case. In the latter case one uses probability tails of the multiple integrals.

We proceed with a more formal setup. Throughout this paper $(S, \sigma(\mathcal{S}))$ will denote a measurable space, where $\mathcal{S}$ is a δ-ring of subsets of S with the property that $S = \bigcup_n S_n$, for some sequence $\{S_n\} \subset \mathcal{S}$. Let M be a symmetric infinitely divisible random measure without Gaussian component on $(S, \mathcal{S})$. That is, $\{M(A), A \in \mathcal{S}\}$ is a stochastic process such that (i) for every $A \in \mathcal{S}$, $M(A)$ is a symmetric infinitely divisible random variable without Gaussian component, (ii) for pairwise disjoint $A_1, A_2, \ldots$ in $\mathcal{S}$ the random variables $M(A_1), M(A_2), \ldots$ are independent ("M is independently scattered"), and if $\cup_{n=1}^\infty A_n \in \mathcal{S}$ then $M(\cup_{n=1}^\infty A_n) = \sum_{n=1}^\infty M(A_n)$ a.s. ("M is σ-additive").

The characteristic function of $M(A)$, $A \in \mathcal{S}$, can be written in the form

$$(1.1) \qquad E\exp\{i\theta M(A)\} = \exp\left\{\int_A \int_0^\infty (1 - \cos\theta x) \rho(dx, s) \lambda(ds)\right\},$$

where λ is a probability measure on $(S, \mathcal{S})$ and $\{\rho(\cdot, s), \ s \in S\}$ is a measurable family of Lévy measures on $(0, \infty)$, and

$$(1.2) \qquad \lambda\{s \in S: \ \rho\big((0, \infty), s\big) = 0\} = 0$$

(see, e.g., Rajput and Rosinski (1989)). Condition (1.2) implies that $M(A) = 0$ a.s. if and only if $\lambda(A) = 0$, that is λ is a control measure of M. Throughout this paper we will assume that λ is atomless.

We will now describe the series expansion of M. For $s \in S$ let

$$R(u, s) = \inf\{x > 0: \ \rho\big((x, \infty), s\big) \leq u\}, \quad u > 0.$$

Let $\{\epsilon_i, i \geq 1\}$, $\{V_i, i \geq 1\}$ and $\{\Gamma_i, i \geq 1\}$ be independent sequences of random variables, the first one being a sequence of i.i.d. Rademacher random variables, the second one being a sequence of i.i.d. S-valued random variables with common distribution λ, while the last one being a sequence of arrival times of a Poisson process on $(0, \infty)$ with unit rate. Then, for every $A \in \mathcal{S}$, the series

$$(1.3) \qquad M_0(A) = \sum_{i=1}^{\infty} \epsilon_i R(\Gamma_i, V_i) \delta_{V_i}(A)$$

converges a.s. and, $\{M_0(A), A \in \mathcal{S}\} =^d \{M(A), A \in \mathcal{S}\}$, where $=^d$ reads "equal in distribution" (see LePage (1980), Rosiński (1990)). Without loss of generality we will identify M with its version M_0 given by (1.3), and we will write

$$(1.4) \qquad M(A) = \sum_{i=1}^{\infty} \xi_i \delta_{V_i}(A), \quad A \in \mathcal{S}$$

where $\xi_i = \epsilon_i R(\Gamma_i, V_i)$. Let $f : S^p \to \mathbf{R}$ be a product measurable symmetric function (that is, f is invariant under permutations of its arguments) and such that f vanishes on the diagonals (that is, $f(\mathbf{s}) = 0$ if two or more of the coordinates of $\mathbf{s}$ coincide.) The multiple integral $I_p^M(f) = \int_{S^p} f dM^{\otimes p}$ can be defined on one hand as the integral with respect to the product random measure $M^{\otimes p}$ (see Kwapień and Woyczyński (1992)) and, on the other hand, it can equivalently be defined as a multilinear random form

$$(1.5) \qquad I_p^M(f) = \sum_{k_1, \ldots, k_p} \xi_{k_1} \cdots \xi_{k_p} f(V_{k_1}, \ldots, V_{k_p})$$

(see Kallenberg and Szulga (1989)). $I_p^M(f)$ exists if and only if

$$(1.6) \qquad \sum_{k_1, \ldots, k_p} \xi_{k_1}^2 \cdots \xi_{k_p}^2 f^2(V_{k_1}, \ldots, V_{k_p}) < \infty \quad \text{a.s.}$$

Finally, we note that the multiple integral can be immediately extended to random integrands $F : \Omega \times S^p \to \mathbf{R}$ such that

(i) there is a σ-field $\mathcal{G}$ independent of the Rademacher sequence $\{\epsilon_j\}$ such that for every $\mathbf{s} \in S^p$, $F(\cdot, \mathbf{s})$ is measurable $\mathcal{G}$
(ii) for every $\omega \in \Omega$, $F(\omega, \cdot)$ is symmetric and vanishes on diagonals.

Indeed, conditions (i) and (ii) imply that

$$(1.7) \qquad I_p^M(F) := \sum_{k_1, \ldots, k_p} \xi_{k_1} \cdots \xi_{k_p} F(\cdot, V_{k_1}, \ldots, V_{k_p})$$

is a conditional Rademacher multilinear form which converges unconditionally a.s. if and only if

$$(1.8) \qquad \sum_{k_1, \ldots, k_p} \xi_{k_1}^2 \cdots \xi_{k_p}^2 F^2(\cdot, V_{k_1}, \ldots, V_{k_p}) < \infty \quad \text{a.s.}$$

Hence (1.8) is a necessary and sufficient condition for the existence of $I_p^M(F)$ provided (i) and (ii) hold. Further, (1.8) is a sufficient condition of the existence of $I_p^M(F)$ for nonsymmetric F's, but it is not necessary. Clearly, $I_p^M(F) = I_p^M(\hat{F})$, where $\hat{F}$ is the symmetrization of F.

Now we will concentrate on multiple symmetric α-stable integrals.

The characteristic function of $M(A)$ in this case has the form

$$E \exp\{i\theta M(A)\} = \exp\{-|\theta|^{\alpha} m(A)\},$$

where m is a σ-finite measure on $\sigma(\mathcal{S})$ that is finite on $\mathcal{S}$. Let $\psi : S \to (0, \infty)$ be such that $\int_S \psi(s) m(ds) = 1$; ψ is used to define the probability control measure λ of M as follows

$$\lambda(A) = \int_A \psi(s) m(ds).$$

Then $R(u, s) = c_\alpha u^{-1/\alpha} \psi(s)^{-1/\alpha}$, where c_α is a numerical constant (see Samorodnitsky and Taqqu (1994), p. 23). Hence

$$(1.9) \qquad M(A) = c_\alpha \sum_{i=1}^{\infty} \epsilon_i \Gamma_i^{-1/\alpha} \psi(V_i)^{-1/\alpha} \delta_{V_i}(A).$$

Samorodnitsky and Taqqu (1990) (see also Samorodnitsky and Szulga (1989)) proved that if
(1.10)

$$N_p(f) := \int_{S^p} |f(s_1, \ldots, s_p)|^\alpha \left[1 + \ln_+ \frac{|f(s_1, \ldots, s_p)|}{\prod_{i=1}^p \psi(s_i)^{1/\alpha}}\right]^{p-1} m(ds_1) \cdots m(ds_p) < \infty,$$

then $I_p^M(f)$ exists and is by (1.5) given by

$$(1.11) \qquad I_p^M(f) = c_\alpha^p \sum_{k_1, \ldots, k_p} [\prod_{i=1}^p \epsilon_{k_i} \Gamma_{k_i}^{-1/\alpha} \psi(V_{k_i})^{-1/\alpha}] f(V_{k_1}, \ldots V_{k_p})$$

(recall that $\log_+ a = \log a$ if $a > 1$ and $= 0$ otherwise). Note that condition (1.10) depends in general on the choice of ψ.

This paper is organized as follows. In the next section we develop the main tool in our treatment of multiple stable integrals, the *product formula*. In Section 3 we derive the asymptotic tail behavior of the distribution of a general term in the product formula, a mixed stable integral. It turns out that different terms in the product formula have different rates of decay of the tail distribution function. This fact allows us to determine the asymptotic tail behavior for the product of multiple stable integrals (Theorem 4.1) and derive from this necessary and sufficient conditions for independence of multiple stable integrals (Theorem 4.3). Finally, in Section 5 we establish an interesting criterion for the independence of multiple Wiener–Itô integrals that follows from the work of Üstünel and Zakai (1989). The other purpose of this section is to emphasize differences and similarities between Gaussian and stable cases. The problem of finding necessary and sufficient conditions for independence of multiple integrals with respect to a general infinitely divisible random measure is still open, even in the compensated Poisson case. We conjecture that the same conditions are required here for independence as in the stable case, hence no cancellation of Poissonian jumps can occur.

2. The product formula for multiple integrals with respect to a symmetric infinitely divisible random measure

In this section we will develop a product formula for multiple integrals with respect to symmetric infinitely divisible random measures M described in Section 1. Let

$$(2.1) \qquad D^p = \{\mathbf{i} \in \mathbf{N}^p : \text{all } i_1, \ldots, i_p \text{ are different}\},$$

and

$$(2.2) \qquad D_n^p = D^p \cap \{1, \ldots, n\}^p.$$

LEMMA 2.1. *Let $F : D_n^p \to \mathbf{R}$, $G : D_n^q \to \mathbf{R}$ be symmetric functions. Then*

$$\sum_{\mathbf{i} \in D_n^p} F(\mathbf{i}) \sum_{\mathbf{j} \in D_n^q} G(\mathbf{j}) = \sum_{r=0}^{p \wedge q} r! \binom{p}{r} \binom{q}{r} \sum_{\mathbf{k} \in D_n^{p+q-r}} H_r(\mathbf{k}), \qquad (2.3)$$

where

$$H_r(\mathbf{k}) = F(k_1, \ldots, k_r, k_{r+1}, \ldots, k_p) G(k_1, \ldots, k_r, k_{p+1}, \ldots, k_{p+q-r}).$$

PROOF. Write the left hand side of (2.3) as

$$\sum_{r=0}^{p \wedge q} \sum_{\{(\mathbf{i},\mathbf{j}) : c(\mathbf{i},\mathbf{j}) = r\}} F(\mathbf{i}) G(\mathbf{j}), \qquad (2.4)$$

where

$$c(\mathbf{i}, \mathbf{j}) = \text{card}(\{i_1, \ldots, i_p\} \cap \{j_1, \ldots, j_q\}).$$

Consider the inner sum in (2.4). If $r = 0$, then

$$\sum_{\{(\mathbf{i},\mathbf{j}) : c(\mathbf{i},\mathbf{j}) = 0\}} F(\mathbf{i}) G(\mathbf{j}) = \sum_{\mathbf{k} \in D_n^{p+q}} H_0(\mathbf{k}).$$

If $r = c(\mathbf{i}, \mathbf{j}) \geq 1$, then $\mathbf{i}$ contains a subsequence $\mathbf{s} \in D_n^r$ such that the elements of $\mathbf{s}$ appear in certain order in $\mathbf{j}$. By the symmetry of F, $F(\mathbf{i}) = F(\mathbf{s}, \mathbf{i}')$, where $\mathbf{i}' \in D_n^{p-r}$ are the remaining elements of $\mathbf{i}$ put together after $\mathbf{s}$ was removed from $\mathbf{i}$. Similarly, $G(\mathbf{j}) = G(\mathbf{s}, \mathbf{j}')$, and $(\mathbf{s}, \mathbf{i}', \mathbf{j}') \in D_n^{p+q-r}$. Hence

$$\sum_{\{(\mathbf{i},\mathbf{j}) : c(\mathbf{i},\mathbf{j}) = r\}} F(\mathbf{i}) G(\mathbf{j}) = r! \binom{p}{r} \binom{q}{r} \sum_{\mathbf{k} \in D_n^{p+q-r}} H_r(\mathbf{k}),$$

which ends the proof. □

Let $f : S^p \to \mathbf{R}$ and $g : S^q \to \mathbf{R}$ be measurable symmetric and vanishing on diagonals functions. Define

$$\begin{aligned}&h_r(s_1, \ldots, s_{p+q-r})\\ &= f(s_1, \ldots, s_r, \ldots, s_p) g(s_1, \ldots, s_r, s_{p+1}, \ldots, s_{p+q-r}).\end{aligned} \qquad (2.5)$$

Note that h_r is obtained from the tensor product $f \otimes g$ by coupling the first r variables of f and g. As a function of the first r variables only h_r is symmetric and vanishes on diagonals, but in general, h_r is neither symmetric nor vanishes on the diagonals. Let M be a random measure given by (1.4). Since λ is atomless, the quadratic variation $[M]$ of M is given by

$$[M](A) = \sum_{i=1}^{\infty} \xi_i^2 \delta_{V_i}(A), \quad A \in \mathcal{S}. \qquad (2.6)$$

Let $I_r^{[M]}(\cdot)$ denote r-tuple integral with respect to $[M]$.

LEMMA 2.2. *Suppose $I_p^M(f)$ and $I_q^M(g)$ exist (see (1.5)-(1.6)). Then, for every $1 \leq r \leq p \wedge q$ and for $\lambda^{\otimes(p+q-2r)}$-almost all $(s_{r+1}, \ldots, s_{p+q-r}) \in S^{p+q-2r}$, the series*

$$H_r(\cdot, s_{r+1}, \ldots, s_{p+q-r}) = \sum_{\mathbf{k} \in D^r} \xi_{k_1}^2 \cdots \xi_{k_r}^2 h_r(V_{k_1}, \ldots, V_{k_r}, s_{r+1}, \ldots, s_{p+q-r}) \qquad (2.7)$$

converges absolutely with probability one. Hence
$$I_r^{[M]}(h_r(\cdot, s_{r+1}, \ldots, s_{p+q-r})) = H_r(\cdot, s_{r+1}, \ldots, s_{p+q-r}).$$

PROOF. Applying Th. 4.7 in Kallenberg and Szulga (1989) to the marked Poisson point process $\{(|\xi_i|, V_i)\}_{i \in \mathbf{N}}$ we get that for $\lambda^{\otimes(p-r)}$-almost all $(s_{r+1}, \ldots, s_p) \in S^{p-r}$ the iterated integral
$$I_r^M(f(\cdot, s_{r+1}, \ldots, s_p))$$
exists and $I_{p-r}^M(I_r^M(f)) = I_p^M(f)$. The analogous statement holds also for g. Hence

$$\sum_{\mathbf{k} \in D^r} \xi_{k_1}^2 \cdots \xi_{k_r}^2 |h_r(V_{k_1}, \ldots, V_{k_r}, s_{r+1}, \ldots, s_{p+q-r})|$$
$$\leq \big(\sum_{\mathbf{k} \in D^r} \xi_{k_1}^2 \cdots \xi_{k_r}^2 f^2(V_{k_1}, \ldots, V_{k_r}, s_{r+1}, \ldots, s_p) \big)^{1/2}$$
(2.8)
$$\times \big(\sum_{\mathbf{k} \in D^r} \xi_{k_1}^2 \cdots \xi_{k_r}^2 g^2(V_{k_1}, \ldots, V_{k_r}, s_{p+1}, \ldots, s_{p+q-r}) \big)^{1/2} < \infty \text{ a.s.}$$

The finiteness of the last two series comes from (1.6) applied to the iterated integrals $I_r^M(f(\cdot, s_{r+1}, \ldots, s_p))$ and $I_r^M(g(\cdot, s_{p+1}, \ldots, s_{p+q-r}))$. □

THEOREM 2.3. *Suppose* $I_p^M(f)$ *and* $I_q^M(g)$ *exist. Then*

(2.9)
$$I_p^M(f) I_q^M(g) = \sum_{r=0}^{p \wedge q} r! \binom{p}{r} \binom{q}{r} (I_{p+q-2r}^M \otimes I_r^{[M]})(h_r).$$

Here $(I_{p+q-2r}^M \otimes I_r^{[M]})(h_r) := I_{p+q-2r}^M(J_r)$, where J_r is given by (2.7).

PROOF. The existence of $I_{p+q-2r}^M(J_r)$ follows from (1.8); one needs to apply (2.8) and the conditions on the existence of $I_p^M(f)$ and of $I_q^M(g)$. The rest of the proof is a direct application of Lemma 2.1. □

Next we will examine the product of symmetric multiple stable integrals.

From (1.9) the quadratic variation $[M]$ of M is a positive α-stable random measure with the series representation
$$[M](A) = c_\alpha^2 \sum_{i=1}^\infty \Gamma_i^{-2/\alpha} \psi(V_i)^{-2/\alpha} \delta_{V_i}(A).$$

Hence the product $I_p^M(f) I_q^M(g)$ is the sum of the α-stable multiple integral $I_{p+q}^M(h_0)$ and mixed, α and $\alpha/2$-stable, multiple integrals $(I_{p+q-2r}^M \otimes I_r^{[M]})(h_r)$, where

$$(I_{p+q-2r}^M \otimes I_r^{[M]})(h_r) = c_\alpha^{p+q} \sum_{\mathbf{k} \in D^{p+q-r}} \big[\prod_{i=1}^r \Gamma_{k_i}^{-2/\alpha} \psi(V_{k_i})^{-2/\alpha} \big]$$
(2.10)
$$\big[\prod_{j=1}^{p+q-2r} \epsilon_{k_{r+j}} \Gamma_{k_{r+j}}^{-1/\alpha} \psi(V_{k_{r+j}})^{-1/\alpha} \big] h_r(V_{k_1}, \ldots V_{k_{p+q-r}}).$$

We will deduce the conditions for independence of $I_p^M(f)$ and $I_q^M(g)$ by computing the tail behavior of the product $I_p^M(f) I_q^M(g)$. We will show that the tails of (2.10) have different rates of decay for different r's and exactly one of them determines the tail behavior of the product (2.9).

3. Tail behavior of the distribution of mixed stable integrals

The product formula established in the previous section shows that for two symmetric stable integrals $I_p^M(f)$ and $I_q^M(g)$ the product $I_p^M(f)I_q^M(g)$ is a linear combination of a $p+q$-tuple symmetric stable integral $I_{p+q}^M(h_0)$ and mixed stable integrals. The tail behavior of the distribution of $I_{p+q}^M(h_0)$ is known:

$$(3.0) \quad \lim_{\lambda \to \infty} \lambda^\alpha (\log \lambda)^{-(p+q)+1} P\Big(|I_{p+q}^M(h_0)| > \lambda\Big) = k_{0,\alpha} \|\widehat{h_0}\|_{L^\alpha(S^{p+q})}^\alpha$$

where $\widehat{h_0}$ is the symmetrization of h_0 and $k_{0,\alpha}$ is positive constant, see Samorodnitsky and Szulga (1989) and Samorodnitsky and Taqqu (1990). To establish the tail behavior of the distribution of the product $I_p^M(f)I_q^M(g)$ we need to understand the tail behavior of the distribution of mixed stable integrals, and this is our goal in this section. Let, therefore, p and q be two positive integers, and let $r \leq p \wedge q$ be another positive integer. We consider the mixed stable integral $(I_{p+q-2r}^M \otimes I_r^{[M]})(h_r)$ with the kernel h_r given by (2.5). The following is the main result of this section.

THEOREM 3.1. *Suppose that $N_p(f) < \infty$ and $N_q(g) < \infty$ (see (1.10)). Then*

$$(3.1) \quad \lim_{\lambda \to \infty} \lambda^{\alpha/2} (\log \lambda)^{-r+1} P\Big(|(I_{p+q-2r}^M \otimes I_r^{[M]})(h_r)| > \lambda\Big) = k_{r,\alpha} C_r(f,g),$$

where $k_{r,\alpha} = c_\alpha^{r\alpha}(r!)^{\alpha/2-1}/(r-1)!$ and

$$(3.2) \quad C_r(f,g) = \int_{S^r} E|I_{p+q-2r}^M(\hat{h}_r(s_1,\ldots,s_r,\cdot))|^{\alpha/2} m(ds_1)\ldots m(ds_r) < \infty.$$

Here $\hat{h}_r$ is the symmetrization of h_r with respect to variables $s_{r+1},\ldots,s_{p+q-r}$, i.e.,

$$\hat{h}_r(s_1,\ldots,s_{p+q-r}) = ((p+q-2r)!)^{-1} \sum_\pi h_r(s_1,\ldots,s_r,s_{\pi(r+1)},\ldots,s_{\pi(p+q-r)}),$$

where π runs over all permutations of the numbers $\{r+1,\ldots,p+q-r\}$. Finally, $I_{p+q-2r}(\hat{h}_r(s_1,\ldots,s_r,\cdot))$ is the ordinary (i.e. not mixed) multiple integral of order $p+q-2r$ with respect to the symmetric α-stable random measure M of $\hat{h}_r$ taken as a function of its last $p+q-2r$ coordinates.

PROOF. We use the series representation of $(I_{p+q-2r}^M \otimes I_r^{[M]})(h_r)$ given by (2.10). Recall that $V_1, V_2, \ldots$ are i.i.d. with common distribution $\psi(s)m(ds)$, $\epsilon_1, \epsilon_2, \ldots$ are i.i.d. Rademacher random variables and $\Gamma_1, \Gamma_2, \ldots$ are the ordered points of a standard Poisson process on $(0,\infty)$, with all three sequences being independent. Let

$$M_r := r! [\prod_{i=1}^r \Gamma_i^{-2/\alpha} \psi(V_i)^{-2/\alpha}] \sum_{\mathbf{j} \in D^{p+q-2r}} [\prod_{k=1}^{p+q-2r} \epsilon_{j_k+r} \Gamma_{j_k+r}^{-1/\alpha} \psi(V_{j_k+r})^{-1/\alpha}]$$

$$(3.3) \qquad h_r(V_1,\ldots,V_r,V_{j_1+r},\ldots,V_{j_{p+q-2r}+r}).$$

We think of M_r as the "main term" of $c_\alpha^{-p-q}(I_{p+q-2r}^M \otimes I_r^{[M]})(h_r)$. The proof is divided into two main steps. The first one shows that

$$(3.4) \quad \lim_{\lambda \to \infty} \lambda^{\alpha/2}(\log \lambda)^{-r+1} P(|M_r| > \lambda) = \frac{c_\alpha^{-(p+q-2r)\alpha/2}(r!)^{\alpha/2-1}}{(r-1)!} C_r(f,g)$$

In the second step we show that the tail of the remainder

$$(3.5) \qquad R_r := c_\alpha^{-p-q}(I_{p+q-2r}^M \otimes I_r^{[M]})(h_r) - M_r$$

is of the smaller order than M_r. Specifically, we will prove that

$$\lim_{\lambda \to \infty} \lambda^{\alpha/2} (\log \lambda)^{-r+1} P(|R_r| > \lambda) = 0. \tag{3.6}$$

Clearly, (3.4)–(3.6) give (3.1) and so conclude the proof. Despite a clear structure of this proof, the actual computations leading to (3.4) and (3.6) are intricate and lengthly; particularly (3.6) involves several truncation arguments and decoupling.

We begin by proving (3.4). Recall that, conditional on Γ_{r+1}, the joint distribution of $\Gamma_1/\Gamma_{r+1}, \ldots, \Gamma_r/\Gamma_{r+1}$ is that of order statistics from i.i.d. uniform sample of size r in $(0, 1)$. Hence

$$P(\frac{1}{r!}|M_r| > \lambda) = \int_0^\infty e^{-x} \frac{x^r}{r!} P_x(\lambda)\, dx, \tag{3.7}$$

where

$$P_x(\lambda) := P\bigg([\prod_{i=1}^r (xU_i)^{-2/\alpha} \psi(V_i)^{-2/\alpha}]$$

$$\Big| \sum_{\mathbf{j} \in D^{p+q-2r}} [\prod_{k=1}^{p+q-2r} \epsilon_{j_k+r}(x + \Gamma_{j_k-1})^{-1/\alpha} \psi(V_{j_k+r})^{-1/\alpha}]$$

$$h_r(V_1, \ldots, V_r, V_{j_1+r}, \ldots, V_{j_{p+q-2r}+r})\Big| > \lambda\bigg). \tag{3.8}$$

The sequences $\epsilon_1, \epsilon_2, \ldots$, $V_1, V_2, \ldots$ and $\Gamma_1, \Gamma_2, \ldots$ in (3.8) are as before, and $U_1, \ldots, U_r$ are independent of them i.i.d. uniform in $(0, 1)$ random variables. Further, $\Gamma_0 = 0$. □

To establish the asymptotic behavior of $P_x(\lambda)$ we need the following lemma.

LEMMA 3.2. *Let Y be a nonnegative random variable with $EY < \infty$ independent of i.i.d. uniform in $(0, 1)$ random variables $U_1, \ldots, U_r$, $r \geq 1$. Then*

$$\lim_{\lambda \to \infty} \lambda (\log \lambda)^{-(r-1)} P(\prod_{j=1}^r U_j^{-1} Y > \lambda) = \frac{1}{(r-1)!} EY. \tag{3.9}$$

Moreover, there is a finite positive constant $K = K(r)$ depending only on r such that for all $\lambda > 0$,

$$P(\prod_{j=1}^r U_j^{-1} Y > \lambda) \leq K\lambda^{-1}(1 + \log_+ \lambda)^{r-1}(1 + EY). \tag{3.10}$$

PROOF OF LEMMA 3.2. Let F_Y be the law of Y. We have

$$P(\prod_{j=1}^r U_j^{-1} Y > \lambda) = P(Y > \lambda) + \int_0^\lambda P(\prod_{j=1}^r U_j^{-1} > \frac{\lambda}{y}) F_Y(dy). \tag{3.11}$$

The assumption $EY < \infty$ implies that

$$P(Y > \lambda) \leq \lambda^{-1} EY, \tag{3.12}$$

$\lambda > 0$. Furthermore,

$$\int_0^\lambda P(\prod_{j=1}^r U_j^{-1} > \frac{\lambda}{y}) F_Y(dy) = \int_0^\lambda P(\Gamma_r > \log \frac{\lambda}{y}) F_Y(dy)$$

(3.13)
$$= \lambda^{-1} \sum_{i=0}^{r-1} \frac{1}{i!} \int_0^\lambda y(\log \frac{\lambda}{y})^i F_Y(dy).$$

Next we observe that for every $i \geq 0$

$$\int_0^\lambda y(1 + \log \frac{\lambda}{y})^i F_Y(dy) \leq \sup_{0 < y \leq \lambda^{-1/2}} y(1 + \log \frac{\lambda}{y})^i + \int_{\lambda^{-1/2}}^\lambda y(1 + \log \frac{\lambda}{y})^i F_Y(dy)$$

$$\leq \lambda \sup_{x \geq \lambda^{3/2}} \frac{(1 + \log x)^i}{x} + (1 + \frac{3}{2} \log \lambda)^i \int_{\lambda^{-1/2}}^\lambda y F_Y(dy)$$

(3.14)
$$\leq c\lambda^{-1/2} (\log \lambda)^i + cEY (\log \lambda)^i$$

for all $\lambda > 3$ (say). Here c is a positive constant that depends only on i. Now the bound (3.10) follows immediately from (3.12), (3.13) and (3.14). Furthermore, one can see by (3.11) and (3.13) that to derive (3.9) one has only to establish that

(3.15)
$$\lim_{\lambda \to \infty} (\log \lambda)^{-j} \int_0^\lambda y(\log \frac{\lambda}{y})^j F_Y(dy) = EY.$$

However, (3.15) follows upon observing that by the argument in (3.14)

$$\lim_{\lambda \to \infty} (\log \lambda)^{-j} \int_0^\lambda y(\log \frac{\lambda}{y})^j F_Y(dy) = \lim_{\lambda \to \infty} (\log \lambda)^{-j} \int_{\lambda^{-1/2}}^\lambda y(\log \frac{\lambda}{y})^j F_Y(dy)$$

and using a dominated convergence argument. □

Let

$$Y_x := \prod_{i=1}^r \psi(V_i)^{-1} \Big| \sum_{\mathbf{j} \in D^{p+q-2r}} [\prod_{k=1}^{p+q-2r} \epsilon_{j_k+r}(x + \Gamma_{j_k-1})^{-1/\alpha} \psi(V_{j_k+r})^{-1/\alpha}]$$

$$h_r(V_1, \ldots, V_r, V_{j_1+r}, \ldots, V_{j_{p+q-2r}+r}) \Big|^{\alpha/2},$$

where $x > 0$ is arbitrary but fixed. We have

(3.16)
$$P_x(\lambda) = P(\prod_{i=1}^r U_i^{-1} Y_x > \lambda^{\alpha/2} x^r).$$

We will show how the first step of the proof can be completed when we have

(3.17)
$$E\Big[\prod_{i=1}^r \psi(V_i)^{-1} \Big| \sum_{\mathbf{j} \in D^{p+q-2r}} [\prod_{k=1}^{p+q-2r} \epsilon_{j_k+r} \Gamma_{j_k}^{-1/\alpha} \psi(V_{j_k+r})^{-1/\alpha}]$$

$$h_r(V_1, \ldots, V_r, V_{j_1+r}, \ldots, V_{j_{p+q-2r}+r}) \Big|^{\alpha/2} \Big] < \infty$$

and

$$E\left[\left(\prod_{i=1}^{r}\psi(V_i)^{-1}\right)\psi(V_{r+1})^{-1/2}\Big|\sum_{\mathbf{j}\in D^{p+q-2r-1}}[\prod_{k=1}^{p+q-2r-1}\epsilon_{j_k+r+1}\Gamma_{j_k}^{-1/\alpha}\psi(V_{j_k+r+1})^{-1/\alpha}]\right.$$

(3.18)
$$\left. h_r(V_1,\ldots,V_r,V_{r+1},V_{j_1+r+1},\ldots,V_{j_{p+q-2r-1}+r+1})\Big|^{\alpha/2}\right] < \infty.$$

(of course, (3.18) is meaningful only if $2r < p+q$). First, using Khinchine's inequality for tetrahedral sums in Rademacher random variables we get that there is a finite $M > 0$ such that for every $x > 0$

$$E\left[\prod_{i=1}^{r}\psi(V_i)^{-1}\Big|\sum_{\mathbf{j}\in D^{p+q-2r}}[\prod_{k=1}^{p+q-2r}\epsilon_{j_k+r}(x+\Gamma_{j_k})^{-1/\alpha}\psi(V_{j_k+r})^{-1/\alpha}]\right.$$

(3.19)
$$\left. h_r(V_1,\ldots,V_r,V_{j_1+r},\ldots,V_{j_{p+q-2r}+r})\Big|^{\alpha/2}\right] \leq M$$

and

$$E\left[\left(\prod_{i=1}^{r}\psi(V_i)^{-1}\right)\psi(V_{r+1})^{-1/2}\right.$$

$$\Big|\sum_{\mathbf{j}\in D^{p+q-2r-1}}[\prod_{k=1}^{p+q-2r-1}\epsilon_{j_k+r+1}(x+\Gamma_{j_k})^{-1/\alpha}\psi(V_{j_k+r+1})^{-1/\alpha}]$$

(3.20)
$$\left. h_r(V_1,\ldots,V_r,V_{r+1},V_{j_1+r+1},\ldots,V_{j_{p+q-2r-1}+r+1})\Big|^{\alpha/2}\right] \leq M.$$

Splitting the inner sum in Y_x accordingly to $j_k \geq 2$ for all k and $j_k = 1$ for some k we get
$$Y_x \leq Y'_x + rx^{-1/2}Y''_x,$$
where Y'_x and Y''_x are the integrands in (3.19) and (3.20), respectively. Hence $EY_x < \infty$ and, by (3.16) and Lemma 3.2,
$$\lim_{\lambda\to\infty}\lambda^{\alpha/2}(\log\lambda)^{-(r-1)}P_x(\lambda) = \frac{x^{-r}}{(r-1)!}EY_x.$$

Hence, making a formal passage to the limit in (3.7),

$$\lim_{\lambda\to\infty}\lambda^{\alpha/2}(\log\lambda)^{-(r-1)}P(|M_r|>\lambda)$$

$$= \lim_{\lambda\to\infty}\lambda^{\alpha/2}(\log\lambda)^{-(r-1)}\int_0^\infty e^{-x}\frac{x^r}{r!}P_x(\frac{\lambda}{r!})\,dx$$

$$= \int_0^\infty e^{-x}\frac{x^r}{r!}(r!)^{\alpha/2}\frac{x^{-r}}{(r-1)!}EY_x\,dx$$

$$= \frac{(r!)^{\alpha/2-1}}{(r-1)!}E\left[\prod_{i=1}^{r}\psi(V_i)^{-1}\Big|\sum_{\mathbf{j}\in D^{p+q-2r}}[\prod_{k=1}^{p+q-2r}\epsilon_{j_k}\Gamma_{j_k}^{-1/\alpha}\psi(V_{j_k+r})^{-1/\alpha}]\right.$$

$$\left. h_r(V_1,\ldots,V_r,V_{j_1+r},\ldots,V_{j_{p+q-2r}+r})\Big|^{\alpha/2}\right]$$

$$= \frac{c_\alpha^{-(p+q-2r)\alpha/2}(r!)^{\alpha/2-1}}{(r-1)!}\int_{S^r}E|I_{p+q-2r}(\hat{h}_r(s_1,\ldots,s_r,\cdot))|^{\alpha/2}m(ds_1)\ldots m(ds_r),$$

which is exactly (3.4). To complete this first step of the proof we need to justify the interchange of the limit and integral in the above derivation. Using (3.10) of Lemma 3.2 with (3.19) and (3.20) we get

$$P_x(\lambda) \leq P\Big((\prod_{i=1}^{r} U_i^{-1})(Y_x' + rx^{-1/2}Y_x'') > \lambda^{\alpha/2} x^r\Big)$$

$$\leq P\Big((\prod_{i=1}^{r} U_i^{-1})Y_x' > \frac{1}{2}\lambda^{\alpha/2} x^r\Big) + P\Big((\prod_{i=1}^{r} U_i^{-1})Y_x'' > \frac{1}{2r}\lambda^{\alpha/2} x^{r+1/2}\Big)$$

$$\leq c\lambda^{-\alpha/2}(1 + \log_+ \lambda)^{r-1}(x^{-r} \vee x^{-(r+1/2)})(1 + \log_+ x)^{r-1},$$

for all $\lambda > 0$ and $x > 0$, where c is a finite positive constant. Therefore, the interchange the limit and the integral is legitimate by the dominated convergence theorem. This proves (3.4) modulo proving (3.17) and (3.18), which we do presently. Since the required argument is similar in both cases, we give the details only for (3.17).

We start with checking the following statement. For every $J = 0, 1, \ldots, p+q-2r$,

$$(3.21) \quad E\Bigg[\bigg(\prod_{i=1}^{r}\psi(V_i)^{-1} \prod_{d=r+1}^{r+J}\psi(V_d)^{-1/2}\Gamma_{d-r}^{-1/2}\bigg)\bigg|\sum_{\mathbf{j} \in D^{p+q-2r-J}}[\prod_{k=1}^{p+q-2r-J}\epsilon_{j_k+r+J}\Gamma_{j_k+J}^{-1/\alpha}\psi(V_{j_k+r+J})^{-1/\alpha}]h_r(V_1,\ldots,V_r,\mathbf{W_j})\bigg|^{\alpha/2}\Bigg] < \infty,$$

where for every $\mathbf{j} \in D^{p+q-2r-J}$, $\mathbf{W_j}$ is a $(p+q-2r)$-dimensional random vector in which J coordinates are $V_{r+1},\ldots,V_{r+J}$ and the remaining $p+q-2r-J$ coordinates are $V_{j_1+r+J},\ldots,V_{j_{p+q-2r-J}+r+J}$.

If $J = p+q-2r$ then (3.21) is trivially true, so we will prove (3.21) by downward induction in J. Assume that (3.21) holds for all $J > J_0$, for some J_0 such that $1 \leq J_0 + 1 \leq p+q-2r$, and let us prove it for $J = J_0$. Note that the assumption of the induction and the Khinchine inequality for tetrahedral sums in Rademacher random variables implies that the finiteness of the expectation in (3.21) for $J > J_0$ will be preserved if we restrict the sum to a nonrandom subset of $D^{p+q-2r-J}$. The first step in the inductive procedure is to note that it is enough to prove that for some $m \geq 1$

$$(3.22) \quad E\Bigg[\bigg(\prod_{i=1}^{r}\psi(V_i)^{-1} \prod_{d=r+1}^{r+J_0}\psi(V_d)^{-1/2}\Gamma_{d-r}^{-1/2}\bigg)\bigg|\sum_{\substack{\mathbf{j} \in D^{p+q-2r-J_0} \\ j_k \geq m, k \leq p+q-2r-J_0}}[\prod_{k=1}^{p+q-2r-J_0}\epsilon_{j_k+r+J_0}\Gamma_{j_k+J_0}^{-1/\alpha}\psi(V_{j_k+r+J_0})^{-1/\alpha}]h_r(V_1,\ldots,V_r,\mathbf{W_j})\bigg|^{\alpha/2}\Bigg] < \infty,$$

because the absolute difference between the random variables in the left hand sides of (3.21) and (3.22) is bounded above by the sum of a finite number of random variables each one of which is of the type given in the left hand side of (3.21) with $J > J_0$ and the sum taken over a nonrandom subset of $D^{p+q-2r-J}$. For such random variables we use the assumption of the induction to establish finiteness of the corresponding moment. Let us, therefore, show that for some $m \geq 1$ (3.22) holds.

Using once again the Khinchine inequality for tetrahedral sums in Rademacher random variables we see that for some positive constant c (that is allowed to change from one place to another) the expectation in the left hand side of (3.22) is bounded above by

$$cE\left[\left(\prod_{i=1}^{r}\psi(V_i)^{-1}\prod_{d=r+1}^{r+J_0}\psi(V_d)^{-1/2}\Gamma_{d-r}^{-1/2}\right)\Big|\sum_{\substack{\mathbf{j}\in D^{p+q-2r-J_0} \\ j_k\geq m,\,k\leq p+q-2r-J_0}}\right.$$

$$\left.\left[\prod_{k=1}^{p+q-2r-J_0}\Gamma_{j_k+J_0}^{-2/\alpha}\psi(V_{j_k+r+J_0})^{-2/\alpha}\right]|h_r(V_1,\ldots,V_r,V_{r+1},\mathbf{W_j})|^2\Big|^{\alpha/4}\right]$$

$$\leq cE\left[\left(\prod_{i=1}^{r}\psi(V_i)^{-1}\prod_{d=r+1}^{r+J_0}\psi(V_d)^{-1/2}\right)\Big|\sum_{\substack{\mathbf{j}\in D^{p+q-2r-J_0} \\ j_k\geq m,\,k\leq p+q-2r-J_0}}\right.$$

$$\left.\left[\prod_{k=1}^{p+q-2r-J_0}\Gamma_{j_k}^{-2/\alpha}\psi(V_{j_k+r+J_0})^{-2/\alpha}\right]|h_r(V_1,\ldots,V_r,V_{r+1},\mathbf{W_j})|^2\Big|^{\alpha/4}\right]$$

$$\leq c\sum_{K=0\vee J_0-q-r}^{J_0\wedge(p-r)}E\left[\prod_{i=1}^{r}\psi(V_i)^{-1}\left(\prod_{d=r+1}^{r+K}\psi(V_d)^{-2/\alpha}\sum_{j_1=m}^{\infty}\cdots\sum_{j_{p-r-K}=m}^{\infty}\right.\right.$$

$$\left[\prod_{k=1}^{p-r-K}\Gamma_{j_k}^{-2/\alpha}\psi(V_{j_k+r+J_0})^{-2/\alpha}\right]|f(V_1,\ldots,V_r,V_{r+1},\ldots,V_{r+K},V_{j_1+r+J_0},$$

$$\ldots,V_{j_{p-r-K}+r+J_0})|^2\Bigg)^{\alpha/4}\left(\prod_{d=r+K+1}^{r+J_0}\psi(V_d)^{-2/\alpha}\sum_{j_1=m}^{\infty}\cdots\sum_{j_{q-r-J_0+K}=m}^{\infty}\right.$$

$$\left[\prod_{k=1}^{q-r-J_0+K}\Gamma_{j_k}^{-2/\alpha}\psi(V_{j_k+r+J_0})^{-2/\alpha}\right]|g(V_1,\ldots,V_r,V_{r+K+1},\ldots,V_{r+J_0},V_{j_1+r+J_0},$$

$$\left.\left.\ldots,V_{j_{q-r-J_0+K}+r+J_0})|^2\right)^{\alpha/4}\right]$$

$$\leq c\sum_{K=0\vee J_0-q-r}^{J_0\wedge(p-r)}\left[E\left(\prod_{d=1}^{r+K}\psi(V_d)^{-2/\alpha}\right.\right.$$

$$\sum_{j_1=m}^{\infty}\cdots\sum_{j_{p-r-K}=m}^{\infty}\left[\prod_{k=1}^{p-r-K}\Gamma_{j_k}^{-2/\alpha}\psi(V_{j_k+r+J_0})^{-2/\alpha}\right]$$

$$\left.|f(V_1,\ldots,V_r,V_{r+1},\ldots,V_{r+K},V_{j_1+r+J_0},\ldots,V_{j_{p-r-K}+r+J_0})|^2\right)^{\alpha/2}\Bigg]^{1/2}$$

$$\left[E\left(\prod_{d=1}^{r+J_0-K}\psi(V_d)^{-2/\alpha}\sum_{j_1=m}^{\infty}\cdots\sum_{j_{q-r-J_0+K}=m}^{\infty}\left[\prod_{k=1}^{q-r-J_0+K}\Gamma_{j_k}^{-2/\alpha}\psi(V_{j_k+r+J_0})^{-2/\alpha}\right]\right.\right.$$

$$\left.\left.|g(V_1,\ldots,V_r,V_{r+1},\ldots,V_{r+J_0-K},V_{j_1+r+J_0},\ldots,V_{j_{q-r-J_0+K}+r+J_0})|^2\right)^{\alpha/2}\right]^{1/2}$$

$$:= c \sum_{K=0 \vee J_0-q-r}^{J_0 \wedge (p-r)} (ES_{1,K}^{\alpha/2})^{1/2} (ES_{2,K}^{\alpha/2})^{1/2}.$$

Hence, to establish (3.22) it is enough to prove that for some $m \geq 1$ $ES_{1,K}^{\alpha/2} < \infty$ and $ES_{2,K}^{\alpha/2} < \infty$. Since the two statement are of the same kind, we only check that

(3.23) $$ES_{1,K}^{\alpha/2} < \infty.$$

For a $\mathbf{j} = (j_1, \ldots, j_{p-r-K})$ with $j_k \geq m$, $1 \leq k \leq p-r-K$ we define

$$X_{\mathbf{j}} = \prod_{d=1}^{r+K} \psi(V_d)^{-2/\alpha} \prod_{k=1}^{p-r-K} \psi(V_{j_k+r+J_0})^{-2/\alpha}$$
(3.24) $$|f(V_1, \ldots, V_r, V_{r+1}, \ldots, V_{r+K}, V_{j_1+r+J_0}, \ldots, V_{j_{p-r-K}+r+J_0})|^2.$$

Then the distribution of $X_{\mathbf{j}}$ does not depend on $\mathbf{j}$, and

$$ES_{1,K}^{\alpha/2} = E\left(\sum_{j_1=m}^{\infty} \cdots \sum_{j_{p-r-K}=m}^{\infty} \left(\prod_{k=1}^{p-r-K} \Gamma_{j_k}^{-2/\alpha}\right) X_{\mathbf{j}}\right)^{\alpha/2}$$

$$\leq E\left(\sum_{j_1=m}^{\infty} \cdots \sum_{j_{p-r-K}=m}^{\infty} \left(\prod_{k=1}^{p-r-K} \Gamma_{j_k}^{-2/\alpha}\right) X_{\mathbf{j}} \mathbf{1}(X_{\mathbf{j}}^{\alpha/2} \leq [\mathbf{j}])\right)^{\alpha/2}$$

$$+ E\left(\sum_{j_1=m}^{\infty} \cdots \sum_{j_{p-r-K}=m}^{\infty} \left(\prod_{k=1}^{p-r-K} \Gamma_{j_k}^{-2/\alpha}\right) X_{\mathbf{j}} \mathbf{1}(X_{\mathbf{j}}^{\alpha/2} > [\mathbf{j}])\right)^{\alpha/2}$$

(3.25) $$:= ES_{11}^{\alpha/2} + ES_{12}^{\alpha/2}.$$

Here $[\mathbf{j}] = \prod_{k=1}^{p-r-K} j_k$. Choose now

$$m > \frac{2}{\alpha}(p \vee q).$$

We now use Proposition 5.1 of Samorodnitsky and Szulga (1989) to conclude that

$$ES_{11}^{\alpha/2} \leq c \left(E\left(X^{\alpha/2}(1+\log_+ X)^{p-r-K-1}\right)\right)^{\alpha/2}.$$

Here X is a generic random variable with the same law as $X_{\mathbf{j}}$, any $\mathbf{j}$. Now (3.1) immediately gives us

$$ES_{11}^{\alpha/2} \leq c \left(\int_T \cdots \int_T |f(x_1, \ldots, x_p)|^{\alpha} \right.$$
(3.26) $$\left. \left(1+\log_+ \frac{|f(x_1,\ldots,x_p)|^2}{\prod_{i=1}^p \psi(x_i)^{2/\alpha}}\right)^{p-r-K-1} m(dx_1)\ldots m(dx_p) \right)^{\alpha/2} < \infty.$$

Another application of Proposition 5.1 of Samorodnitsky and Szulga (1989) shows that
(3.27)
$$ES_{12}^{\alpha/2} \leq cE\Big(X^{\alpha/2}(1+\log_+ X)^{p-r-K}\Big)$$
$$\leq c\int_{S^p} |f(x_1,\ldots,x_p)|^\alpha \Big(1+\log_+ \frac{|f(x_1,\ldots,x_p)|^2}{\prod_{i=1}^p \psi(x_i)^{2/\alpha}}\Big)^{p-r-K} m(dx_1)\ldots m(dx_p)$$
$$< \infty.$$

The relations (3.25), (3.26) and (3.27) now establish that $ES_{1,K}^{\alpha/2} < \infty$, and so we have completed the inductive argument, thus proving (3.22) and, hence, (3.21) as well.

Having obtained (3.21) we immediately see that (3.17) holds as a particular case of (3.21) with $J = 0$. Therefore, (3.4) has been completely proved.

Now we will prove (3.6). Observe that

(3.28)
$$R_r = \sum_{r_0=0}^{r-1} c(r_0) R_{r,r_0}$$

for some constants $c(r_0)$, $r_0 = 0, \ldots, r-1$, where

$$R_{r,r_0} = [\prod_{i=1}^{r_0} \Gamma_i^{-2/\alpha} \psi(V_i)^{-2/\alpha}]$$

$$\sum_{m_1=r_0+2}^{\infty} \sum_{m_2=m_1+1}^{\infty} \cdots \sum_{m_{r-r_0}=m_{r-r_0-1}+1}^{\infty} [\prod_{d=1}^{r-r_0} \Gamma_{m_d}^{-2/\alpha} \psi(V_{m_d})^{-2/\alpha}]$$

$$\sum_{\substack{\mathbf{j} \in D^{p+q-2r}, j_k+r_0 \neq m_d \\ k=1,\ldots,p+q-2r, d=1,\ldots r-r_0}} [\prod_{k=1}^{p+q-2r} \epsilon_{j_k+r_0} \Gamma_{j_k+r_0}^{-1/\alpha} \psi(V_{j_k+r_0})^{-1/\alpha}]$$

$$h_r(V_1,\ldots,V_{r_0}, V_{m_1},\ldots,V_{m_{r-r_0}}, V_{j_1+r_0},\ldots,V_{j_{p+q-2r}+r_0}),$$

$r_0 = 0, \ldots, r-1$. That is, (3.6) will follow once we check that for every such r_0

(3.29)
$$\lim_{\lambda \to \infty} \lambda^{\alpha/2} (\log \lambda)^{-(r-1)} P(|R_{r,r_0}| > \lambda) = 0.$$

We start with "decoupling" $R_{r,r-0}$. We know by Theorem 1 of de la Peña and Montgomery–Smith (1995) that for some constant $c > 0$ one has

(3.30)
$$P(|R_{r,r_0}| > \lambda) \leq cP(|R_{r,r_0}^{(1)}| > \lambda/c)$$

for all $\lambda > 0$, where

$$R_{r,r_0}^{(1)} = [\prod_{i=1}^{r_0} \Gamma_i^{-2/\alpha} \psi(V_i)^{-2/\alpha}]$$

$$\sum_{m_1=r_0+2}^{\infty} \sum_{m_2=m_1+1}^{\infty} \cdots \sum_{m_{r-r_0}=m_{r-r_0-1}+1}^{\infty} [\prod_{d=1}^{r-r_0} \Gamma_{m_d}^{-2/\alpha} \psi(V_{m_d}^{(d)})^{-2/\alpha}]$$

$$\sum_{\substack{\mathbf{j} \in D^{p+q-2r}, j_k+r_0 \neq m_d \\ k=1,\ldots,p+q-2r, d=1,\ldots r-r_0}} [\prod_{k=1}^{p+q-2r} \epsilon_{j_k+r_0}^{(k)} \Gamma_{j_k+r_0}^{-1/\alpha} \psi(V_{j_k+r_0}^{(r-r_0+k)})^{-1/\alpha}]$$

$$h_r(V_1,\ldots,V_{r_0}, V_{m_1}^{(1)}, \ldots, V_{m_{r-r_0}}^{(r-r_0)}, V_{j_1+r_0}^{(r-r_0+1)}, \ldots, V_{j_{p+q-2r}+r_0}^{(p+q-r-r_0)}),$$

$r_0 = 0, \ldots, r-1$. Here $(V_j^{(m)}, j \geq 1)$, $m = 1, \ldots, p+q-r-r_0$ are i.i.d. copies of $(V_j, j \geq 1)$ independent of i.i.d. copies $(\epsilon_j^{(m)}, j \geq 1)$, $m = 1, \ldots, p+q-2r$ of $(\epsilon_j, j \geq 1)$. Furthermore all these new sequences of random variables are independent of the sequences $(\Gamma_j, j \geq 1)$ and $(V_j, j \geq 1)$. Applying now the contraction principle for Rademacher series (e.g. Theorem 3.4. in Ledoux and Talagrand (1991)) we conclude further that there is a constant $c > 0$ such that

(3.31) $$P(|R_{r,r_0}^{(1)}| > \lambda) \leq cP(|R_{r,r_0}^{(2)}| > \lambda/c)$$

for all $\lambda > 0$, where

$$R_{r,r_0}^{(2)} = [\prod_{i=1}^{r_0} \Gamma_i^{-2/\alpha} \psi(V_i)^{-2/\alpha}]$$

$$\sum_{m_1=r_0+2}^{\infty} \sum_{m_2=m_1+1}^{\infty} \cdots \sum_{m_{r-r_0}=m_{r-r_0-1}+1}^{\infty} [\prod_{d=1}^{r-r_0} \Gamma_{m_d}^{-2/\alpha} \psi(V_{m_d}^{(d)})^{-2/\alpha}]$$

$$\sum_{\substack{\mathbf{j} \in D^{p+q-2r}, j_k + r_0 \neq m_d \\ k=1,\ldots,p+q-2r, d=1,\ldots r-r_0}} [\prod_{k=1}^{p+q-2r} \epsilon_{j_k}^{(k)} (\Gamma_{j_k}^{(1)})^{-1/\alpha} \psi(V_{j_k}^{(r-r_0+k)})^{-1/\alpha}]$$

$$h_r(V_1, \ldots, V_{r_0}, V_{m_1}^{(1)}, \ldots, V_{m_{r-r_0}}^{(r-r_0)}, V_{j_1}^{(r-r_0+1)}, \ldots, V_{j_{p+q-2r}}^{(p+q-r-r_0)}),$$

$r_0 = 0, \ldots, r-1$, where the new sequence of random variables, $(\Gamma_j^{(1)}, j \geq 1)$ is a copy of $(\Gamma_j, j \geq 1)$ and is independent of all other random variables around. It follows from (3.30) and (3.31) that (3.6) will follow once we prove that

(3.32) $$\lim_{\lambda \to \infty} \lambda^{\alpha/2} (\log \lambda)^{-(r-1)} P(|R_{r,r_0}^{(2)}| > \lambda) = 0$$

for every $r_0 = 0, \ldots r-1$.

We start by proving a similar statement for a "truncated" version of $R_{r,r_0}^{(2)}$. For an $M \geq 1$ define

$$R_{r,r_0}^{(2)}(M) = [\prod_{i=1}^{r_0} \Gamma_i^{-2/\alpha} \psi(V_i)^{-2/\alpha}]$$

$$\sum_{m_1=M}^{\infty} \sum_{m_2=m_1+1}^{\infty} \cdots \sum_{m_{r-r_0}=m_{r-r_0-1}+1}^{\infty} [\prod_{d=1}^{r-r_0} \Gamma_{m_d}^{-2/\alpha} \psi(V_{m_d}^{(d)})^{-2/\alpha}]$$

$$\sum_{\substack{\mathbf{j} \in D^{p+q-2r}, j_k \geq M, j_k + r_0 \neq m_d \\ k=1,\ldots,p+q-2r, d=1,\ldots r-r_0}} [\prod_{k=1}^{p+q-2r} \epsilon_{j_k}^{(k)} (\Gamma_{j_k}^{(1)})^{-1/\alpha} \psi(V_{j_k}^{(r-r_0+k)})^{-1/\alpha}]$$

$$h_r(V_1, \ldots, V_{r_0}, V_{m_1}^{(1)}, \ldots, V_{m_{r-r_0}}^{(r-r_0)}, V_{j_1}^{(r-r_0+1)}, \ldots, V_{j_{p+q-2r}}^{(p+q-r-r_0)}).$$

We claim that for all M big enough,

(3.33) $$\lim_{\lambda \to \infty} \lambda^{\alpha/2} (\log \lambda)^{-(r-1)} P(|R_{r,r_0}^{(2)}(M)| > \lambda) = 0.$$

To proceed, we need the following lemma.

LEMMA 3.3. *Let Z be a nonnegative random variable defined on the same probability space as the first r points $\Gamma_1, \ldots, \Gamma_r$, $r \geq 1$, of a unit rate Poisson process on $(0, \infty)$. Assume that there is a $K < \infty$ such that*

(3.34) $$E(Z|\Gamma_1, \ldots, \Gamma_r) \leq K \text{ a.s..}$$

Then

$$(3.35) \quad \limsup_{t \downarrow 0} t^{-1} \left(\log \frac{1}{t}\right)^{-(r-1)} P(\prod_{i=1}^{r} \Gamma_i \leq tZ) < \infty.$$

PROOF OF LEMMA 3.3. Let F_r stand for the law of the product $\prod_{i=1}^{r} \Gamma_i$. We have by the Markov inequality

$$P(\prod_{i=1}^{r} \Gamma_i \leq tZ) \leq P(\prod_{i=1}^{r} \Gamma_i \leq t) + Kt \int_{t}^{\infty} y^{-1} F_r(dy)$$

$$\leq P(\prod_{i=1}^{r} \Gamma_i \leq t) + Kt \int_{t}^{\infty} \frac{P(\prod_{i=1}^{r} \Gamma_i \leq x)}{x^2} dx,$$

and (3.35) follows immediately from the fact that

$$\lim_{t \downarrow 0} t^{-1} \left(\log \frac{1}{t}\right)^{-(r-1)} P(\prod_{i=1}^{r} \Gamma_i \leq t) < \infty$$

(see e.g. Lemma 3.1 of Samorodnitsky and Szulga (1989)). □

Back, once again, to the proof of Theorem 3.1, we notice that in order to use Lemma 3.3 to prove (3.33) it is enough to show that for all M big enough,

$$(3.36) \quad E|R_{r,r_0}^{(3)}(M)|^{\alpha/2} < \infty,$$

where

$$R_{r,r_0}^{(3)}(M) = \prod_{i=1}^{r_0} \psi(V_i)^{-2/\alpha}$$

$$\sum_{m_1=M}^{\infty} \sum_{m_2=m_1+1}^{\infty} \cdots \sum_{m_{r-r_0}=m_{r-r_0-1}+1}^{\infty} [\prod_{d=1}^{r-r_0} \Gamma_{m_d-r_0}^{-2/\alpha} \psi(V_{m_d}^{(d)})^{-2/\alpha}]$$

$$\sum_{\substack{\mathbf{j} \in D^{p+q-2r}, j_k \geq M, j_k+r_0 \neq m_d \\ k=1,\ldots,p+q-2r, d=1,\ldots r-r_0}} [\prod_{k=1}^{p+q-2r} \epsilon_{j_k}^{(k)} (\Gamma_{j_k}^{(1)})^{-1/\alpha} \psi(V_{j_k}^{(r-r_0+k)})^{-1/\alpha}]$$

$$h_r(V_1, \ldots, V_{r_0}, V_{m_1}^{(1)}, \ldots, V_{m_{r-r_0}}^{(r-r_0)}, V_{j_1}^{(r-r_0+1)}, \ldots, V_{j_{p+q-2r}}^{(p+q-r-r_0)}).$$

Using first the Khinchine inequality for tetrahedral sums in Rademacher random variables and then twice the Cauchy–Schwartz inequality we have

$$E|R_{r,r_0}^{(3)}(M)|^{\alpha/2}$$

$$\leq cE \Bigg[\prod_{i=1}^{r_0} \psi(V_i)^{-4/\alpha} \sum_{\substack{\mathbf{j} \in D^{p+q-2r} \\ j_k \geq M, k=1,\ldots,p+q-2r}} [\prod_{k=1}^{p+q-2r} (\Gamma_{j_k}^{(1)})^{-2/\alpha} \psi(V_{j_k}^{(r-r_0+k)})^{-2/\alpha}]$$

$$\Bigg(\sum_{\substack{\mathbf{m} \in D^{r-r_0} \\ m_d \geq M, d=1,\ldots,r-r_0}} [\prod_{d=1}^{r-r_0} \Gamma_{m_d-r_0}^{-2/\alpha} \psi(V_{m_d}^{(d)})^{-2/\alpha}]$$

$$h_r(V_1, \ldots, V_{r_0}, V_{m_1}^{(1)}, \ldots, V_{m_{r-r_0}}^{(r-r_0)}, V_{j_1}^{(r-r_0+1)}, \ldots, V_{j_{p+q-2r}}^{(p+q-r-r_0)})\Bigg)^2\Bigg]^{\alpha/4}$$

$$\leq cE\left[\prod_{i=1}^{r_0}\psi(V_i)^{-1}\left(\sum_{\substack{\mathbf{j}\in D^{p+q-2r}\\j_k\geq M, k=1,\ldots,p+q-2r}}[\prod_{k=1}^{p+q-2r}(\Gamma_{j_k}^{(1)})^{-2/\alpha}\psi(V_{j_k}^{(r-r_0+k)})^{-2/\alpha}]\right.\right.$$

$$\left(\sum_{\substack{\mathbf{m}\in D^{r-r_0}\\m_d\geq M, d=1,\ldots,r-r_0}}[\prod_{d=1}^{r-r_0}\Gamma_{m_d-r_0}^{-2/\alpha}\psi(V_{m_d}^{(d)})^{-2/\alpha}]\right.$$

$$\left.f(V_1,\ldots,V_{r_0},V_{m_1}^{(1)},\ldots,V_{m_{r-r_0}}^{(r-r_0)},V_{j_1}^{(r-r_0+1)},\ldots,V_{j_{p-r}}^{(p-r_0)})\right)^2$$

$$\left(\sum_{\substack{\mathbf{m}\in D^{r-r_0}\\m_d\geq M, d=1,\ldots,r-r_0}}[\prod_{d=1}^{r-r_0}\Gamma_{m_d-r_0}^{-2/\alpha}\psi(V_{m_d}^{(d)})^{-2/\alpha}]\right.$$

$$\left.\left.\left.g(V_1,\ldots,V_{r_0},V_{m_1}^{(1)},\ldots,V_{m_{r-r_0}}^{(r-r_0)},V_{j_{p-r+1}}^{(p-r_0+1)},\ldots,V_{j_{p+q-2r}}^{(p+q-r-r_0)})\right)^2\right)^{\alpha/4}\right]$$

$$\leq c\left\{E\left[\prod_{i=1}^{r_0}\psi(V_i)^{-1}\left(\sum_{\substack{\mathbf{j}\in D^{p-r}\\j_k\geq M, k=1,\ldots,p-r}}[\prod_{k=1}^{p-r}(\Gamma_{j_k}^{(1)})^{-2/\alpha}\psi(V_{j_k}^{(r-r_0+k)})^{-2/\alpha}]\right.\right.\right.$$

$$\left(\sum_{\substack{\mathbf{m}\in D^{r-r_0}\\m_d\geq M, d=1,\ldots,r-r_0}}[\prod_{d=1}^{r-r_0}\Gamma_{m_d-r_0}^{-2/\alpha}\psi(V_{m_d}^{(d)})^{-2/\alpha}]\right.$$

$$\left.\left.\left.f(V_1,\ldots,V_{r_0},V_{m_1}^{(1)},\ldots,V_{m_{r-r_0}}^{(r-r_0)},V_{j_1}^{(r-r_0+1)},\ldots,V_{j_{p-r}}^{(p-r_0)})\right)^2\right)^{\alpha/2}\right]\right\}^{1/2}$$

$$\left\{E\left[\prod_{i=1}^{r_0}\psi(V_i)^{-1}\left(\sum_{\substack{\mathbf{j}\in D^{q-r}\\j_k\geq M, k=1,\ldots,q-r}}[\prod_{k=1}^{q-r}(\Gamma_{j_k}^{(1)})^{-2/\alpha}\psi(V_{j_k}^{(r-r_0+k)})^{-2/\alpha}]\right.\right.\right.$$

$$\left(\sum_{\substack{\mathbf{m}\in D^{r-r_0}\\m_d\geq M, d=1,\ldots,r-r_0}}[\prod_{d=1}^{r-r_0}\Gamma_{m_d-r_0}^{-2/\alpha}\psi(V_{m_d}^{(d)})^{-2/\alpha}]\right.$$

$$\left.\left.\left.g(V_1,\ldots,V_{r_0},V_{m_1}^{(1)},\ldots,V_{m_{r-r_0}}^{(r-r_0)},V_{j_1}^{(r-r_0+1)},\ldots,V_{j_{q-r}}^{(q-r_0)})\right)^2\right)^{\alpha/2}\right]\right\}^{1/2}.$$

Since both expectations above are of the same nature, our statement (3.36) will follow once we prove that

$$(3.37)\quad E\left[\prod_{i=1}^{r_0}\psi(V_i)^{-1}\left(\sum_{\substack{\mathbf{j}\in D^{p-r}\\j_k\geq M, k=1,\ldots,p-r}}[\prod_{k=1}^{p-r}(\Gamma_{j_k}^{(1)})^{-2/\alpha}\psi(V_{j_k}^{(r-r_0+k)})^{-2/\alpha}]\right.\right.$$

$$\left(\sum_{\substack{\mathbf{m}\in D^{r-r_0}\\m_d\geq M, d=1,\ldots,r-r_0}}[\prod_{d=1}^{r-r_0}\Gamma_{m_d-r_0}^{-2/\alpha}\psi(V_{m_d}^{(d)})^{-2/\alpha}]\right.$$

$$\left.\left.f(V_1,\ldots,V_{r_0},V_{m_1}^{(1)},\ldots,V_{m_{r-r_0}}^{(r-r_0)},V_{j_1}^{(r-r_0+1)},\ldots,V_{j_{p-r}}^{(p-r_0)})\right)^2\right)^{\alpha/2}\right]<\infty.$$

However, the fact that for M large enough (3.37) follows from $N_p(f) < \infty$ can be established by exactly the same argument as that used in proving (3.21). Hence, (3.33) follows.

Recall that (3.33) is a statement analogous to (3.32) but for a "truncated" version of $R_{r,r_0}^{(2)}$. Our next step is to prove another such statement, this time for a "less truncated" version of $R_{r,r_0}^{(2)}$. Specifically, we claim that for all M big enough,

(3.38) $$\lim_{\lambda \to \infty} \lambda^{\alpha/2}(\log \lambda)^{-(r-1)} P(|R_{r,r_0}^{(4)}(M)| > \lambda) = 0,$$

where for $M \geq 1$

$$R_{r,r_0}^{(4)}(M) = [\prod_{i=1}^{r_0} \Gamma_i^{-2/\alpha}\psi(V_i)^{-2/\alpha}]$$

$$\sum_{m_1=r_0+2}^{\infty} \sum_{m_2=m_1+1}^{\infty} \cdots \sum_{m_{r-r_0}=m_{r-r_0-1}+1}^{\infty} [\prod_{d=1}^{r-r_0} \Gamma_{m_d}^{-2/\alpha}\psi(V_{m_d}^{(d)})^{-2/\alpha}]$$

$$\sum_{\substack{J \in D^{p+q-2r}, j_k \geq M, j_k + r_0 \neq m_d \\ k=1,\ldots,p+q-2r, d=1,\ldots r-r_0}} [\prod_{k=1}^{p+q-2r} \epsilon_{j_k}^{(k)}(\Gamma_{j_k}^{(1)})^{-1/\alpha}\psi(V_{j_k}^{(r-r_0+k)})^{-1/\alpha}]$$

$$h_r(V_1,\ldots,V_{r_0},V_{m_1}^{(1)},\ldots,V_{m_{r-r_0}}^{(r-r_0)},V_{j_1}^{(r-r_0+1)},\ldots,V_{j_{p+q-2r}}^{(p+q-r-r_0)}).$$

Observe that for all $M \geq r_0 + 2$,

$$R_{r,r_0}^{(4)}(M) = R_{r,r_0}^{(2)}(M) + \sum_{n=r_0+2}^{M-1} R_{r,r_0}(M,n)$$

where for $n = r_0 + 2, \ldots, M-1$

$$R_{r,r_0}(M,n) = [\prod_{i=1}^{r_0} \Gamma_i^{-2/\alpha}\psi(V_i)^{-2/\alpha}]\Gamma_n^{-2/\alpha}\psi(V_n^{(1)})^{-2/\alpha}$$

$$\sum_{m_2=n+1}^{\infty} \cdots \sum_{m_{r-r_0}=m_{r-r_0-1}+1}^{\infty} [\prod_{d=2}^{r-r_0} \Gamma_{m_d}^{-2/\alpha}\psi(V_{m_d}^{(d)})^{-2/\alpha}]$$

$$\sum_{\substack{J \in D^{p+q-2r}, j_k \geq M, j_k + r_0 \neq m_d \\ k=1,\ldots,p+q-2r, d=1,\ldots r-r_0}} [\prod_{k=1}^{p+q-2r} \epsilon_{j_k}^{(k)}(\Gamma_{j_k}^{(1)})^{-1/\alpha}\psi(V_{j_k}^{(r-r_0+k)})^{-1/\alpha}]$$

$$h_r(V_1,\ldots,V_{r_0},V_n^{(1)},V_{m_2}^{(2)},\ldots,V_{m_{r-r_0}}^{(r-r_0)},V_{j_1}^{(r-r_0+1)},\ldots,V_{j_{p+q-2r}}^{(p+q-r-r_0)}).$$

Since (3.33) holds for M big enough, (3.38) will follow once we prove that for all M big enough and all $n = r_0 + 2, \ldots, M-1$

(3.39) $$\lim_{\lambda \to \infty} \lambda^{\alpha/2}(\log \lambda)^{-(r-1)} P(|R_{r,r_0}(M,n)| > \lambda) = 0.$$

Applying the same downward induction as the one used in the proof of (3.21) we see that (3.39) will follow if we prove that for all M large enough,

(3.40) $$\lim_{\lambda \to \infty} \lambda^{\alpha/2}(\log \lambda)^{-(r-1)} P(|R_{r,r_0}^{(5)}(M)| > \lambda) = 0,$$

where

$$R_{r,r_0}^{(5)}(M,n) = [\prod_{i_1=1}^{r_0} \Gamma_{i_1}^{-2/\alpha}\psi(V_{i_1})^{-2/\alpha}][\prod_{i_2=r_0+2}^{r-1} \Gamma_{i_2}^{-2/\alpha}\psi(V_{i_2}^{(i_2-r_0+1)})^{-2/\alpha}]$$

$$\sum_{\substack{\mathbf{j} \in D^{p+q-2r} \\ j_k \geq M, k=1,\ldots,p+q-2r}} \Big[\prod_{k=1}^{p+q-2r} \epsilon_{j_k}^{(k)} (\Gamma_{j_k}^{(1)})^{-1/\alpha} \psi(V_{j_k}^{(r-r_0+k)})^{-1/\alpha} \Big]$$

$$h_r(V_1, \ldots, V_{r_0}, V_{r_0+2}^{(1)}, \ldots, V_{r+1}^{(r-r_0)}, V_{j_1}^{(r-r_0+1)}, \ldots, V_{j_{p+q-2r}}^{(p+q-r-r_0)}).$$

However, the fact that (3.40) holds for all M large enough follows immediately from the fact that (3.33) holds for all M large enough. Therefore, (3.38) follows.

We are now in a position to complete the proof of (3.32). Fix an M for which (3.38) holds. It is clear that (3.32) will follow once we establish that for every $n = 1, M-1$

(3.41) $$\lim_{\lambda \to \infty} \lambda^{\alpha/2} (\log \lambda)^{-(r-1)} P(|R_{r,r_0}^{(6)}(n)| > \lambda) = 0,$$

where

$$R_{r,r_0}^{(6)}(n) = \Big[\prod_{i=1}^{r_0} \Gamma_i^{-2/\alpha} \psi(V_i)^{-2/\alpha} \Big] (\Gamma_n^{(1)})^{-1/\alpha} \psi(V_n^{(r-r_0+1)})^{-1/\alpha}$$

$$\sum_{\substack{m_1 \geq r_0+2 \\ m_1 \neq n}} \sum_{\substack{m_2 \geq m_1+1 \\ m_2 \neq n}} \cdots \sum_{\substack{m_{r-r_0} \geq m_{r-r_0-1}+1 \\ m_{r-r_0} \neq n}} \Big[\prod_{d=1}^{r-r_0} \Gamma_{m_d}^{-2/\alpha} \psi(V_{m_d}^{(d)})^{-2/\alpha} \Big]$$

$$\sum_{\substack{\mathbf{j} \in D^{p+q-2r-1}, j_k > n, j_k+r_0 \neq m_d \\ k=1,\ldots,p+q-2r-1, d=1,\ldots r-r_0}} \Big[\prod_{k=1}^{p+q-2r-1} \epsilon_{j_k}^{(k)} (\Gamma_{j_k}^{(1)})^{-1/\alpha} \psi(V_{j_k}^{(r-r_0+k+1)})^{-1/\alpha} \Big]$$

$$h_r(V_1, \ldots, V_{r_0}, V_{m_1}^{(1)}, \ldots, V_{m_{r-r_0}}^{(r-r_0)}, V_n^{(r-r_0+1)}, V_{j_2}^{(r-r_0+1)}, \ldots, V_{j_{p+q-2r-1}}^{(p+q-r-r_0)}).$$

Repeating the truncation and downward induction argument used a number of times before, we see that (3.41) will be a consequence of the following statement.

(3.42) $$\lim_{\lambda \to \infty} \lambda^{\alpha/2} (\log \lambda)^{-(r-1)} P(|R_{r,r_0}^{(7)}| > \lambda) = 0,$$

where

$$R_{r,r_0}^{(7)} = \prod_{i_1=1}^{r_0} \Gamma_{i_1}^{-2/\alpha} \prod_{i_2=r_0+2}^{r+1} \Gamma_{i_2}^{-2/\alpha} \prod_{j=1}^{p+q-2r} (\Gamma_j^{(1)})^{-1/\alpha}$$

$$\prod_{k_1=1}^{r} \psi(V_{k_1})^{-2/\alpha} \prod_{k_2=r+1}^{p+q-r} \psi(V_{k_2})^{-1/\alpha}$$

$$f(V_1, \ldots, V_r, V_{r+1}, \ldots, V_p) g(V_1, \ldots, V_r, V_{p+1}, \ldots, V_{p+q-r}).$$

We need yet another lemma.

LEMMA 3.4. *Let Z be a nonnegative random variable independent of a sequence $(\Gamma_j, j \geq 1)$ of the ordered points of a unit rate Poisson process on $(0, \infty)$, with $EZ < \infty$. Then for every $r \geq 1$,*

(3.43) $$\lim_{t \downarrow 0} t^{-1} (\log \tfrac{1}{t})^{-(r-1)} P\Big(\Big(\prod_{i=1}^{r-1} \Gamma_i \Big) \Gamma_{r+1} \leq tZ \Big) = 0.$$

PROOF OF LEMMA 3.4. Let $\Gamma_2^{(1)}$ be a copy of Γ_2 independent both of the original Poisson points and Z. We have

$$P\Big((\prod_{i=1}^{r-1}\Gamma_i)\Gamma_{r+1} \leq tZ\Big) \leq P\Big((\prod_{i=1}^{r-1}\Gamma_i) \leq t\frac{Z}{\Gamma_2^{(1)}}\Big)$$

$$\leq O\Big(t(\log\frac{1}{t})^{r-2}\Big)$$

by Lemma 3.3 since $E(\Gamma_2)^{-1} < \infty$. □

We now complete the proof of Theorem 3.1. It follows from Lemma 3.4 that (3.42) will be established if

$$E\Big|\prod_{k_1=1}^{r}\psi(V_{k_1})^{-2/\alpha}\prod_{k_2=r+1}^{p+q-r}\psi(V_{k_2})^{-1/\alpha}f(V_1,\ldots,V_r,V_{r+1},\ldots,V_p)$$

(3.44) $$g(V_1,\ldots,V_r,V_{p+1},\ldots,V_{p+q-r})\Big|^{\alpha/2} < \infty.$$

However, (3.44) follows from $N_p(f) < \infty$ and $N_q(g) < \infty$ by Hölder's inequality. This proves (3.41), and hence (3.32). Therefore, the proof of Theorem 3.1 is now complete.

4. Tail of the product and independence of multiple stable integrals

Theorems 2.3 and 3.1 allow us to determine the asymptotic behavior of the tail of the product of two multiple stable integrals.

THEOREM 4.1. *Assume $N_p(f) < \infty$ and $N_q(g) < \infty$, where $f : S^p \to \mathbf{R}$ and $g : S^q \to \mathbf{R}$ are symmetric vanishing on diagonals functions. Let $r = \max\{i : C_i(f,g) \neq 0\}$, where $C_i(f,g)$ is given by (3.2) and $C_0(f,g) = \|\widehat{f \otimes g}\|_{L^\alpha(S^q)}^\alpha$. Then, as $\lambda \to \infty$,*

$$P(|I_p^M(f)I_q^M(g)| > \lambda) \sim \begin{cases} k_{0,\alpha}\lambda^{-\alpha}(\ln\lambda)^{p+q-1}C_0(f,g) & \text{if } r=0 \\ \\ k_{r,\alpha}\lambda^{-\alpha/2}(\ln\lambda)^{r-1}C_r(f,g) & \text{if } r=1,\ldots,p\wedge q, \end{cases}$$

where $k_{r,\alpha}$ are finite numerical constants in (3.0) and (3.1).

PROOF. The proof follows from Theorems 2.3 and 3.1 by a standard application of the inequality

$$P(|X| > (1+\epsilon)\lambda) - P(|Y| > \epsilon\lambda) \leq P(|X+Y| > \lambda)$$
$$\leq P(|X| > (1-\epsilon)\lambda) + P(|Y| > \epsilon\lambda),$$

which is valid for any random variables X and Y, $\epsilon \in (0,1)$, and $\lambda > 0$. □

Now we are ready to establish conditions for independence of $I_p^M(f)$ and $I_q^M(g)$. The following lemma will be used below and in Section 5. Its proof uses a method from Üstünel and Zakai (1989).

LEMMA 4.2. *Let $u : S^k \to \mathbf{R}$ and $v : S^n \to \mathbf{R}$ be measurable symmetric functions such that $u \in L^2(S^k, \nu^{\otimes k})$ and $v \in L^2(S^n, \nu^{\otimes n})$, where ν is a σ-finite measure on S. Then*

$$(4.1) \quad \|\widehat{u \otimes v}\|^2_{L^2(S^{k+n})} = \frac{k!n!}{(k+n)!} \|u \otimes v\|^2_{L^2(S^{k+n})} + \sum_{i=1}^{k \wedge n} c_i \|u \otimes_i v\|^2_{L^2(S^{k+n-2i})},$$

where

$$(4.2) \quad u \otimes_i v(s_1, \ldots, s_{k+n-2i})$$

$$= \int_{S^i} u(s_1, \ldots, s_{k-i}, x_1, \ldots, x_i) v(x_1, \ldots, x_i, s_{k-i+1}, \ldots, s_{k+n-2i}) \nu(dx_1) \ldots \nu(dx_i).$$

and $c_i = \binom{k}{i}\binom{n}{i}/\binom{k+n}{k}$.

PROOF. We have

$$\|\widehat{u \otimes v}\|^2_{L^2(S^{k+n})}$$

$$= ((k+n)!)^{-2} \int_{S^{k+n}} |\sum_\pi u(s_{\pi_1}, \ldots, s_{\pi_k}) v(s_{\pi_{k+1}}, \ldots, s_{\pi_{k+n}})|^2 \, d\nu^{\otimes(k+n)}$$

$$= ((k+n)!)^{-2} \int_{S^{k+n}} D_{\pi,\mu}(s_1, \ldots, s_{k+n}) \, d\nu^{\otimes(k+n)},$$

where π, μ run over all permutations of $1, \ldots, k+n$ and

$$D_{\pi,\mu} = u(s_{\pi_1}, \ldots, s_{\pi_k}) v(s_{\mu_{k+1}}, \ldots, s_{\mu_{k+n}}) u(s_{\mu_1}, \ldots, s_{\mu_k}) v(s_{\pi_{k+1}}, \ldots, s_{\pi_{k+n}}).$$

Suppose that $k \leq n$ and consider the sets $J_1(\pi, \mu) = \{\pi_1, \ldots, \pi_k\} \cap \{\mu_1, \ldots, \mu_k\}$ and $J_2(\pi, \mu) = \{\pi_{k+1}, \ldots, \pi_{k+n}\} \cap \{\mu_{k+1}, \ldots, \mu_{k+n}\}$. If $\mathrm{Card}(J_1(\pi, \mu)) = k - i$, then $\mathrm{Card}(J_2(\pi, \mu)) = n + i$, $i = 0, \ldots, k$. If $i = 0$ then $D_{\pi,\mu} = |u \otimes v|^2$. Suppose $i \geq 1$. Then by Fubini's theorem

$$\int_{S^{k+n}} D_{\pi,\mu} \, d\nu^{k+n} = \|u \otimes_i v\|^2_{L^2(S^{k+n-2i})}.$$

Since there are $\binom{k}{i}\binom{n}{i} k! n! (k+n)!$ terms $D_{\pi,\mu}$ with $\mathrm{Card}(J_1(\pi, \mu)) = k - i$, $i \geq 0$, the proof is complete. □

THEOREM 4.3. *Let $f_i : S^{p_i} \to \mathbf{R}$ and $g_i : S^{q_i} \to \mathbf{R}$ be symmetric vanishing on diagonals functions, $p_i, q_i \geq 1$, and such that $N_{p_i}(f_i) < \infty$ and $N_{q_i}(g_i) < \infty$, $i = 1, \ldots, n$. The sequences $\{I^M_{p_i}(f_i)\}_{i=1}^n$ and $\{I^M_{q_i}(g_i)\}_{i=1}^n$ are independent of each other if and only if there exist disjoint measurable sets $A, B \subset S$ such that $\mathrm{supp}\{f_i\} \subset A^{p_i}$ modulo $m^{\otimes p_i}$ and $\mathrm{supp}\{g_i\} \subset B^{q_i}$ modulo $m^{\otimes q_i}$, $i = 1, \ldots, n$.*

PROOF. *Sufficiency.* Under the assumption of disjointness of A and B, the random measures $M_1(E) := M(E \cap A)$ and $M_2(E) := M(E \cap B)$, $E \in \mathcal{S}$, are independent. Since $I^M_{p_i}(f_i) = I^{M_1}_{p_i}(f_i)$ and $I^M_{q_i}(g_i) = I^{M_2}_{q_i}(g_i)$, the independence of the sequences $\{I^M_{p_i}(f_i)\}_{i=1}^n$ and $\{I^M_{q_i}(g_i)\}_{i=1}^n$ is obvious.
Necessity. First we will give the proof for $n = 1$. Put $p = p_1$, $q = q_1$, $f = f_1$, and $g = g_1$. Assume that $I^M_p(f)$ and $I^M_q(g)$ are independent. Then the tail of $|I^M_p(f) I^M_q(g)|$ is of the order $\lambda^{-\alpha}(\log \lambda)^{p+q-1}$. From Theorem 4.1 we have $C_1(f,g) = \cdots = C_{p \wedge q}(f,g) = 0$. Hence for $r = 1, \ldots, p \wedge q$

$$(4.3) \quad \hat{h}_r(s_1, \ldots, s_r, \cdot) = 0 \quad m^{\otimes(p+q-r)} - \text{a.e.}$$

(see the definition of $\hat{h}_r$ in Theorem 3.1). Choose a sequence $s_1, \ldots, s_{p+q-1}$ of points from S and define a measure $\nu = \sum_{i=2}^{p+q-1} \delta_{s_i}$. Consider the multiple integral

$$(4.4) \quad L(s_1, \ldots, s_{p+q-1}) := \int_{S^{p+q-1}} |\hat{h}_1(s_1, x_2, \ldots, x_{p+q-1})|^2 \, \nu(dx_2) \cdots \nu(dx_{p+q-1}).$$

Notice that for each s_1, $h_r(s_1, \cdot)$ is a product of two symmetric measurable functions, so that from Lemma 4.2 we get

$$L(s_1, \ldots, s_{p+q-1}) \geq \frac{p!q!}{(p+q)!} \int_{S^{p-1}} |f(s_1, \cdot)|^2 d\nu^{\otimes(p-1)} \int_{S^{q-1}} |g(s_1, \cdot)|^2 d\nu^{\otimes(q-1)}$$

$$(4.5) \quad \geq \frac{p!q!(p-1)!(q-1)!}{(p+q)!} |f(s_1, \ldots, s_p))|^2 |g(s_1, s_{p+1}, \ldots, s_{p+q-1})|^2.$$

Now observe that

$$L(s_1, \ldots, s_{p+q-1}) = \sum_{r=1}^{p \wedge q} a_r \sum_{1 = i_1 < i_2 < \cdots < i_r \leq p+q-1} |\hat{h}_r(s_{i_1}, \ldots, s_{i_r}, \cdot)|^2,$$

for some constants $a_r \geq 0$; the symmetrization in $\hat{h}_r(s_{i_1}, \ldots, s_{i_r}, \cdot)$ is taken with respect to variables s_j with $j \in \{1, \ldots, p+q-1\} \setminus \{i_1, \ldots, i_r\}$. In view of (4.3), $L = 0$ $m^{\otimes(p+q-1)}$–almost everywhere; (4.5) implies that

$$f(s_1, \ldots, s_p) g(s_1, s_{p+1}, \ldots, s_{p+q-1}) = 0 \quad m^{\otimes(p+q-1)} - a.e.$$

Consequently,

$$0 = m^{\otimes(p+q-1)} \{(s_1, \ldots, s_{p+q-1}) : f(s_1, \ldots, s_p) \neq 0, \ g(s_1, s_{p+1}, \ldots, s_{p+q-1}) \neq 0\}$$

$$= \int_S m^{\otimes(p-1)} \{(s_2, \ldots, s_p) : f(s_1, \cdot) \neq 0\}$$

$$m^{\otimes(q-1)} \{(s_{p+1}, \ldots, s_{p+q-1}) : g(s_1, \cdot) \neq 0\} m(ds_1).$$

Hence the sets

$$A = \{s_1 : m^{\otimes(p-1)} \{(s_2, \ldots, s_p) : f(s_1, \cdot) \neq 0\} > 0\}$$

and

$$B = \{s_1 : m^{\otimes(q-1)} \{(s_{p+1}, \ldots, s_{p+q-1}) : g(s_1, \cdot) \neq 0\} > 0\}$$

satisfy $m(A \cap B) = 0$ and from the symmetry of f and g, $\text{supp}\{f\} \subset A^p$ modulo $m^{\otimes p}$ and $\text{supp}\{g\} \subset B^q$ modulo $m^{\otimes q}$. Finally we may replace A by $A \cap B^c$ to have A and B disjoint. This completes the proof for $n = 1$.

Let now $n > 1$. Using our theorem for $n = 1$, for every pair $(i, j) \in \{1, \ldots, n\}^2$ there exist disjoint $E_{i,j}, F_{i,j} \in \mathcal{S}$ such that $\text{supp}\{f_i\} \subset E_{i,j}^{p_i}$ modulo $m^{\otimes p_i}$ and $\text{supp}\{f_{q_j}\} \subset F_{i,j}^{q_j}$ modulo $m^{\otimes q_j}$. Since

$$A = \bigcup_{i=1}^{n} \bigcap_{j=1}^{n} E_{i,j} \quad \text{and} \quad B = \bigcup_{j=1}^{n} \bigcap_{i=1}^{n} F_{i,j}$$

satisfy the conditions of the theorem, the proof is complete. $\square$

COROLLARY 4.4. *Let $f_i : S^{p_i} \to \mathbf{R}$ be symmetric vanishing on diagonals functions such that $N_{p_i}(f_i) < \infty$, $p_i \geq 1$, $i \geq 1$. Then $I_{p_i}^M(f_i)$, $i \geq 1$, are independent if and only if there exist disjoint sets $A_i \in \mathcal{S}$, $i \geq 1$, such that for every i, $\text{supp}\{f_i\} \subset A_i^{p_i}$ modulo $m^{\otimes p_i}$.*

5. Independence of multiple integrals with respect to Gaussian and Poisson random measures

The purpose of this section is to exhibit the similarities and differences between the independence criteria for stable and Gaussian multiple integrals. We will also establish a surprising criterion for the independence of multiple Wiener–Itô integrals that follows from the work of Üstünel and Zakai (1989). Finally, we will give an example that can be used to disprove the claim made in the conclusion of the proof of independence of multiple integrals in Privault (1996). Consequently, the problem of necessary and sufficient conditions for independence of multiple Poisson integrals remains open. We expect the conditions be the same as in Theorem 4.3.

Let W be a Wiener random measure on $(S, \mathcal{S})$ with atomless σ-finite control measure m. Let $I_p^W(f)$ denote the p-tuple Wiener-Itô integral of $f \in L^2(S^p) (= L^2(S^p, m^{\otimes p}))$ with respect to W. We refer the reader to Major (1980) for the necessary facts on Wiener-Itô integrals. In what follows, $f \in L^2(S^p)$, $g \in L^2(S^q)$ are assumed to be symmetric and vanishing on the diagonals. $I_p^W(f)$ and $I_q^W(g)$ satisfy the *product formula*

$$(5.1) \qquad I_p^W(f) I_q^W(g) = \sum_{i=0}^{p \wedge q} i! \binom{p}{i} \binom{q}{i} I_{p+q-2i}^W(f \otimes_i g)$$

where $f \otimes_i g$ is defined analogously to (4.2),

$$(5.2) \quad f \otimes_i g(s_1, \ldots, s_{p+q-2i})$$
$$= \int_{S^i} f(s_1, \ldots, s_{k-i}, x_1, \ldots, x_i) g(x_1, \ldots, x_i, s_{k-i+1}, \ldots, s_{k+n-2i})$$
$$\times m(dx_1) \ldots m(dx_i).$$

We have the *isometry property* $E[I_k^W(h)^2] = k! \|\hat{h}\|_{L^2(S^k)}^2$, where $\hat{h}$ denotes the symmetrization of h.

THEOREM 5.1 (Üstünel and Zakai (1989)). *Let $f_i \in L^2(S^{p_i})$, $g_i \in L^2(S^{q_i})$, $p_i, q_i \geq 1$, $i = 1, \ldots, n$. The sequences $\{I_{p_i}^W(f_i)\}_{i=1}^n$ and $\{I_{q_i}^W(g_i)\}_{i=1}^n$ are independent of each other if and only if for every $i, j = 1, \ldots, n$,*

$$f_i \otimes_1 g_j = 0 \qquad m^{\otimes(p_i + q_j - 2)} - a.e.$$

From the proof of this result given in Üstünel and Zakai (1989) we deduce an interesting criterion for the independence of multiple Wiener-Itô integrals. We combine their proof with that in Kallenberg (1991) to obtain the following.

THEOREM 5.2. *Let $f_i \in L^2(S^{p_i})$, $g_i \in L^2(S^{q_i})$, $p_i, q_i \geq 1$, $i = 1, \ldots, n$. The sequences $\{I_{p_i}^W(f_i)\}_{i=1}^n$ and $\{I_{q_i}^W(g_i)\}_{i=1}^n$ are independent of each other if and only if for every $i, j = 1, \ldots, n$,*

$$(5.3) \qquad Cov(I_{p_i}^W(f_i)^2, I_{q_j}^W(g_j)^2) = 0.$$

PROOF. The necessity is obvious, so we will prove the sufficiency.

First we will obtain a lower bound for the covariance of squares of multiple Wiener-Itô integrals. Let $f \in L^2(S^p)$ and $g \in L^2(S^q)$ be symmetric vanishing on diagonals functions. Using product formula (5.1), orthogonality of multiple integrals of different orders, and the isometry property,

$$E[(I_p^W(f)I_q^W(g))^2] = \sum_{i=0}^{p\wedge q}[i!\binom{p}{i}\binom{q}{i}]^2(p+q-2i)!\|\widetilde{(f\otimes_i g)}\|^2_{L^2(S^{p+q-2i})}.$$

Dropping terms with $i \geq 1$ and then applying Lemma 4.2 gives

$$E[(I_p^W(f)I_q^W(g))^2] \geq (p+q)!\|\widetilde{(f\otimes g)}\|^2_{L^2(S^{p+q})}$$
$$\geq p!q!\|f\otimes g\|^2_{L^2(S^{p+q})} + p!q!pq\|f\otimes_1 g\|^2_{L^2(S^{p+q-2})}.$$

Hence

$$\text{Cov}([I_p^W(f)]^2, [I_q^W(g)]^2) = E[(I_p^W(f)I_q^W(g))^2] - E[I_p^W(f)^2]E[I_q^W(g)^2]$$
(5.4)
$$\geq p!q!pq\|f\otimes_1 g\|^2_{L^2(S^{p+q-2})}.$$

Let H_f denote the Hilbert space in $L^2(S)$ spanned by all functions

$$s \to \int_{S^{p-1}} f(x_1,\ldots,x_{p-1},s)h(x_1,\ldots,x_{p-1})\,m(dx_1)\cdots m(dx_{p-1}),$$

where $h \in L^2(S^{p-1})$, and similarly define H_g. If $\text{Cov}([I_p^W(f)]^2, [I_q^W(g)]^2) = 0$, then by (5.4), $\|f\otimes_1 g\|^2_{L^2(S^{p+q-2})} = 0$ and this in turn implies that H_f and H_g are orthogonal. Under the assumption of our theorem, for every f_i and g_j, $i,j = 1,\ldots,n$, the Hilbert spaces H_{f_i} and H_{g_j} are orthogonal. Let $\{\phi_{i,k}\}_k$ be an orthonormal basis for H_{f_i} and let $\{\psi_{j,k}\}_k$ be an orthonormal basis for H_{g_j}. f_i and g_j admit orthogonal expansions in the tensor product bases,

$$f_i = \sum_{k_1,\ldots,k_{p_i}} a_{k_1,\ldots,k_{p_i}} \phi_{i,k_1} \otimes \cdots \otimes \phi_{i,k_{p_i}},$$

$$g_j = \sum_{k_1,\ldots,k_{q_j}} b_{k_1,\ldots,k_{q_j}} \psi_{j,k_1} \otimes \cdots \otimes \psi_{j,k_{q_j}}.$$

Recall now Itô formula (see Major (1980), Theorem 4.2)

(5.5) $\quad I_r^W(\phi_1^{\otimes r_1} \otimes \cdots \otimes \phi_m^{\otimes r_m}) = \prod_{k=1}^m H_{r_k}(I_1^W(\phi_k)), \quad r_1 + \cdots + r_m = r,$

which holds for any orthonormal sequences $\{\phi_k\}$ in $L^2(S)$. Here H_n is the Hermite polynomial of degree n. From Itô's formula, $I_{p_i}^W(f_i)$ is a function of the Gaussian sequence $\{I_1^W(\phi_{i,k})\}_k$ and $I_{q_j}^W(g_j)$ is a function of the Gaussian sequence $\{I_1^W(\psi_{j,k})\}_k$. Because the Gaussian sequences

$$\{I_1^W(\phi_{i,k}) : k \geq 1, i = 1,\ldots n\} \quad \text{and} \quad \{I_1^W(\psi_{j,k}) : k \geq 1, j = 1,\ldots n\}$$

are uncorrelated, thus independent, the sequences $\{I_{p_i}^W(f_i)\}_{i=1}^n$ and $\{I_{q_i}^W(g_i)\}_{i=1}^n$ are independent. $\square$

COROLLARY 5.3. $I_{p_1}^W(f_1),\ldots,I_{p_n}^W(f_n)$ are independent if and only if

$$\text{Cov}(I_{p_i}^W(f_i)^2, I_{p_j}^W(f_j)^2) = 0, \quad \text{for every } i \neq j.$$

Here $f_i \in L^2(S^{p_i})$, $p_i \geq 1, i = 1,\ldots,n$.

It is interesting to compare Corollary 5.3 with the basic fact that jointly normal random variables are independent if and only if they are uncorrelated. Such statement is false for multiple Itô–Wiener integrals because multiple integrals of different orders are always uncorrelated. In particular, the terms in the chaos expansion of an L^2 functional of a Wiener process are uncorrelated and by Corollary 5.3 they are independent if and only if their squares are uncorrelated.

COROLLARY 5.4. $Cov(I_p^W(f)^2, I_q^W(g)^2) \geq 0$, for any $f \in L^2(S^p)$ and $g \in L^2(S^q)$.

PROOF. Follows from (5.4). □

Finally we will show that $Cov(I_p^N(f)^2, I_q^N(g)^2) = 0$ does not imply independence of multiple integrals $I_p^N(f)$ and $I_q^N(g)$ with respect to a compensated Poisson random measure N. This will also serve as a counterexample to the claim made in the conclusion of the proof of Theorem 1 in Privault (1996).

EXAMPLE 5.5. Let $S = [-1, 1]$, m the Lebesgue measure and let N be a compensated Poisson random measure on $[-1, 1]$, i.e., $E \exp i\theta N(A) = \exp[m(A)(e^{i\theta} - 1 - i\theta)]$. Let $f : [-1, 1]^2 \to \mathbf{R}$ be given by

$$f(s_1, s_2) = \begin{cases} 0 & \text{if } s_1 s_2 \geq 0 \\ -1 & \text{on } [-1, -1/2) \times [1/2, 1] \cup [-1/2, 0) \times (0, 1/2] \\ & \cup (0, 1/2) \times [-1/2, 0) \cup [1/2, 1] \times [-1, -1/2) \\ 1 & \text{elsewhere.} \end{cases}$$

Define $g : [-1, 1] \to \mathbf{R}$ by

$$g(s) = \begin{cases} -1 & \text{if } s < 0 \\ 1 & \text{if } s \geq 0. \end{cases}$$

Therefore we have $p = 2$ and $q = 1$. Elementary but tedious computations give $E[I_2^N(f)^2 I_1^N(g)^2] = 8$, $E[I_2^N(f)^2] = 4$, and $E[I_1^N(g)^2] = 2$. Consequently, $Cov(I_2^N(f)^2, I_1^N(g)^2) = 0$. To prove that $I_2^N(f)$ and $I_1^N(g)$ are not independent we notice that $I_2^N(f)$ and $I_1^N(g)$ take on only integer values. Moreover, if $I_2^N(f) = 2$ then $I_1^N(g)$ must be even. Since the events $\{I_2^N(f) = 2\}$ and $\{I_1^N(g) = 1\}$ have positive probabilities, $I_2^N(f)$ and $I_1^N(g)$ are dependent.

References

[1] de la Peña, V.H. and Montgomery–Smith, S.J. (1995) *Decoupling inequalities for the tail probabilities of multivariate U-statistics*, Annals of Probability **23** 817-851.

[2] Kallenberg, O. (1991) *On an independence criterion for multiple Wiener integrals*, Annals of Probability **19** 483-485.

[3] Kallenberg, O. and Szulga, J. (1989) *Multiple integration with respect to Poisson and Lévy processes*, Probability Theory and Related Fields **83** 101-134.

[4] Kwapień, S. and Woyczyński, W. (1992) *Random Series and Stochastic Integrals: Single and Multiple*, Birkhäuser.

[5] Ledoux, M. and Talagrand, M. (1991) *Probability in Banach Spaces: Isoperimetry and Processes*, Springer Verlag.

[6] Le Page, R. (1980) *Multidimensional infinitely divisible variables and processes*, Part 2, Lectures Notes in Mathematics **860** Springer–Verlag 279-284.

[7] Major, P. (1980) *Multiple Wiener–Itô integral*, Lecture Notes in Mathematics **849** Springer-Verlag.

[8] Privault, N. (1996) *On the independence of multiple stochastic integrals with respect to a class of martingales*, C. R. Acad. Sci. Paris **323** 515-520.

[9] Rajput, B. and Rosiński, J. (1989) *Spectral representations of infinitely divisible processes*, Probability Theory and Related Fields **82** 451–458.

[10] Rosiński, J. (1990) *On series representations of infinitely divisible random vectors*, Annals of Probability **18** 405–430.

[11] Samorodnitsky, G. and Szulga, J. (1989) *An asymptotic evaluation of the tail of a multiple symmetric α-stable integral*, Annals of Probability **17** 1503–1520.

[12] Samorodnitsky, G. and Taqqu, M.S. (1990) *Multiple stable integrals of Banach-valued functions*, Journal of Theoretical Probability **3** 267–287.

[13] Samorodnitsky, G. and Taqqu, M.S. (1994) *Non-Gaussian Stable Processes*, Chapman & Hall.

[14] Urbanik, K. (1967) *Some prediction problems for strictly stationary processes*, Proceedings of the Fifth Berkeley Symposium **2** 235–258.

[15] Üstünel, A.S. and Zakai, M. (1989) *On the independence and conditioning on Wiener space*, Annals of Probability **17** 1441–1453.

University of Tennessee, Department of Mathematics, Ayres Hall, Knoxville, TN 37996 USA

E-mail address: `rosinski@math.utk.edu`

Cornell University, School of Operations Research and Industrial Engineering, ETC Building, Ithaca, NY 14853 USA

E-mail address: `gennady@orie.cornell.edu`

A domination inequality for martingale polynomials

Jerzy Szulga

1. Introduction

A stochastic process $(X_t, t > 0)$ and a partition $t_0 < t_1 < t_2 < \ldots$ of the half-line define a sequence of increments $\theta_k = X_{t_k} - X_{t_{k-1}}$. For a d-variate function f, a *chaos polynomial of degree d*,

$$\sum_{j_1 < \cdots < j_d} f(t_{j_1}, \ldots, t_{j_d}) \theta_{j_1} \cdots \theta_{j_d}, \tag{1.1}$$

is an approximation of the multiple stochastic integral, and also the factual multiple integral of a simple function,

$$\int \cdots \int_{t_1 < \cdots < t_d} f(t_1, \ldots, t_d) \, dX_{t_1} \cdots X_{t_d}.$$

A reasonable integration theory with refined partitions leading to a legitimate limit necessitates restraints on potential integrators. Wiener's and Itô's works (e.g., [**Wie30, Itô51**]) laid the foundations for such theory. A Brownian motion and a square integrable Borel function serve as a winner example of an integrator-integrand couple. In the last two decades the theory has grown to include semimartingales and multiple stochastic integrals (cf., e.g., [**Pro90, KW92, Sz98**] for a historic angle).

Increments of a typical integrator are often 'orthogonal' in some sense, also assuming an exclusive temporal structure and preventing higher powers like θ_j^2 to appear in polynomial (1.1). On the other hand, many applications rely heavily on such non-chaos polynomials, particularly in statistics. Actually, (1.1) could be an arbitrary polynomial. This happens, for example, when a process has constant trajectories and repeats itself on some intervals. In other words, a general polynomial is an element of the algebra spanned by increments (θ_k).

A diagonal in the cube $[0, t]^d$ is a set of points $(x_1, \ldots, x_d)$ such that at least two coordinates are equal. For $d = 2$, there is one diagonal, for $d = 3$, there are 4 diagonals, for $d = 4$, 14 diagonals, etc. A formula for the number of diagonals (so called, Bell's number) is not quite simple. The product X_t^d can be seen as a

---

1991 *Mathematics Subject Classification.* Primary 60G44, 60F05; Secondary 60H05, 60H25, 05E35, 15A69, 15A63.

© 1999 American Mathematical Society

random measure of the cube. The corresponding random measure of a diagonal could be approximated by a random polynomial

$$\sum_{j_1<\cdots<j_d} a_{j_1,\ldots,j_k} \theta_{j_1}^{m_1} \cdots \theta_{j_d}^{m_d},$$

where the coefficients and exponents would depend on this diagonal. Consequently, the random measure on the diagonal would give rise to a 'diagonal' stochastic integral. A combination of the off-diagonal integral and suitably normalized diagonal integrals is known in the literature as the Stratonovich integral.

Let us consider a Brownian motion for the sake of illustration, and let us recall that the quadratic variation of Brownian motion is deterministic and higher order variations vanish. Then, a quadratic form $\sum_k f(t_k)\theta_k^2$ leads to the deterministic integral $\int f dt$ but a cubic form $\sum_k a_k \theta_k^3$ and higher order forms disappear in the limit as partitions become finer. Thus, a limit of a general Brownian polynomial is a mixture of stochastic Wiener's and deterministic integrals. The mixture is intricate algebraically yet simple analytically. If higher order variations do not vanish, like in the case of a Poisson process, then combinatorics and notation could overshadow the analytical subtlety. At least, the visage of the objects might be fairly appalling. Notwithstanding, [**Cad97**] proved some limit theorems for such stochastic polynomials.

We intend to show how a domination inequality could alleviate the combinatorial and notational burden that is imminent in construction procedures and limit theorems for non-chaos polynomials and integrals.

The method applies to a class of martingales that includes Brownian and Poisson motions, and Azéma martingales. The study of such martingales was originated in [**AY89**], and continued in [**Eme89, Mey89**], to mention but a sample. In particular, Azéma and related martingales are solutions of Meyer's structure equation

$$d[X,X]_t = dt - f(X_{t-})dX_t.$$

An exhaustive bibliography would exceed the limitations of this note. Some references can be found in, e.g., [**Sz98**], and more information - in the volumes of Séminaire des Probabilités over the last decade (see References). For general martingale and integration matters we refer to [**LS86, JS87**].

2. Notation

$\mathbf{1}_{\{\cdot\}}$ denotes the indicator function, and $|A|$ marks the number of elements in a finite set A. We write $\mathbb{N} = \{1, 2, \ldots\}$, $\overline{N} = \{0, 1, 2, \ldots\}$, and $[j, k] = \{j, j+1, \ldots, k\}$, for $j, k \in \mathbb{N}$, $j < k$. Also, $\mathbb{N}_\infty$ denotes the set of sequences of nonnegative integers that become zero eventually.

A boldface character will denote a sequence. There is no need to distinguish between a finite or infinite sequence, for the former may be augmented by zeros. For example,

$$\boldsymbol{x} = (x_1, x_2, \ldots), \quad \boldsymbol{m} = (m_1, \ldots, m_k), \quad \boldsymbol{1} = (1, \ldots, 1), \quad \boldsymbol{0} = (0, \ldots, 0), \text{ etc.}$$

Despite a possible ambiguity, the length of a sequence will be always known from the context. Let $\boldsymbol{m}, \boldsymbol{k} \in \mathbb{N}_\infty$ and $\boldsymbol{x}, \boldsymbol{t}$ be sequences of scalars (i.e., of real or complex numbers). Denote

$$\boldsymbol{x^m} = \prod_j x_j^{m_j}, \text{ (by convention, } 0^0 \stackrel{df}{=} 1);$$

$$|\boldsymbol{m}| = \sum_j |m_j|, \quad \boldsymbol{x}^* = \max |x_j|, \quad \boldsymbol{m}_* = \min\{m_j : m_j > 0\};$$

$$\operatorname{supp} \boldsymbol{m} = \{j : m_j \neq 0\}, \quad \operatorname{size} \boldsymbol{m} = |\operatorname{supp} \boldsymbol{m}| = \sum_j \mathbb{1}_{\{m_j \neq 0\}};$$

$$\boldsymbol{xt} = (x_1 t_1, t_2 t_2, \ldots), \quad \boldsymbol{k} \wedge \boldsymbol{m} = (k_1 \wedge m_1, k_2 \wedge m_2, \ldots), \quad \boldsymbol{m}_+ = \boldsymbol{m} \vee \boldsymbol{0}.$$

$\boldsymbol{m}_*, \operatorname{supp} \boldsymbol{m}, \operatorname{size} \boldsymbol{m}$ are specifically chosen to accommodate an infinite sequence.

A polynomial in $\boldsymbol{x}$ of degree $d = \deg(P) = \max\{|\boldsymbol{m}| : F(\boldsymbol{m}) \neq 0\}$ can be written as

$$(2.1) \qquad P(\boldsymbol{x}) = \sum_{\boldsymbol{m} \in \mathbb{N}_\infty} F(\boldsymbol{m}) \boldsymbol{x^m},$$

where a function F with a finite support takes values in a vector space. For some homogeneous polynomials, a reader might prefer a more lucid notation. For example, for just one choice of $\boldsymbol{m} \in \mathbb{N}^k$, one may write explicitly

$$(2.2) \qquad P_{\boldsymbol{m}}(\boldsymbol{x}) = \sum_{1 \leq j_1 < \cdots < j_k \leq n} f_{\boldsymbol{m}}(j_1, \ldots, j_k) x_{j_1}^{m_1} \cdots x_{j_k}^{m_k}.$$

The sum's constraint may be replaced by a seemingly more general phrase "$j : j_k$ are different", subject to redefining coefficients $f_{\boldsymbol{m}}(j_1, \ldots, j_k)$. In the current form it will be called a *tetrahedral polynomial*, for the index $\boldsymbol{j}$ runs through the main tetrahedron in $\mathbb{N}^k$. If the coefficients are scalar, we write

$$\|f_{\boldsymbol{m}}\|_2^2 = \sum_{1 \leq j_1 < \cdots < j_k \leq n} |f_{\boldsymbol{m}}(j_1, \ldots, j_k)|^2.$$

The notation in (2.1) is more than just a shorthand (or its opposite). It will prove to be useful in the next section. Actually, the notation may become a burden, for example, if one tries to write down a general polynomial, even to simply augment the following list of examples by writing "explicitly" $P_3, P_4, \ldots, P_d$:

$$P_0(\boldsymbol{x}) = f_0$$

$$P_1(\boldsymbol{x}) = P_0(\boldsymbol{x}) + \sum_{j=1}^n f_1(j) x_j,$$

$$P_2(\boldsymbol{x}) = P_1(\boldsymbol{x}) + \sum_{j<k \leq n} f_{1,1}(j,k) x_j x_k + \sum_{j=1}^n f_2(j) x_j^2,$$

$$\cdots\cdots \quad \cdots\cdots$$

$$P_{17}(\boldsymbol{x}) = \sum_{j<k<l} f_{3,6,1}(j,k,l) x_j^3 x_k^6 x_l,$$

Two types of notation are illustrated by the following table:

coordinate:		j^{th}	k^{th}	l^{th}
$f_0 =$	$F(0,\ldots,$	$0,0,\ldots,$	$0,\ldots,$	$0,\ldots,0)$
$f_1(j) =$	$F(0,\ldots,$	$1,0,\ldots,$	$0,\ldots,$	$0,\ldots,0)$
$f_{1,1}(j,k) =$	$F(0,\ldots,$	$1,0,\ldots,$	$1,\ldots,$	$0,\ldots,0)$
$f_2(j) =$	$F(0,\ldots,$	$2,0,\ldots,$	$0,\ldots,$	$0,\ldots,0)$
$f_{3,6,1}(j,k,l) =$	$F(0,\ldots,$	$3,0,\ldots,$	$6,\ldots,$	$1,\ldots,0)$

A polynomial P is called k-*homogeneous*, $k \leq \deg(P)$, if $F(\boldsymbol{m}) = 0$, whenever $|\boldsymbol{m}| \neq k$ (*homogeneous*, if it is homogeneous for some k). P is called a *chaos polynomial*, if $F(\boldsymbol{m}) = 0$, whenever $m^* \geq 2$. $P_{17}(\boldsymbol{x})$ is an example of a 10-homogeneous polynomial with $\deg(P) = 10$. The sum $\sum_{j,k} f_{1,1}(j,k) x_j x_k$ in $P_2(\boldsymbol{x})$ is an example of a chaos polynomial of degree 2. Of course, from the plain algebraic point of view there is no difference between the notions of 'polynomial' and 'chaos polynomial'. However, our scalars are factually linearly independent random variables. A chaos polynomial is a linear combination of the *multiplicative family* spanned by $\boldsymbol{x}$,

(2.3)
$$\mathcal{X} \stackrel{\mathrm{df}}{=} \{ \boldsymbol{x^m} : \boldsymbol{m}^* \leq 1 \} = \{ x_{j_1} \cdots x_{j_k} : k = 1, \cdots, n; 1 \leq j_1 < \cdots < j_k \leq n \} \cup \{ 1 \}.$$

3. Reduction to chaos

3.1. Illustration. Consider a simple random chaos polynomial $P = X_1 \cdots X_n$, formed by square integrable random variables. Under plain Hölder's inequality $\mathsf{E}|P|^{2/n} < \infty$, and even $\mathsf{E}|P|^2 < \infty$. when X_k are independent. Our special processes will still have this property in spite of dependent increments.

For $\boldsymbol{m} \in \mathbb{N}_\infty$, we define the index

(3.1)
$$H = H(\boldsymbol{m}) = |\boldsymbol{m}| - \text{size } \boldsymbol{m},$$

For example, $H(2) = 1, H(2,3) = 3, H(3,6,1) = 7$, etc. Alternatively,

(3.2)
$$H(\boldsymbol{m}) = \sum_k (m_k - 1) \mathbb{1}_{\{m_k \geq 2\}}.$$

Then, we define the *chaos index* of a polynomial (2.1) with the help of the formula

(3.3)
$$\chi(P) \stackrel{\mathrm{df}}{=} \max \{ H(\boldsymbol{m}) : F(\boldsymbol{m}) \neq 0 \}$$

P is a chaos polynomial if and only if $\chi(P) = 0$ (i.e., iff $\boldsymbol{m}^* < 2$ for all $\boldsymbol{m}$). Let $\boldsymbol{\varepsilon} = (\varepsilon_j)$ be a Rademacher sequence. Note the identities

(3.4)
$$\sum_j f(j) x_j^2 = \mathsf{E}\Big(\sum_j f(j) x_j \varepsilon_j \Big)\Big(\sum_j x_j \varepsilon_j \Big)$$
$$\sum_{j<k} g(j,k) x_j^2 x_k^3 = \mathsf{E}\Big(\sum_{j<k} g(j,k) x_j x_k \varepsilon_j \varepsilon_k \Big)\Big(\sum_{j<k} x_j x_k^2 \varepsilon_j \varepsilon_k \Big), \text{ etc.}$$

Put $V = (\sum_j |x_j|^2)^{1/2}$. The Cauchy-Schwartz inequality implies that

$$\Big| \sum_j f(j) x_j^2 \Big| \leq \Big(\mathsf{E}\Big| \sum_j f(j) x_j \varepsilon_j \Big|^2 \Big)^{1/2} \Big(\mathsf{E}\Big| \sum_j x_j \varepsilon_j \Big|^2 \Big)^{1/2}$$
$$\leq \Big(\mathsf{E}\Big| \sum_j f(j) x_j \varepsilon_j \Big|^2 \Big)^{1/2} V;$$

(3.5)
$$\Big| \sum_{j<k} g(j,k) x_j^2 x_k^3 \Big| \leq \Big(\mathsf{E}\Big| \sum_{j<k} g(j,k) x_j x_k \varepsilon_j \varepsilon_k \Big|^2 \Big)^{1/2} \Big(\mathsf{E}\Big| \sum_{j<k} x_j x_k^2 \varepsilon_j \varepsilon_k \Big|^2 \Big)^{1/2}$$
$$\leq \Big(\mathsf{E}\Big| \sum_{j<k} g(j,k) x_j x_k \varepsilon_j \varepsilon_k \Big|^2 \Big)^{1/2} V^3.$$

In a more general context, put $\mu_j = \mathbb{1}_{\{m_j \geq 2\}}$, and let ι_1 be the first index j for which $m_j \geq 2$, ι_2 be the second index j, for which $m_j \geq 2$, etc., and let $p = p(\boldsymbol{m})$

be the length of the obtained vector, $(\imath_1, \ldots, \imath_p)$. With $P_{\boldsymbol{m}}$ we associate a chaos polynomial

$$P_\chi(\boldsymbol{x}; \boldsymbol{t}; \boldsymbol{m}) = \sum_{1 \le j_1 < \cdots < j_k} f_{j_1, \ldots, j_k} t_{j_1}^{\mu_1} \cdots t_{j_k}^{\mu_k} x_{j_1} \cdots x_{j_k},$$

where $\boldsymbol{t}$ is a scalar sequence of length of $\boldsymbol{x}$. Of course, $P(\boldsymbol{x}) = P_\chi(\boldsymbol{x}, \boldsymbol{t})$, if P is already a chaos polynomial, i.e., if $m^* < 2$. It follows, like it did for (3.4) and (3.5), that

$$(3.6) \quad P_{\boldsymbol{m}}(\boldsymbol{x}) = \mathsf{E}\left(P_\chi(\boldsymbol{x}; \boldsymbol{\varepsilon}; \boldsymbol{m}) \sum_{1 \le l_1 < \cdots < l_p} x_{l_1}^{m_{\imath_1}-1} \varepsilon_{l_1} \cdots x_{l_p}^{m_{\imath_p}-1} \varepsilon_{l_p}\right), \quad m^* \ge 2,$$

and

$$(3.7) \quad |P_{\boldsymbol{m}}(\boldsymbol{x})| \le V^H \left(\mathsf{E}|P_\chi(\boldsymbol{x}; \boldsymbol{\varepsilon}; \boldsymbol{m})|^2\right)^{1/2}.$$

$H = \chi(P)$, because we chose just one sequence $\boldsymbol{m}$.

LEMMA 3.1. *Let $\boldsymbol{X} = (X_j, j = 1, \ldots, n)$ be a sequence of square integrable real random variables. Choose one $\boldsymbol{m} \in \mathbb{N}_\infty$. Let $P(\boldsymbol{X}) = P_{\boldsymbol{m}}(\boldsymbol{X})$ be a homogeneous polynomial whose coefficients take values in a Banach space. Then, for a Rademacher sequence $\boldsymbol{\varepsilon}$ independent of $\boldsymbol{X}$,*

$$(3.8) \quad \mathsf{E}|P(\boldsymbol{X})|^{2/(H+1)} \le \left(\mathsf{E}|P_\chi(\boldsymbol{X}; \boldsymbol{\varepsilon}; \boldsymbol{m})|^2\right)^{1/(H+1)} \left(\mathsf{E}V^2\right)^{H/(H+1)}.$$

PROOF. Starting with (3.6) and (3.7), we apply Hölder's inequality (also, tacitly, Fubini's theorem),

$$\mathsf{E}A^{\frac{1}{H+1}} B^{\frac{H}{H+1}} \le \left(\mathsf{E}A^{p\frac{1}{H+1}}\right)^{1/p} \left(\mathsf{E}B^{q\frac{H}{H+1}}\right)^{1/q},$$

where $p = H + 1$, $q = (H + 1)/H$, and $A = \mathsf{E}_\varepsilon |P_\chi(X; \boldsymbol{\varepsilon}; \boldsymbol{m})|^2$, $B = V^2$. □

Of course, the hypothesis is trivially true for a chaos polynomial P.

COROLLARY 3.2. *In addition to assumptions of the lemma, let the family $\mathcal{X}$ (2.3) consist of orthogonal random variables. If the coefficients $f_{\boldsymbol{m}}(\cdot)$ are real, then*

$$(3.9) \quad \mathsf{E}|P(\boldsymbol{X})|^{2/(H+1)} \le \|f_{\boldsymbol{m}}\|_2^{2/(H+1)} \left(\mathsf{E}V^2\right)^{H/(H+1)}.$$

PROOF. A Rademacher polynomial chaos, implicit in $P_\chi(\boldsymbol{x}; \boldsymbol{\varepsilon}; \boldsymbol{m})$, is already spanned by a multiplicative family of orthogonal random variables (so called, Walsh functions). The estimate follows in virtue of Fubini's theorem. □

3.2. Non-homogeneous polynomials. Using Steinhaus variables σ_j (i.e., $\sigma_j = \exp\{2\pi i U_j\}$, where U_j are independent random variables uniformly distributed on $[0, 1]$), we associate two chaos polynomials with a non-chaos polynomial P (2.1)

$$(3.10) \quad \begin{aligned} P^\chi(\boldsymbol{x}) &= \sum_{\boldsymbol{m}^* \le 1} F(\boldsymbol{m}) \boldsymbol{x}^{\boldsymbol{m}}, \\ P_\chi(\boldsymbol{x}, \boldsymbol{\sigma}) &= \sum_{\boldsymbol{m}^* \ge 2} F(\boldsymbol{m}) \boldsymbol{x}^{\boldsymbol{m} \wedge \boldsymbol{1}} \boldsymbol{\sigma}^{(\boldsymbol{m}-\boldsymbol{1})_+}. \end{aligned}$$

LEMMA 3.3. *Let $H = \chi(P)$. The chaos decomposition holds:*

$$(3.11) \quad P(\boldsymbol{x}) = P^\chi(\boldsymbol{x}) + \mathsf{E}\left(P_\chi(\boldsymbol{x}, \boldsymbol{\sigma}) \sum_{|\boldsymbol{n}| \le H} \boldsymbol{x}^{\boldsymbol{n}} \boldsymbol{\sigma}^{-\boldsymbol{n}}\right),$$

PROOF. The second term of the decomposition is but the sum over the multi-indices m such that $m^* \geq 2$. First, we observe that $\{(x\sigma^{-1})^n : |n| \leq H\}$ is a family of orthogonal random variables. Secondly, every term $F(m)x^{m \wedge 1}\sigma^{(m-1)_+}$ from the sum forming P_χ admits exactly one non-orthogonal random variable $(x\sigma^{-1})^n$ such that $n = (m-1)_+$. Thus, the expectation of the corresponding product recovers $F(m)x^m$, where $m^* \geq 2$. □

THEOREM 3.4. *Let $P(x)$ be a non-chaos polynomial in scalar variables $x = x_1, x_2, \ldots$ and $H = \chi(P) \geq 1$, with coefficients in a complex Banach space. Let X consist of square integrable random variables. Then*

(3.12)
$$\mathsf{E}|P(X)|^{2/(H+1)} \leq \left(\mathsf{E}|P^\chi(X)|^2\right)^{1/(H+1)} + \left(\mathsf{E}|P_\chi(X,\sigma)|^2(1+\mathsf{E}V^2)^H\right)^{1/(H+1)}$$

PROOF. We apply the Cauchy-Schwartz inequality to the second summand in (3.11), and continue estimation as follows:

$$\left(\mathsf{E}_\sigma |P_\chi(x,\sigma)|^2\right)^{1/2} \left(\sum_{|n|\leq H} |x|^{2n}\right)^{1/2} \leq \left(\mathsf{E}_\sigma |P_\chi(x,\sigma)|^2\right)^{1/2} (1+V^2)^{H/2},$$

where $|x|^2 = (|x_1|^2, |x_2|^2, \ldots)$. In fact, considering nonnegative numbers $a_k = |x_k|^2$ forming the sequence a, we have applied the inequality

(3.13)
$$\sum_{|n|\leq H} a^n \leq \sum_{k=0}^H |a|^k \leq (1+|a|)^H.$$

Here, $|a|$ is the sum of coordinates, not the absolute value, so $|a| = V^2$.

In view of the subadditivity of the function $x^{2/(H+1)}$, (3.11) and the obtained estimate of the second summand lead to the inequality

$$|P(X)|^{2/H+1} \leq |P^\chi(X)|^{2/(H+1)} + \left(\mathsf{E}_\sigma |P_\chi(X,\sigma)|^2\right)^{1/(H+1)} (1+V^2)^{H/(H+1)}.$$

Now, we apply the expectation $\mathsf{E}_X(\cdot)$ to both sides of the obtained inequality. Like in the homogeneous case, we use Hölder's inequality for $A = \left(\mathsf{E}_\sigma |P_\chi(X,\sigma)|^2\right)^{1/(H+1)}$ and $B = 1+V^2$ with $p = H+1, q = (H+1)/H$. Finally, Jensen's inequality applied to the term containing P^χ completes the proof. □

COROLLARY 3.5. *If the multiplicative family $\mathcal{X}$ (2.3) consists of orthogonal random variables, then*

$$\mathsf{E}|P(X)|^{2/(H+1)} \leq (2+\mathsf{E}V^2)^{H/(H+1)} \|F\|_2^{2/(H+1)}$$

PROOF. The argument is similar to the one used in Corollary 3.2 but relies on the orthogonality of Steinhaus polynomials instead of Walsh functions. Put

$$a = \mathsf{E}|P^\chi(X)|^2 = \sum_{m^*<2} |F(m)|^2, \quad b = \mathsf{E}|P_\chi(X,\sigma)|^2 = \sum_{m^*\geq 2} |F(m)|^2.$$

It remains to verify the inequality

$$a^{1/(H+1)} + (c^H b)^{1/(H+1)} \leq (1+c)^{H/(H+1)} (a+b)^{1/(H+1)}.$$

□

4. Integrals with respect to piecewise constant processes

4.1. Variations and integrals. A scalar sequence $(a_1, a_2, \dots)$ gives rise to a step function $f = \sum_j a_j \mathbb{1}_{(j,j+1)}$ on $[0, \infty)$ with discontinuities in $\overline{N}$. Conversely, a step function with a possible discontinuities in $\overline{N}$ defines $a_j = f(j-)$. Let f_- denote the left-continuous extension of f on $[0, \infty)$. The following function is right-continuous:

$$\sum_{j=1}^{[t]} a_j = \int_0^t f(x_-)d[x].$$

By replacing $t \mapsto nt$ we condense the time scale, so all possible discontinuities fall into $D_n = \{j/n : j = 0, 1, \dots\}$.

Since in each instance we deal with a bounded time interval $[0, t]$, we may and do assume that $t \leq T = 1$, without loss of generality. Then, we replace D_n by $D_n \cap [0, 1]$, and assume that $t \leq 1$, without further comments.

For a positive integer m, a sequence $X_1, X_2, \dots$ of random variables defines right-continuous step variation processes

$$M_n^m(t) = \sum_{j=1}^{[nt]} X_j^m$$

that satisfy the identities

$$\sum_{j=1}^{[nt]} a_j X_j^m = \int_0^t f(x-)dM_n^m(x).$$

Let us distinguish the cases $m = 1$ and $m = 2$, and denote $M_n = M_n^1$ and $V_n = (M_n^2)^{1/2}$ (the quadratic variation of M_n).

Since the integrals are step functions with discontinuities in D_n, the procedure can be iterated, giving rise to the multiple integrals

(4.1)
$$I_t^{\boldsymbol{m}}(f\boldsymbol{m}) = \int_{\Delta_k(t)} f\boldsymbol{m}\, d\boldsymbol{M}^{\boldsymbol{m}} = \int \cdots \int_{\Delta_k(t)} f\boldsymbol{m}(x_1, \dots, x_k) dM^{m_1}(x_1) \cdots dM^{m_k}(x),$$

where $\boldsymbol{m}$ is a finite sequence with no 0 element and $k = \text{size } \boldsymbol{m}$,

$$\Delta_k(t) = \{\, \boldsymbol{x} = (x_1, \dots, x_k) : 1 \leq x_1 < \dots < x_k \leq t \,\}, \quad \Delta_k = \Delta_k(1),$$

$f\boldsymbol{m} : [0,1]^k \to R$ is constant on each of the rectangles $(a_1, b_1) \times \cdots \times (a_k, b_k)$, where $a_j, b_j \in D_n$, and its support is contained in $\Delta_k(T)$. In other words, $f\boldsymbol{m}$ is measurable with respect to the ring $\mathcal{A}_n$ induced by the rectangles contained in Δ_k.

REMARK 4.1. If $\boldsymbol{m}^* = 1$, then $I_t^{\boldsymbol{m}}$ is a right-continuous martingale. We will call it a *chaos integral*.

We restrict the integrands to functions vanishing on diagonals in $[0,1]^k$ even though at this moment the restriction is unnecessary. However, it will become important later on.

Let $\mathcal{A}$ be the ring induced by rectangles $(a_1, b_1) \times \cdots \times (a_k, b_k)$ with rational a_j, b_j's. If an $\mathcal{A}$-measurable function f takes a finite number of values and has the support in Δ_k, then one can find an n such that integrals (4.1) are well defined. Denote the vector space of such functions by $\mathbb{E}_{k,0}$.

Integral (4.1) corresponds to polynomial (2.2). A finite linear combination of integrals (4.1) produces an analog of polynomial (2.1), but with notation compatible with (4.1) (no infinite sequences $\boldsymbol{m}$, no 0 among indices).

For step functions, there is no difference between polynomials and integrals, because one can be represented by the other. Therefore, we carry over the concepts of homogeneity, degree, and the chaos index to general integrals. The integrals can be written as

$$(4.2) \qquad I_t(f) = \int f^t \, d\boldsymbol{M} = \sum_{\boldsymbol{m}} \int_{\Delta^k(t)} f_{\boldsymbol{m}} \, d\boldsymbol{M}^{\boldsymbol{m}},$$

where $f^t = (f_{\boldsymbol{m}} \mathbb{1}_{\Delta^k(t)})$, and each $f_{\boldsymbol{m}}$ belongs to $\mathbb{E}_{k,0}$. Like before, $k = $ size $\boldsymbol{m}$. Denote by $\mathbb{E}_0$ the collection of such f's. Also, let us write

$$(4.3) \qquad \|f^t\|_2^2 = \sum_{\boldsymbol{m}} \int_{\Delta^k(t)} |f_{\boldsymbol{m}}(\boldsymbol{x})|^2 d\boldsymbol{x}.$$

We say that $f = (f_{\boldsymbol{m}}) \in L^2(\boldsymbol{\Delta})$, if each $f_{\boldsymbol{m}} \in L^2(\Delta^k)$.

4.2. Special martingales. We say that a square-integrable random variable X is orthogonal to a sigma-field $\mathcal{F}$, if, for every square-integrable $\mathcal{F}$-measurable random variable A

$$(4.4) \qquad Cov(X, A) = 0, \quad \text{or, assuming } \mathbb{E} X = 0, \quad \mathbb{E} X A = 0.$$

Let $(X_j, \mathcal{F}_j : j \in \mathbb{N})$'s be an adapted sequence ($\mathcal{F}_0$ is the trivial sigma-field and $X_0 = 0$, by convention). For the sake of convenience but without loss of generality we may and do assume that the random variables have zero mean. Assume that

$$(4.5) \qquad \begin{array}{l} 1. \text{ for every } k \in \mathbb{N}, X_k \text{ is orthogonal to } \mathcal{F}_{k-1}; \\ 2. \text{ for every } k \in \mathbb{N}, X_k^2 \text{ is orthogonal to } \mathcal{F}_{k-1}; \\ 3. \ \mathbb{E} X_k^2 = 1/n \end{array}$$

The first condition turns X_k's into martingale differences. Then $M_n(t)$, defined in the preceding subsection, is a martingale on $[0,1]$, with respect to the filtration $\mathcal{G}_t^{(n)} = \mathcal{F}_{[nt]}$), and $V_n^2 = [M_n, M_n]$ is its variation process. The second condition makes $X_k^2 - \mathbb{E} X_k^2$ martingale differences, too. Thus, the predictable compensator of $|M_n|^2$ satisfies the identity $\langle M_n, M_n \rangle_t = \mathbb{E} M_n^2(t) = [nt]/n$.

EXAMPLE 4.2 ([**Eme89**]). A martingale differences sequence solving Emery's discrete structure equation

$$X_k^2 = 1/n + f_k(X_1, \ldots, X_{k-1}) X_k, \quad f_k \text{ are Borel functions}, \quad k = 1, \ldots, n,$$

also satisfies (4.5). The probability distribution of a solution is unique.

Further, because of (4.5), the multiplicative family $\mathcal{X}$ (2.3) consist of orthogonal random variables. Now, it suffices to interpret the results from the previous section in the current language. Recall that the chaos index is now given by the formula

$$H = \max \{ H(\boldsymbol{m}) : f_{\boldsymbol{m}} \neq 0 \}.$$

For example, we obtain

THEOREM 4.3. *Assume* (4.5). *Then, for* $t \leq 1$,

(4.6)
$$\mathsf{E}|\sup_{s\leq t} \int f^s \, d\mathbf{M}|^{2/(H+1)} \leq 3\|f^t\|_2^{2/(H+1)}.$$

PROOF. First, we recall Corollary 3.5. The sum in $\|F\|_2$ can be written now as the integral of a step function.

Secondly, integral (4.1) is a right-continuous martingale when $m^* \leq 1$, by Remark 4.1. Hence, we apply Doob's inequality

$$\mathsf{E}|\sup_{s\leq t} Z_s|^2 \leq 2\mathsf{E}|Z_t|^2 \qquad \text{(for a martingale } Z_t\text{)},$$

to chaos integrals induced by polynomials P^χ and P_χ (3.10). Finally, since $\mathsf{E}V^2 \leq 1$, hence $2^{1/(H+1)}(2+\mathsf{E}V^2)^{H/(H+1)} \leq 3$. □

Let $M_t = M_t^1$ be a square integrable martingale cadlag (right continuous with left limits) martingale on $[0,1]$. Let $V^2 = X^2 = [X,X]$ denote its quadratic variation. Higher order variations can be defined by induction, $M^m = [M^{m-1}, M]$, $m \geq 2$. Integrals (4.1), hence the integral (4.2), are well defined for caglad functions (left continuous with right limits). Moreover, they are limits in the topology of convergence in probability, uniform on intervals, of their counterparts for step functions (cf., e.g., [**Pro90**]).

A cadlag adapted process X_t is called a *special martingale*, if, for every n, the increments $X_k^n = X_{k/n} - X_{(k-1)/n}$ satisfy (4.5). The quadratic variation of a special martingale satisfies $d\mathsf{E}V_t^2 \leq dt$.

THEOREM 4.4. *Let M be a special martingale.*
1. *Then, the integral* (4.2) *exists and inequality* (4.6) *continues to hold, for $f = (f\mathbf{m}) \in L^2(\mathbf{\Delta})$.*
2. *Further, for every $\epsilon > 0$ and every stopping time $\tau \leq 1-\epsilon$,*

(4.7) $$\mathsf{E}\sup_{t\leq \epsilon} |I_{\tau+t}(f) - I_t(f)|^{2/(H+1)} \leq 3 \sup_{u \leq 1-\epsilon} \|f^{u+\epsilon} - f^u\|_2^{2/(H+1)}.$$

PROOF. Inequality (4.6) means that the linear mappings

$$L^2(\mathbf{\Delta}) \supset \mathbb{E}_0 \ni f \mapsto I_t(f) = \int f^t \, d\mathbf{M} \in L^{2/(M+1)}$$

is continuous, hence the mapping can be extended from $\mathbb{E}_0$ onto $L^2(\mathbf{\Delta})$, and the inequality will be preserved.

In the second statement we use Doob's optional sampling theorem (or, Markov's property) of the square integrable martingales corresponding to polynomials P_χ and P^χ (i.e., if Y_t is a square integrable martingale, then $Y_{\tau+t} - Y_\tau$ is such, too). The upper estimate involves

$$3\left(\mathsf{E}\int |f|^2 \mathbf{1}_{\mathbf{\Delta}(\tau+\epsilon)\backslash \mathbf{\Delta}(\tau)}\right)^{1/(H+1)},$$

and we just use the uniform bound. □

COROLLARY 4.5. *For a special martingale and $f \in L^2(\mathbf{\Delta})$,*

(4.8) $$\mathsf{P}\left(\sup_{t\leq 1} |I(f^t)| \geq a\right) \leq a^{-2/(H+1)} 3\|f\|_2^{1/(H+1)}.$$

If $\rho_f(\epsilon) = \sup\{\|f^{u+\epsilon} - f^u\|_2^2 : u \leq 1-\epsilon\}$, then

(4.9) $\quad \mathsf{P}\left(|\sup_{t \leq \epsilon} |I(f^{\tau+t} - f^\tau)| \geq b\right) \leq b^{-2/(H+1)} \rho_f(\epsilon)^{1/(H+1)}.$

PROOF. Both (4.8) and (4.9) follow from the corresponding statements of the theorem by Chebyshev's inequality.

If $f \in L^2(\mathbf{\Delta})$, then $t \mapsto \|f^t\|^2$ is a non-decreasing continuous, hence uniformly continuous function. $\square$

EXAMPLE 4.6 ([**Mey89**]). A martingale that is a solution of Meyer's structure equation

$$[X,X]_t = t + \int_0^t f(X_-)dX, \quad f \text{ is continuous}, \quad k=1,\ldots,n,$$

is special, for example (cf., [**Eme89, Pro90, Sz98**]):

1. for $f(x) = 0$, a Brownian motion B_t, (it is the unique continuous solution);
2. for $f(x) = c \neq 0$, a compensated Poisson process $c\xi_t^{1/c^2} - t$ (the superscript marks the intensity);
3. for $f(x) = -x$, the original Azéma martingale [**AY89**]

$$\text{sign}(B_t)\sqrt{2(t - G_t)} \quad (\text{where } G_t \stackrel{df}{=} \sup\{s \leq t : B_s = 0\});$$

4. for $f(x) = -2x$, Parthasarathy's process $\sqrt{t}(-1)^{\eta_t}$, where η_t is a non-homogeneous Poisson process with the density $t^{-1}/4$;

5. Limit theorems

The discussed processes can be seen as random elements with values in $D = D[0,1]$, Skorohod's space of cadlag functions. We write $\mathcal{L}(M)$ for the probability distribution of a process X. The restriction to the interval $[0,1]$ is unnecessary, though convenient, for a simple continuous change of time will carry all results to $D[0,T]$ or $D[0,\infty)$. D becomes a Polish (separable complete metric) space under Skorohod's metric δ. Thus, Prohorov-Parthasarathy-Billingsley- (plus many other authors') theory of probability measures on Polish spaces becomes felicitous. D could have been endowed with the supremum norm but then it would become non-separable.

Let us address two issues associated with Skorohod's metric. First, it requires a great deal of explanations, too much for this note. So, we refer to [**LS86**] or [**JS87**] for the missing details.

Secondly, in contrast to D being endowed with the supremum norm, the vector structure is not compatible with Skorohod's topology. That is, even if $\mathcal{L}(Y(n)) \to \mathcal{L}(Y)$ and $\mathcal{L}(Z(n)) \to \mathcal{L}(0)$, then not necessarily $\mathcal{L}(Y(n) + Z(n)) \to \mathcal{L}(Y)$.

REMARK 5.1. In order to explain the importance of this issue, consider martingales $Y(n), Y$ such that $\mathcal{L}(Y(n)) \to \mathcal{L}(Y)$. Let $C_Y = \{t > 0 : Y_t - Y_{t-} = 0\} \cup \{0\}$ denote the set of continuity points (which is dense in $[0,1]$). Then, finite dimensional distributions of the process $(Y_t(n) : t \in C_Y)$ converge weakly to finite dimensional distributions of $(Y_t : t \in C_Y)$. By a continuity argument, for every simple function g with discontinuities in C_Y, the finite dimensional distributions of the integral processes $(\int_0^t g dY(n) : t \in C_Y)$ converge weakly to the finite dimensional distributions of the integral processes $(\int_0^t g\,dY : t \in C_Y)$. Having known only this, we cannot strengthen this convergence to the convergence in Skorohod's topology.

However, if we know already that $\mathcal{L}(\int_0^t g\,dY(n)) \to Q$, for some probability law Q on D, then the preceding argument allows the identification of the limit law, $Q = \mathcal{L}(\int_0^t g\,dY)$.

The method used in solving Meyer's structure equation in Example 4.6 illustrates the theory. First, the discrete structure equation (Example 4.2) is solved. Then, the discrete martingale solutions $M(n)$ are extended to continuous time step cadlag martingale that satisfy the structure equation. Next, laws $\mathcal{L}(M(n))$ are showed to be relatively compact in the complete Skorohod space. Finally, a cluster point is proven to be a martingale satisfying the structure equation.

Our situation is as follows. We are given a sequence of matrices $f(n)$ and a sequence of step-martingales $M(n)$ in D, with discontinuity points sitting in $\{j/n\}$. Polynomial (2.2) (or (2.1)) gives rise to a cadlag process (4.1) (or (4.2)).

Then, the question, posed and studied in [**Cad97**], is if and when

$$(5.1) \qquad \mathcal{L}(I_n(f(n))) \to \mathcal{L}(I(f)), \text{ in Skorohod's topology},$$

for some f and a martingale M.

REMARK 5.2. Estimates (4.8) and (4.9) lead to and ensure Aldous' criterion ([**LS86**, page 515] or [**JS87**, page 320]) for the relative compactness of laws of $\int g\,d\mathbf{M}_n$, for a $g \in L^2(\mathbf{\Delta})$. It is important that the upper bounds for the probability tails do not depend on n.

THEOREM 5.3. *Let martingales $M(n)$ satisfy (4.5) and $f(n) \to f$ in $L^2(\mathbf{\Delta})$. If*

$$(5.2) \qquad \mathcal{L}(M(n)) \to \mathcal{L}(M),$$

then (5.1) holds.

PROOF. We write $I(f) = \int g\,d\mathbf{M}, I_n(f) = \int f\,d\mathbf{M}(n)$.

Recall the meaning of $\mathbb{E}_0$, a dense subspace of $\mathbb{E} = L^2(\mathbf{\Delta})$, and let $g_0 \in \mathbb{E}_0$. By virtue of Remark 5.2, $\mathcal{L}(I_n(g_0))$ is relatively compact, hence a limit law Q_0 exists. Without loss of generality, we may even assume that $\mathcal{L}(I_n(g_0)) \to Q_0$. Since (5.2) leads to the convergence of finite dimensional distributions of the processes,

$$(I_n(g_0) : g_0 \in \mathbb{E}_0) \stackrel{wc_f}{\to} (I(g_0) : g_0 \in \mathbb{E}_0),$$

hence, $Q_0 = \mathcal{L}(I(g_0))$, as expected. Returning to Skorohod's metric, we observe that

$$\delta(I_n(f_n), I(f)) \leq \delta(I_n(g_0), I(g_0)) + R_n,$$

where

$$R_n = \delta(I_n(f(n)), I_n(f)) + \delta(I_n(f), I_n(g_0)) + \delta(I(g_0), I(f))$$

Let $\epsilon > 0$. Since the convergence in probability to 0 (all the more, in $L^{2/(M+1)}$), uniform on intervals, implies the convergence in Skorohod's metric to $\mathcal{L}(0)$, we can choose $g_0 \in \mathbb{E}_0$ such that $R_n < \epsilon$.

Finally, we take $\limsup_{n \to \infty}$, and then let $\epsilon \to 0$. $\square$

6. Concluding remarks

1. The method used in (3.6), (3.7), or (3.11), (3.12) remains valid when the coefficients $a_{j_1,\ldots,j_k}$ are functions on $[0,T]$ and the absolute value is replaced by a monotone functional acting on these functions. For example, for bounded functions one may work with the supremum norm. Formulae (3.6) and (3.11) need just a vector space. It must be a complex vector space when Steinhaus variables are used. A similar decomposition and a domination inequality can be still derived in a real normed space. Then several independent Rademacher sequences would be more appropriate. Yet, the real case is slightly more complex.

EXERCISE 6.1. Let the matrix $[\varepsilon_j^n]$, where the index j follows the length of vector $\boldsymbol{x}$ and $n \geq 0$, consist of independent Rademacher random variables whenever $n \neq 0$, and of ones, when $n = 0$. Show that the entries ε_j^n (still remembering, that n's are indices not exponents) may replace the powers of Steinhaus random variables in (3.11), by matching superscripts and exponents appropriately.

2. The concluding inequality (3.13) can be proved by, e.g., mathematical induction. Alternatively, since the second inequality in (3.13) is obvious, it suffices to notice that the convex function $F(\boldsymbol{t}) = \sum_{|\boldsymbol{n}|\leq H} \boldsymbol{t}^{\boldsymbol{n}}$ attains its maximum on the vertices of the simplex $|\boldsymbol{t}| = \alpha$. That is, $F(\boldsymbol{t}_0) = \sum_{k=0}^{H} \alpha^k$, when one element of the sequence $\boldsymbol{t}_0$ is equal α and other are zero. Of course, these bounds could be sharpened but the effort seems futile at this time.

3. The issue of Corollary 3.5 is how to dispose of Steinhaus random variables once they did their job. In presence of martingale differences, scalar coefficients constitute no problem. Vector coefficients (say, from a Banach space) require a contraction principle for bounded multipliers. In the case of martingale differences, the range of coefficients would have to be a Burkholder's UMD-space [**Bur81**]. However, we do not know if a UMD-space is sufficient in the multilinear context. For example, a bilinear version of the UMD-property would require the fulfillment of the inequality

$$\mathsf{E}\|\sum_{j<k} u_{j,k}s_j s_k X_j X_k\|^2 \leq C \sup_j |s_j|^2 \mathsf{E}\|\sum_{j<k} u_{j,k} X_j X_k\|^2,$$

for every square integrable martingale differences, vector matrix $[u_{j,k}]$, and bounded multipliers (s_j). Even though a large class of Banach spaces (e.g., L^p-spaces, when $1 < p < \infty$; see also MCP-spaces in [**Sz98**]) admits this property we do not know its complete scope.

4. For non-chaos integrals, $L^2(\boldsymbol{\Delta})$ is a proper subset of integrable functions. For example, for a Brownian motion, a mixture of L^1- and L^2-spaces is the domain of the integral ([**Sz98**]). In other words, we should not expect an inequality inverse to the (4.6).

5. Theorem 5.3 can be imitated in many ways, by considering various modes of convergence "$\stackrel{*}{\to}$" and "$\stackrel{**}{\to}$". That is, its new outline will look as follows:
Let $M(n)$ be martingales satisfying (4.5) and $f(n) \to f$ in $L^2(\boldsymbol{\Delta})$. Then,

$$M(n) \stackrel{*}{\to} M \quad \Rightarrow \quad \int f(n)\,d\boldsymbol{M}(n) \stackrel{**}{\to} \int f\,d\boldsymbol{M}.$$

Whenever modes of convergence are compatible with the vector structure in a space of random variables, the proof of a statement becomes an easy edition of the proof of Theorem 5.3. For example, this happens for the assumed convergence in L^p and the convergence in $L^{p \wedge 2/(M+1)}$ in the hypothesis (which include the convergence in probability, for $p = 0$). Another example involves convergence of finite dimensional distributions, even of processes $(I(g) : g \in L^2)$ with parameters being square integrable functions.

However, all these results utilize only square integrable functions (cf. the preceding point).

6. The metric convergence, compatible with the vector structure, allows one to work with one homogeneous polynomial or integral at a time. Since the addition is not continuous in Skorohod's metric, thus one needs to handle all polynomials and integrals simultaneously. Yet, some summands, perhaps homogeneous random chaoses or special chaoses of this note, might defy this restraint.

References

[AY89] J. Azéma and M. Yor. Étude d'une martingale remarquable, Sém. Prob. XXIII, Lecture Notes in Math. 1372:88–130 Springer, Berlin, 1989.

[Bur81] D.L. Burkholder. A geometrical characterization of Banach spaces in which martingale difference sequences are unconditional. *Ann. Probab.*, 9:997–1011, 1981.

[Cad97] B. Cadre. Functional asymptotic behavior of some random multilinear forms. Stoch. Process. Appl. 68 (1997), no. 1, 49–64.

[Eme89] M. Emery. *On the Azéma martingales*, Sém. Prob. XXIII, Lecture Notes in Math. 1372:66–87. Springer, 1989.

[Itô51] K. Itô. Multiple Wiener integral. *J. Math. Soc. Japan*, 3:157–169, 1951.

[JS87] J. Jacod and A. Shiryaev. *Limit Theorems for Stochastic Processes*. Springer-Verlag, 1987.

[KW92] S. Kwapień and W.A. Woyczyński. *Random Series and Stochastic Integrals*. Bürkhauser, Boston, 1992.

[LS86] R. Sh. Liptser and A. Shiryaev. *Theory of Martingales*. Kluwer Academic Publishers, Dordrecht-Boston-London, 1986.

[Mey89] P.-A. Meyer. Construction de solutions d'"équations de structure". *Sém. Prob. XXIII*, Lecture Notes in Math. 1372:142–145 Springer, Berlin, 1989.

[Pro90] P. Protter. *Stochastic Integration and Differential Equations*. Springer, Berlin, 1990.

[Sz98] J. Szulga. *Introduction to Random Chaos*. Chapman & Hall, New York-London, 1998.

[Wie30] N. Wiener. The homogeneous chaos. *Amer. J. Math.*, 60:897–936, 1930.

DEPARTMENT OF MATHEMATICS, AUBURN UNIVERSITY, AUBURN, ALABAMA 36849
Current address: Department of Mathematics, Auburn University, Auburn, Alabama 36849
E-mail address: szulga@mail.auburn.edu

A log–concavity proof for a Gaussian exponential bound

Richard A. Vitale

ABSTRACT. The Prékopa–Leindler result on marginal log–concavity is used to provide an alternate proof of an exponential inequality for a bounded, mean–zero Gaussian process.

1. Introduction

The subject of this note is the following bound:

THEOREM 1. *If $\{X_t, t \in K\}$ is a bounded, mean zero Gaussian process, then*

$$(1) \qquad E \exp\{\sup_{t \in K}(X_t - (1/2)\sigma_t^2)\} \leq \exp\{E \sup_{t \in K} X_t\},$$

where $\sigma_t^2 = E X_t^2$.

This arose in [9] as part of a study of maximum likelihood in infinite dimensions, and in [11] it was shown to imply a deviation bound of Maurey–Pisier type ([5]):

$$P(\sup_{t \in K} X_t - E \sup_{t \in K} X_t \geq a) \leq \exp\{-(a^2/2\sigma^2)\},$$

where $a > 0$ and $\sigma = \sup_{t \in K} \sigma_t$. In view of these appearances, it seems worthwhile to provide another proof of (1), which is the purpose here.

2. Preliminaries

In proving (1), it is enough to take K finite, since the general case then follows from the Monotone Convergence Theorem. On the other hand, it is also convenient to use isonormal indexing, where $K \subset \mathbb{R}^d$, in which case the convex hull of K is rather more natural and yields the same bound. Accordingly, we consider processes $\{X_t, t \in K\}$, where K is a convex body (i.e., compact, convex set) in $\mathbb{R}^d$, and $X_t = <t, Z> = \sum_1^d t_i Z_i$, where $t \in K$ and $Z = (Z_1, Z_2, \ldots, Z_d)$ is a standard Gaussian vector in $\mathbb{R}^d$. The functional on the left–hand side of (1) can then be expressed as

$$W(K) = \int_{\mathbb{R}^d} (1/2\pi)^{d/2} \exp\{\sup_{t \in K}(<t, z> - (1/2)(\|t\|^2 + \|z\|^2))\} \, dz,$$

1991 *Mathematics Subject Classification.* Primary 60G15; Secondary 60E15.

or, equivalently,

$$(2) \qquad W(K) = \int_{\mathbb{R}^d} (1/2\pi)^{d/2} \exp\{-(1/2)\delta^2(z,K)\}\, dz,$$

where $\delta(z,K) = \inf_{t \in K} \|z-t\|$ is the distance from z to K. (Note: (2) differs from the corresponding definition in [11] by a re-scaling of K.)

We will employ a result of Prékopa and Leindler ([4], [6]) that has been used elsewhere in the probability literature, for example, in a recent study of a Gaussian correlation question ([7]). Recall that a function $f: \mathbb{R}^d \times \mathbb{R}^n \to \mathbb{R}^+$ is *log-concave* if for $x, x' \in \mathbb{R}^d$, $c, c' \in \mathbb{R}^n$, and $0 \leq \theta \leq 1$,

$$f(\theta(x,c) + (1-\theta)(x',c')) \geq f^\theta(x,c) f^{1-\theta}(x',c').$$

The Prékopa–Leindler result says that for such an f

$$\int_{\mathbb{R}^d} f(x,c) dx,$$

as a function of c, is log–concave as well.

We also use a classical result on "rounding" of convex bodies that is attributed by Schneider [8] for dimension three to Hadwiger [2, 3] and which has antecedents in the work of Pontryagin; a strengthening appears in [10].

THEOREM 2. *Suppose that K is a convex body in $\mathbb{R}^d$ that is not a singleton. Then there is a sequence of orthogonal transformations such that the sequence of (Minkowski) averages $(1/n)(O_1 K + O_2 K + \cdots + O_n K)$ converges to the ball centered at the origin which shares the same mean width as K.*

Here *mean width* is the average separation of supporting hyperplanes with opposite normals $U, -U, \|U\| = 1$, chosen at random.

3. Proof of the bound

We begin by establishing two lemmas.

LEMMA 1. *W is a continuous functional on the space of convex bodies in $\mathbb{R}^d$ topologized by the Hausdorff metric.*

PROOF. It suffices to verify that $W(K + \varepsilon B) - W(K) \to 0$ as $\varepsilon \searrow 0$, where B is the unit ball in $\mathbb{R}^d$. Assuming that $0 < \varepsilon \leq 1$ and using $0 \leq \delta(z,K) - \delta(z, K + \varepsilon B) \leq \varepsilon$ gives

$$W(K + \varepsilon B) - W(K) = (1/2\pi)^{d/2} \int_{\mathbb{R}^d} \Big[\exp\{-(1/2)\delta^2(z, K + \varepsilon B)\}$$

$$- \exp\{-(1/2)\delta^2(z, K)\}\Big] dz$$

$$\leq (1/2\pi)^{d/2} \int_{\mathbb{R}^d} (1/2) \left[\delta^2(z, K) - \delta^2(z, K + \varepsilon B)\right]$$

$$\cdot \exp\{-(1/2)\delta^2(z, K + \varepsilon B)\} dz$$

$$\leq \varepsilon (1/2\pi)^{d/2} \int_{\mathbb{R}^d} (1/2) \left[\delta(z, K) + \delta(z, K + \varepsilon B)\right]$$

$$\cdot \exp\{-(1/2)\delta^2(z, K + \varepsilon B)\} dz$$

$$\leq \varepsilon (1/2\pi)^{d/2} \int_{\mathbb{R}^d} \delta(z, K) \exp\{-(1/2)\delta^2(z, K + B)\} dz.$$

The second lemma uses a dual "trick" to show a convexity property of δ. □

LEMMA 2. With $z = (z_1, z_2, \ldots, z_d) \in \mathbb{R}^d$, $c = (c_1, c_2, \ldots, c_n) \in \mathbb{R}^n$, and convex bodies $K_1, K_2, \ldots, K_n \subset \mathbb{R}^d$, let

$$g(z,c) = \sup_{\|u\| \leq 1} \left[<z,u> - \sum_{1}^{n} c_i \sup_{y \in K_i} <y,u> \right]. \tag{3}$$

Then

(i) $g(z,c) = \delta(z, c_1 K_1 + c_2 K_2 + \cdots c_n K_n)$ for $c \in (\mathbb{R}^+)^n$.
(ii) $g^2 : \mathbb{R}^d \times \mathbb{R}^n \to \mathbb{R}^+$ is convex.
(iii) $\int_{\mathbb{R}^d} \exp\{-(1/2)g^2(z,c)\} dz$ is a log-concave function of c.

PROOF. For (i), we recall a standard representation in convex geometry ([1])

$$\delta(z,K) = \sup_{\|u\| \leq 1} \left[<z,u> - \sup_{y \in K} <y,u> \right]$$

and the identity

$$\sup_{y \in c_1 K_1 + c_2 K_2 + \cdots + c_n K_n} <y,u> = \sum_{i=1}^{n} c_i \sup_{y \in K_i} <y,u>.$$

For (ii), observe that $g(z,c)$ is always non-negative (take $u = 0$ in (3)), and therefore it is enough to confirm that g is convex. This is immediate from its representation as a supremum of linear functionals. Finally, the Prékopa–Leindler result implies (iii). □

PROOF OF THEOREM 1. Lemma 2 implies that, if $K_1, K_2, \ldots, K_n$ are convex bodies in $\mathbb{R}^d$ and if $0 \leq \theta_i \leq 1$ for $i = 1, 2, \ldots, n$ with $\theta_1 + \theta_2 + \cdots + \theta_n = 1$, then

$$\Pi_1^n W^{\theta_i}(K_i) \leq W(\theta_1 K_1 + \theta_2 K_2 + \cdots + \theta_n K_n).$$

Suppose now that K is a given convex body in $\mathbb{R}^d$ and that $K_i = O_i K$ for certain orthogonal transformations $O_1, O_2, \ldots, O_n$. Then choosing $\theta_i = 1/n$, $i = 1, 2, \ldots, n$ and observing that $W(O_i K) = W(K)$, $i = 1, 2, \ldots, n$, one has

$$W(K) \leq W((1/n)(O_1 K_1 + O_2 K_2 + \cdots + O_n K_n)).$$

For a Hadwiger sequence of orthogonal transformations, the argument on the right tends to αB, a scaled version of the unit ball B in $\mathbb{R}^d$. With Lemma 1, one gets in the limit

$$W(K) \leq W(\alpha B).$$

To determine α, we use the mean width condition of Theorem 2, which is equivalent to

$$E \sup_{t \in K} X_t = E \sup_{t \in \alpha B} X_t = E \sup_{t \in \alpha B} <t, Z>.$$

The last quantity is $\alpha E\|Z\|$, so that

$$\alpha = E \sup_{t \in K} X_t / E\|Z\| \sim E \sup_{t \in K} X_t / \sqrt{d}. \tag{4}$$

We now compute a bound for $W(\alpha B)$. It is direct to see that

$$W(\alpha B) = E \exp\{(1/2)[\|Z\|^2 - \delta^2(Z, \alpha B)]\}.$$

Separating into cases $\|Z\| \leq \alpha$ and $\alpha < \|Z\|$ gives the bound

$$E 1_{(\|Z\| \leq \alpha)} \exp\{(1/2)\|Z\|^2\} + E 1_{(\alpha < \|Z\|)} \exp\{\alpha\|Z\| - (1/2)\alpha^2\},$$

which can be further majorized by

(5) $$E1_{(\|Z\|\leq\alpha)}\exp\{(1/2)\alpha^2\} + E\exp\{\alpha\|Z\|\}.$$

Now let $d \to \infty$ and using (4) conclude that the first term of (5) tends to 0. The second term is $E\exp\{(E\sup_{t\in K} X_t)\cdot\|Z\|/E\|Z\|\}$ which by standard methods (e.g., uniform integrability) tends to $\exp\{\sup_{t\in K} X_t\}$. This concludes the proof. $\square$

Acknowledgments

A referee provided a careful reading. I thank the organizers of the Special Session for their hospitality.

References

[1] Beer, G. (1993). *Topologies on Closed and Closed Convex Sets*. Kluwer, Boston.
[2] Hadwiger, H. (1955). *Altes und Neues über konvexe Körper*. Birkhauser, Basel.
[3] Hadwiger, H. (1957). *Vorlesungen über Inhalt, Oberfläche und Isoperimetrie*. Springer, Berlin.
[4] Leindler, L. (1972). On a certain converse of Hölder's inequality II. *Acta. Sci. Math. Szeged* **33**, 217–223.
[5] Pisier, G. (1986). Probabilistic methods in the geometry of Banach spaces. *Lecture Notes in Mathematics* **1206**, 167–241.
[6] Prékopa, A. (1973). On logarithmic concave measures and functions. *Acta Sci. Math. Szeged* **34**, 335–343.
[7] Schectman, G., Schlumprecht, Th., and Zinn, J. (1998). On the Gaussian measure of the intersection. *Ann. Probab.* **26**, 346–357.
[8] Schneider, R. (1993). *Convex Bodies: the Brunn-Minkowski Theory*. Cambridge University Press, New York.
[9] Tsirel'son, B.S. (1985). A geometric approach to maximum likelihood estimation for infinite-dimensional Gaussian location II. *Theory Prob. Appl.* **30**, 820–828.
[10] Vitale, R.A. (1994). Stochastic smoothing of convex bodies: two examples. *Rend. Circolo Mat. Palermo* **35**, 315–322.
[11] Vitale, R.A. (1996). The Wills functional and Gaussian processes. *Ann. Probab.* **24**, 2172–2178.

DEPARTMENT OF STATISTICS, U-120, UNIVERSITY OF CONNECTICUT, STORRS, CT 06269-3120
E-mail address: rvitale@uconnvm.uconn.edu

Selected Titles in This Series

(Continued from the front of this publication)

207 **Yujiro Kawamata and Vyacheslav V. Shokurov, Editors,** Birational algebraic geometry: A conference on algebraic geometry in memory of Wei-Liang Chow (1911–1995), 1997

206 **Adam Korányi, Editor,** Harmonic functions on trees and buildings, 1997

205 **Paulo D. Cordaro and Howard Jacobowitz, Editors,** Multidimensional complex analysis and partial differential equations: A collection of papers in honor of François Treves, 1997

204 **Yair Censor and Simeon Reich, Editors,** Recent developments in optimization theory and nonlinear analysis, 1997

203 **Hanna Nencka and Jean-Pierre Bourguignon, Editors,** Geometry and nature: In memory of W. K. Clifford, 1997

202 **Jean-Louis Loday, James D. Stasheff, and Alexander A. Voronov, Editors,** Operads: Proceedings of Renaissance Conferences, 1997

201 **J. R. Quine and Peter Sarnak, Editors,** Extremal Riemann surfaces, 1997

200 **F. Dias, J.-M. Ghidaglia, and J.-C. Saut, Editors,** Mathematical problems in the theory of water waves, 1996

199 **G. Banaszak, W. Gajda, and P. Krasoń, Editors,** Algebraic K-theory, 1996

198 **Donald G. Saari and Zhihong Xia, Editors,** Hamiltonian dynamics and celestial mechanics, 1996

197 **J. E. Bonin, J. G. Oxley, and B. Servatius, Editors,** Matroid theory, 1996

196 **David Bao, Shiing-shen Chern, and Zhongmin Shen, Editors,** Finsler geometry, 1996

195 **Warren Dicks and Enric Ventura,** The group fixed by a family of injective endomorphisms of a free group, 1996

194 **Seok-Jin Kang, Myung-Hwan Kim, and Insok Lee, Editors,** Lie algebras and their representations, 1996

193 **Chongying Dong and Geoffrey Mason, Editors,** Moonshine, the Monster, and related topics, 1996

192 **Tomek Bartoszyński and Marion Scheepers, Editors,** Set theory, 1995

191 **Tuong Ton-That, Kenneth I. Gross, Donald St. P. Richards, and Paul J. Sally, Jr., Editors,** Representation theory and harmonic analysis, 1995

190 **Mourad E. H. Ismail, M. Zuhair Nashed, Ahmed I. Zayed, and Ahmed F. Ghaleb, Editors,** Mathematical analysis, wavelets, and signal processing, 1995

189 **S. A. M. Marcantognini, G. A. Mendoza, M. D. Morán, A. Octavio, and W. O. Urbina, Editors,** Harmonic analysis and operator theory, 1995

188 **Alejandro Adem, R. James Milgram, and Douglas C. Ravenel, Editors,** Homotopy theory and its applications, 1995

187 **G. W. Brumfiel and H. M. Hilden,** $SL(2)$ representations of finitely presented groups, 1995

186 **Shreeram S. Abhyankar, Walter Feit, Michael D. Fried, Yasutaka Ihara, and Helmut Voelklein, Editors,** Recent developments in the inverse Galois problem, 1995

185 **Raúl E. Curto, Ronald G. Douglas, Joel D. Pincus, and Norberto Salinas, Editors,** Multivariable operator theory, 1995

184 **L. A. Bokut', A. I. Kostrikin, and S. S. Kutateladze, Editors,** Second International Conference on Algebra, 1995

183 **William C. Connett, Marc-Olivier Gebuhrer, and Alan L. Schwartz, Editors,** Applications of hypergroups and related measure algebras, 1995

182 **Selman Akbulut, Editor,** Real algebraic geometry and topology, 1995

181 **Mila Cenkl and Haynes Miller, Editors,** The Čech Centennial, 1995

(See the AMS catalog for earlier titles)